Accelerated Life Models

Modeling and Statistical Analysis

MONOGRAPHS ON STATISTICS AND APPLIED PROBABILITY

General Editors

D.R. Cox, V. Isham, N. Keiding, T. Louis, N. Reid, R. Tibshirani, and H. Tong

1 Stochastic Population Models in Ecology and Epidemiology *M.S. Barlett* (1960)
2 Queues *D.R. Cox and W.L. Smith* (1961)
3 Monte Carlo Methods *J.M. Hammersley and D.C. Handscomb* (1964)
4 The Statistical Analysis of Series of Events *D.R. Cox and P.A.W. Lewis* (1966)
5 Population Genetics *W.J. Ewens* (1969)
6 Probability, Statistics and Time *M.S. Barlett* (1975)
7 Statistical Inference *S.D. Silvey* (1975)
8 The Analysis of Contingency Tables *B.S. Everitt* (1977)
9 Multivariate Analysis in Behavioural Research *A.E. Maxwell* (1977)
10 Stochastic Abundance Models *S. Engen* (1978)
11 Some Basic Theory for Statistical Inference *E.J.G. Pitman* (1979)
12 Point Processes *D.R. Cox and V. Isham* (1980)
13 Identification of Outliers *D.M. Hawkins* (1980)
14 Optimal Design *S.D. Silvey* (1980)
15 Finite Mixture Distributions *B.S. Everitt and D.J. Hand* (1981)
16 Classification *A.D. Gordon* (1981)
17 Distribution-Free Statistical Methods, 2nd edition *J.S. Maritz* (1995)
18 Residuals and Influence in Regression *R.D. Cook and S. Weisberg* (1982)
19 Applications of Queueing Theory, 2nd edition *G.F. Newell* (1982)
20 Risk Theory, 3rd edition *R.E. Beard, T. Pentikäinen and E. Pesonen* (1984)
21 Analysis of Survival Data *D.R. Cox and D. Oakes* (1984)
22 An Introduction to Latent Variable Models *B.S. Everitt* (1984)
23 Bandit Problems *D.A. Berry and B. Fristedt* (1985)
24 Stochastic Modelling and Control *M.H.A. Davis and R. Vinter* (1985)
25 The Statistical Analysis of Composition Data *J. Aitchison* (1986)
26 Density Estimation for Statistics and Data Analysis *B.W. Silverman* (1986)
27 Regression Analysis with Applications *G.B. Wetherill* (1986)
28 Sequential Methods in Statistics, 3rd edition
G.B. Wetherill and K.D. Glazebrook (1986)
29 Tensor Methods in Statistics *P. McCullagh* (1987)
30 Transformation and Weighting in Regression
R.J. Carroll and D. Ruppert (1988)
31 Asymptotic Techniques for Use in Statistics
O.E. Bandorff-Nielsen and D.R. Cox (1989)
32 Analysis of Binary Data, 2nd edition *D.R. Cox and E.J. Snell* (1989)

Accelerated Life Models
Modeling and Statistical Analysis

Vilijandas Bagdonavičius
Mikhail Nikulin

CHAPMAN & HALL/CRC

A CRC Press Company
Boca Raton   London   New York   Washington, D.C.

Library of Congress Cataloging-in-Publication Data

Bagdonavičius , Vilijandas
    Accelerated life models : modeling and statistical analysis / Vilijandas Bagdonavičius ,
Mikhail Nikulin.
        p. cm. — (Monographs on statistics and applied probability)
    Includes bibliographical references and index.
    ISBN 1-58488-186-0 (alk. paper)
    1. Accelerated life testing—Mathematical models. 2. Accelerated life testing—Statistical
methods. I. Bagdonavičius V. (Vilijandas) II. Title. III. Series.

TA169.3 .N55 2001
620′ .004—dc21
                                        2001043390

Visit the CRC Press Web site at www.crcpress.com

© 2002 by Chapman & Hall/CRC

No claim to original U.S. Government works
International Standard Book Number 1-58488-186-0
Library of Congress Card Number 2001043390
Printed in the United States of America 1 2 3 4 5 6 7 8 9 0
Printed on acid-free paper

To Login N. Bolshev

Contents

Preface

Failure time regression (FTR) data are obtained by observing failure times of units functioning under various values of explanatory variables (called also regressors, stresses, covariables) such as temperature, voltage, load, pressure, humidity, design, manufacture, etc. The purpose of the FTR data analysis is to estimate reliability under specified values of interest of these variables. In the particular case of *accelerated life testing* (ALT) data are collected from experiments under higher than usual stress conditions and reliability under usual (design) stress is estimated.

Models relating reliability characteristics to explanatory variables are called failure time regression models. In ALT such models are called accelerated life models.

Degradation data with explanatory variables are obtained when failure times and quantities characterizing degradation of units under explanatory variables are measured. Models relating degradation to explanatory variables and models relating intensity of failures to degradation and to these variables are used for analysis of such data.

This book gives models and methods of statistical analysis for FTR (ALT) and degradation data with explanatory variables.

Some chapters of books such as *Statistical Reliability Theory* (1989) by L. Gertsbakh, *Statistical Methods for Reliability Data* (1998) by W. Meeker and L. Escobar, *Statistical Analysis of Reliability Data* (2000) by M. Crowder, A. Kimber, R. Smith and T. Sweeting, are also focused on these topics.

The books *Statistical Methods in Accelerated Life Testing* (1988) by R. Viertl and *Accelerated Testing: Statistical Models, Test Plans, and Data Analysis* (1990) by W. Nelson are entirely consecrated to models and statistical analysis of ALT data.

FTR models and methods of their analysis have much in common with regression models in survival analysis. In the well-known books on survival analysis of T. Fleming and D. Harrington, *Counting Processes and Survival Analysis* (1991), and P. K. Andersen, O. Borgan, R. Gill and N. Keiding, *Statistical Models Based on Counting Processes* (1993), data are presented in the form of failure, censoring, and explanatory variable processes. Such data presentation has an important advantage since the data are viewed in dynamics and can be analyzed using the theory of counting processes. In the recent papers on statistical analysis of reliability data such data presentation is found more and more often. We use it in the present book and consequently some results from the theory of counting processes. Nevertheless, we tried to minimize the number of required mathematical notions and avoided technical details by explaining ideas and referring to papers and books where thorough analysis can be found. A short review of used results from the stochastic process theory is given in Appendix.

The most used FTR and ALT models are the parametric accelerated failure time (AFT) and semiparametric proportional hazards (PH) model. Nevertheless, they are quite restricted and do not always agree well with real data. So a number of models, which are alternatives or generalizations of them, are presented. A unified approach to the formulation of the models was used. It gives the possibility to formulate new models and see relations between them and well-known models. We think that most practical situations may be modeled by one or another of the presented models.

For most models, estimation procedures from FTR censored with time-varying and constant explanatory variables are given. In particular, plans of experiments and estimation methods from ALT data are considered. Methods for analysis of ALT data when the production process is unstable are given also.

For the most important models formal goodness-of-fit tests are given.

In the statistical literature degradation and the failure time data are analyzed separately. We give methods of statistical analysis of degradation models with the intensity of failures depending on the level of degradation and explanatory variables.

We consider here only the classes of univariate models. These models can be generalized to the case of multivariate life data as it was done, for example, in our monographs *Semiparametric Models in Accelerated Life Testing* (1995) and *Additive and Multiplicative Semiparametric Models in Accelerated Life Testing and Survival Analysis* (1998), published in *Queen's Papers in Pure and Applied Mathematics* (Kingston, Canada).

Formal exposition of material did not leave a place for numerical examples and diagnostic plots. Many such plots and examples for some of the models can be found in the books of W. Nelson and W. Meeker and L. Escobar. The recent book of T. Therneau and P. Grambsch, *Modeling Survival Data. Extending the Cox Model* (2000), gives graphical analysis and many practical recommendations on application of survival regression data, which can be useful for analysis of FTR data. Many useful results related with our book can be found also in the monographs *Statistical and Probabilistic Models in Reliability* (1999), edited by D. Ionescu and N. Limnios, *Recent Advances in Reliability Theory* (2000), edited by N. Limnios and M. Nikulin, *Goodness-of-fit Tests and Validity Models* (2001), edited by C. Huber, N. Balakrishnan, M. Nikulin and M. Mesbah.

We used preliminary versions of our book to give the basis of courses for MSc and Ph.D students the Vilnius University, the University Bordeaux 1, the University Victor Segalen Bordeaux 2 in the Department Sciences et Modélisation and in the Institut de Santé Publique, d'Epidémiologie et de Développement (l'ISPED) for the students of DESS Statistique Appliquée aux Sciences Sociales et de Santé.

Grants from Le Conseil Régional d'Aquitaine and l'Université Victor Segalen Bordeaux 2 supported our work. We thank these agencies for making our collaboration possible.

We thank our friends and colleagues, A. Alioum, A. Bikelis, C. Cuadras, D. Commenges, D. Cox, K. El Himdi, L. Gerville-Réache, C. Huber, I. Ibragimov, W. Kahle, V. Kazakevičius, N. Keiding, H. Läuter, M. Läuter, J. Lawless, M.-L.T. Lee, N. Limnios, W. Meeker, M. Mesbah, V. Nikoulina, O. Pons, N. Singpurwalla, T. Smith, V. Solev, F. Spizzichino, K. Suzuki, R. Viertl, Z. Ying, A. Zerbet and A. Zoglat for their many useful discussions, advice, and help.

Lastly, we wish to thank the participants of the European Seminar, "Mathematical Methods for Survival Analysis and Reliability," organized by Université René Descartes, Paris 5 (C. Huber), Université de Technologie de Compiègne (N. Limnios), Université de Bretagne Sud, Vannes (M. Mesbah), Université Victor Segalen Bordeaux 2 (M. Nikulin), Université de Lille 1 (J.-L. Bon), Potsdam University (H. Läuter), Otto-von-Guericke University, Magdeburg (W. Kahle), McMaster University, Hamilton (N. Balakrishnan), and Steklov Mathematical Institute, Saint Petersburg (V. Solev) for their helpful comments, recommendations, suggestions, and encouragement during the preparation of the project.

Vilijandas Bagdonavičius and Mikhail Nikulin

CHAPTER 1

Failure time distributions

1.1 Introduction

Suppose that the failure time T is a nonnegative, absolutely continuous random variable. Failure-time distribution can be defined by one of the following functions:

Survival function

The survival function

$$S(t) = \mathbf{P}\{T \geq t\}, \quad t \geq 0; \tag{1.1}$$

for fixed t means the probability of survival up to time t.

Cumulative distribution function

The cumulative distribution function (c.d.f)

$$F(t) = \mathbf{P}\{T < t\} = 1 - S(t) \tag{1.2}$$

for fixed t means the probability of failure before the moment t.

Probability density function

The probability density function is a function $p(t) \geq 0$ such that for any $t \geq 0$

$$F(t) = \int_0^t p(s)ds. \tag{1.3}$$

If the cumulative distribution function has the derivative at the point t then

$$p(t) = \lim_{h \to 0} \frac{\mathbf{P}(t \leq T < t + h)}{h} = F'(t) = -S'(t). \tag{1.4}$$

For fixed t the probability density function characterizes the probability of failure in a small interval after the moment t.

Hazard rate function

The hazard rate function

$$\alpha(t) = \lim_{h \to 0} \frac{\mathbf{P}(t \leq T < t + h \mid T > t)}{h} = \frac{p(t)}{S(t)}, \tag{1.5}$$

for fixed t characterizes the probability of failure in a small interval after the moment t, given survival to time t. So it means the risk of failure of the units survived.

Cumulative hazard function

Cumulative hazard function

$$A(t) = \int_0^t \alpha(u)du = -ln\{S(t)\}. \tag{1.6}$$

The survival function can be found from the hazard rate or the cumulative hazard function using the following relation:

$$S(t) = \exp\{-A(t)\} = \exp\{-\int_0^t \alpha(u)du\}. \tag{1.7}$$

The quantile function

The quantile function: for $0 < p < 1$

$$t_p = inf\{t : F(t) \geq p\}. \tag{1.8}$$

When $F(t)$ is strictly increasing continuous function then

$$t_p = F^{-1}(p), \quad 0 < p < 1. \tag{1.9}$$

For fixed p the quantile t_p means the time at which a specific proportion p of the population fails.

Any of above considered functions can be found from another.

The mean and the variance

Important survival characteristics are the mean failure-time $\mathbf{E}(T)$ and the variance $\mathbf{Var}(T)$:

$$\mathbf{E}(T) = \int_0^\infty S(t)dt, \quad \mathbf{Var}(T) = 2\int_0^\infty tS(t)dt - \{\mathbf{E}(T)\}^2.$$

The mean and the variance can be found from any of the above considered functions but not vice versa.

1.2 Parametric classes of failure time distributions

For many FTR models, failure time distributions under different constant explanatory variables should belong to the same parametric class.

Classes of failure-time distributions can be formulated by specifying the form of one of the considered functions: survival, cumulative distribution, probability density, or hazard rate. In this chapter distributions will be classified by shape of the hazard rate function.

Analyzing failure-time data, the five most common shapes of hazard rates observed are: constant, monotone (increasing or decreasing), ∩-shaped, and ∪-shaped. The last includes three typical periods: burn in (or infant mortality) period, relatively low failure intensity period, and senility period, with progressively increasing risk of failure.

1.2.1 Constant hazard rate

The only continuous distribution with constant hazard rate is the exponential distribution.

Exponential distribution $\mathcal{E}(\theta)$

$$S(t,\theta) = e^{-t/\theta}, \quad t \geq 0 \quad (\theta > 0),$$

$$p(t,\theta) = \frac{1}{\theta}e^{-t/\theta} \quad t \geq 0;$$

$$\alpha(t,\theta) = 1/\theta;$$

$$t_p = -\theta\ln(1-p); \quad 0 < p < 1;$$

$$\mathbf{E}(T) = \theta, \quad \mathbf{Var}(T) = \theta^2.$$

Hazard rates for various values of the parameter θ are given in Figure 1.1.

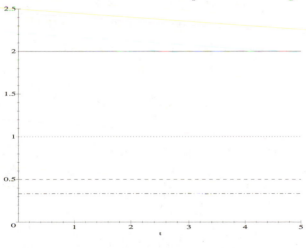

Figure 1.1

1.2.2 Monotone hazard rate

There are many families of survival distributions with a monotone hazard rate.

Weibull distribution $W(\theta, \nu)$

$$S(t, \theta, \nu) = \exp\left\{-(\frac{t}{\theta})^\nu\right\} \quad (\theta, \nu > 0); \quad t \geq 0;$$

$$\alpha(t, \theta, \nu) = \frac{\nu}{\theta^\nu} t^{\nu-1};$$

$$p(t, \theta, \nu) = \frac{\nu}{\theta^\nu} t^{\nu-1} \exp\left\{-(\frac{t}{\theta})^\nu\right\};$$

$$t_p = \theta \left(-\ln(1-p)\right)^{1/\nu}; \quad 0 < p < 1;$$

$$\mathbf{E}(T) = \theta\,\Gamma(1+1/\nu), \quad \mathbf{Var}(T) = \theta^2 \left(\Gamma(1+2/\nu) - \Gamma^2(1+1/\nu)\right).$$

The hazard rate is a *power function*. Note that $W(\theta, 1) = \mathcal{E}(\theta)$.
With $0 < \nu < 1$ the hazard rate is decreasing from ∞ to 0 (Figure 1.2).

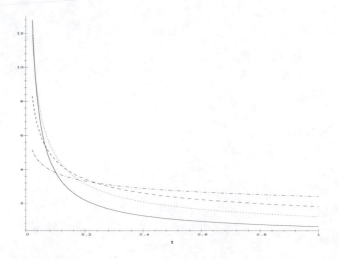

Figure 1.2

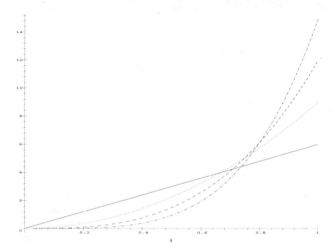

Figure 1.3

With $\nu > 1$ the hazard rate is increasing from 0 to ∞ (Figure 1.3).

Gamma distribution $G(\theta, \nu)$

$$p(t, \theta, \nu) = \frac{1}{\theta^\nu \Gamma(\nu)} t^{\nu-1} e^{-\frac{t}{\theta}} \quad (\theta, \nu > 0); \quad t \geq 0;$$

$$F(t, \theta, \nu) = \frac{1}{\Gamma(\nu)} \int_0^{t/\theta} u^{\nu-1} e^{-u} du;$$

$$\alpha(t, \theta, \nu) = \frac{p(t, \theta, \nu)}{1 - F(t, \theta, \nu)};$$

$$\mathbf{E}(T) = \theta\nu, \quad \mathbf{Var}(T) = \theta^2 \nu.$$

Note that $G(\theta, 1) = \mathcal{E}(\theta)$.

With $\nu > 1$ the hazard rate is increasing from 0 to $\frac{1}{\theta}$ (Figure 1.4).

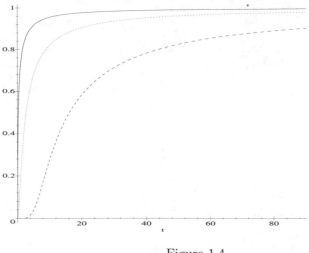

Figure 1.4

With $0 < \nu < 1$ the hazard rate is decreasing from ∞ to $\frac{1}{\theta}$ (Figure 1.5).

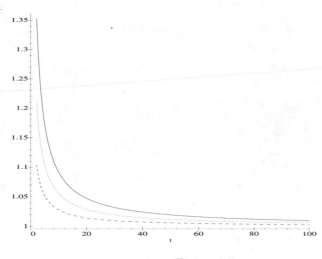

Figure 1.5

Goodness-of-fit tests distinguish Weibull and gamma distributions only when the size of the data is very large.

Gompertz-Makeham distribution $GM(\gamma_0, \gamma_1, \gamma_2)$

$$S(t, \theta) = \exp\{-\gamma_0 t - \frac{\gamma_1}{2}(e^{-\gamma_2 t} - 1)\}, \quad (\gamma_0, \gamma_1 > 0, \gamma_2 \in \mathbf{R});$$

$$p(t, \theta, \nu) = (\gamma_0 + \gamma_1 e^{-\gamma_2 t}) \exp\{-\gamma_0 t - \frac{\gamma_1}{2}(e^{-\gamma_2 t} - 1)\};$$

$$\alpha(t, \theta) = \gamma_0 + \gamma_1 e^{-\gamma_2 t}.$$

Note that $GM(\gamma_0, \gamma_1, 0) = \mathcal{E}(\gamma_0 + \gamma_1)$.

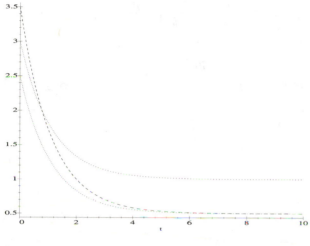

Figure 1.6

With $\gamma_2 > 0$ the hazard rate is decreasing from $\gamma_0 + \gamma_1$ to γ_0 (Figure 1.6).

With $\gamma_2 < 0$ the hazard rate is increasing from $\gamma_0 + \gamma_1$ to ∞ (Figure 1.7).

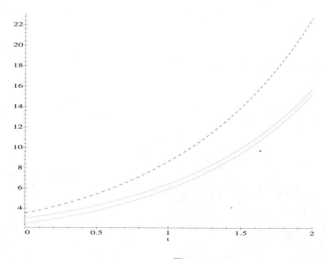

Figure 1.7

Mixture of two exponential distributions $ME(\theta_1, \theta_2, p_1)$

$$S(t, \theta_1, \theta_2, p_1) = p_1 \exp\{-\frac{t}{\theta_1}\} + p_2 \exp\{-\frac{t}{\theta_2}\} \quad (0 < p_1 < 1, \quad p_2 = 1 - p_1, \quad \theta_2 > \theta_1 >$$

$$p(t, \theta_1, \theta_2, p_1) = \frac{p_1}{\theta_1} \exp\{-\frac{t}{\theta_1}\} + \frac{p_2}{\theta_2} \exp\{-\frac{t}{\theta_2}\};$$

$$\alpha(t, \theta_1, \theta_2, p_1) = p(t, \theta_1, \theta_2, p_1)/S(t, \theta_1, \theta_2, p_1);$$

$$\mathbf{E}(T) = p_1\theta_1 + p_2\theta_2.$$

The hazard rate is decreasing from $c_2 = \frac{p_1}{\theta_1} + \frac{p_2}{\theta_2}$ to $c_1 = \frac{1}{\theta_2}$ (Figure 1.8).

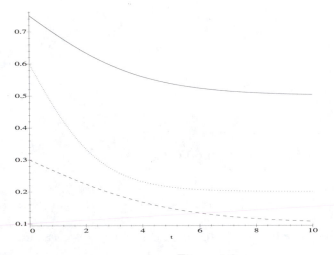

Figure 1.8

Generalized Weibull distribution $GW(\theta, \nu, \gamma)$

$$S(t, \theta, \nu, \gamma) = \exp\left\{1 - \left(1 + (\frac{t}{\theta})^\nu\right)^{1/\gamma}\right\}, \quad (\theta, \nu, \gamma > 0); \quad t \geq 0; \quad (1.10)$$

$$\alpha(t, \theta, \nu, \gamma) = \frac{\nu}{\gamma\theta^\nu} t^{\nu-1}\{1 + (\frac{t}{\theta})^\nu\}^{1/\gamma-1};$$

$$t_p = \theta\{(1 - \ln(1 - p))^\gamma - 1\}^{1/\nu}; \quad 0 < p < 1.$$

Note that $GW(\theta, \nu, 1) = W(\theta, \nu)$, $GW(\theta, 1, 1) = \mathcal{E}(\theta)$.

The generalized Weibull distribution was suggested by accelerated life models considered in the following chapter.

This class of distributions has very nice properties. In dependence of parameter values the hazard rate can be constant, monotone (increasing or decreasing), ∩-shaped, and ∪- shaped. All moments of this distribution are finite.

With $\nu > 1$, $\nu > \gamma$ the hazard rate is increasing from 0 to ∞ (Figure 1.9).

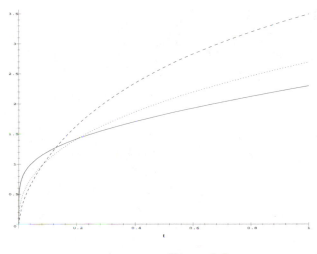

Figure 1.9

With $\nu = 1$, $\gamma < 1$ the hazard rate is increasing from $(\gamma\theta)^{-1}$ to ∞ (Figure 1.10).

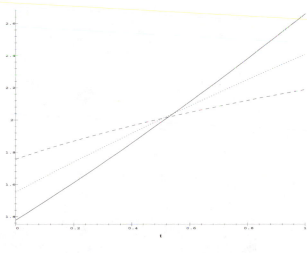

Figure 1.10

With $0 < \nu < 1$, $\nu < \gamma$ the hazard rate is decreasing from ∞ to 0 (Figure 1.11).

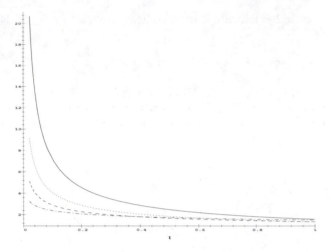

Figure 1.11

With $0 < \nu < 1$, $\nu = \gamma$ the hazard rate is decreasing from ∞ to θ^{-1} (Figure 1.12).

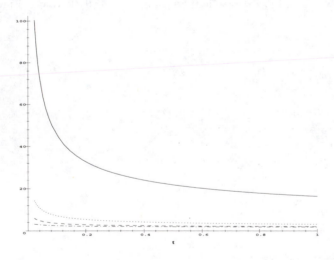

Figure 1.12

Exponentiated Weibull distribution $EW(\theta, \nu, \gamma)$.

$$S(t, \theta, \nu, \gamma) = 1 - \left\{ 1 - \exp[-(\frac{t}{\theta})^{\nu}] \right\}^{1/\gamma} \quad (\theta, \nu, \gamma > 0); \quad t \geq 0; \quad (1.11)$$

$$\alpha(t,\theta,\nu,\gamma) = \frac{\nu\{1 - \exp[-(\frac{t}{\theta})^{\nu}]\}^{(1-\gamma)/\gamma}\exp[-(\frac{t}{\theta})^{\nu}](\frac{t}{\theta})^{\nu-1}}{\gamma\theta\{1 - (1 - \exp[-(\frac{t}{\theta})^{\nu}])\}^{1/\gamma}}$$

$$t_p = \theta[-\ln(1 - p^{\gamma})]^{1/\nu}; \quad 0 < p < 1.$$

Note that $EW(\theta,\nu,1) = W(\theta,\nu)$, $EW(\theta,1,1) = \mathcal{E}(\theta)$.

This distribution was introduced by Efron (1988). Its properties were studed by Mudholkar and Srivastava (1995). All moments of this distribution are finite.

With $\nu > 1$, $\nu \geq \gamma$ the hazard rate is increasing from 0 to ∞.

With $\nu = 1$, $\gamma \leq 1$ the hazard rate is increasing from $(\gamma\theta)^{-1}$ to ∞.

With $0 < \nu < 1$, $\nu < \gamma$ the hazard rate is decreasing from ∞ to 0.

With $0 < \nu < 1$, $\nu = \gamma$ the hazard rate is decreasing from θ^{-1} to 0.

Summary. For the values of parameters where the hazard rate is increasing we have different families of survival distributions:

$W(\theta,\nu)$: $\alpha(t)$ increases from 0 to ∞;

$G(\theta,\nu)$: $\alpha(t)$ increases from 0 to $c > 0$;

$GM(\gamma_0,\gamma_1,\gamma_2)$: $\alpha(t)$ increases from $c > 0$ to ∞;

$GW(\theta,\nu,\gamma)$: $\alpha(t)$ increases from $c \geq 0$ to ∞;

$EW(\theta,\nu,\gamma)$: $\alpha(t)$ increases from $c \geq 0$ to ∞.

For the values of parameters where the hazard rate is decreasing:

$W(\theta,\nu)$: $\alpha(t)$ decreases from ∞ to 0;

$G(\theta,\nu)$: $\alpha(t)$ decreases from ∞ to $c > 0$;

$ME(\theta_1,\theta_2,p_1)$:$\alpha(t)$ decreases from c_2 to c_1, $c_2 > c_1$.

$GM(\gamma_0,\gamma_1,\gamma_2)$: $\alpha(t)$ decreases from $c_1 > 0$ to $c_2 : 0 < c_2 < c_1$;

$GW(\theta,\nu,\gamma)$: $\alpha(t)$ decreases from ∞ to $c \geq 0$;

$EW(\theta,\nu,\gamma)$: $\alpha(t)$ decreases from $0 < c \leq \infty$ to 0.

1.2.3 ∩-shaped hazard rate

Lognormal distribution $LN(\mu,\sigma)$

$$S(t,\mu,\sigma) = 1 - \Phi\left(\frac{\ln t - \mu}{\sigma}\right), \quad (\mu \in \mathbf{R}, \sigma > 0); \quad t \geq 0; \qquad (1.12)$$

$$p(t,\mu,\sigma) = \frac{1}{\sigma t}\varphi\left(\frac{\ln t - \mu}{\sigma}\right);$$

$$\alpha(t,\mu,\sigma) = \frac{p(t,\mu,\sigma)}{S(t,\mu,\sigma)};$$

$$t_p = e^{\sigma\Phi^{-1}(p)+\mu};$$

$$\mathbf{E}(T) = e^{\mu+\sigma^2/2}, \quad \mathbf{Var}(T) = e^{2\mu+\sigma^2/2}(e^{\sigma^2} - 1).$$

Here Φ is the distribution function of the standard normal law,

$$\varphi(t) = \frac{1}{\sqrt{2\pi}}e^{-t^2/2} = \Phi'(x).$$

The hazard rate increases from 0 to its maximum value and then decreases to 0, i.e., it is $\cap$-shaped (Figure 1.13).

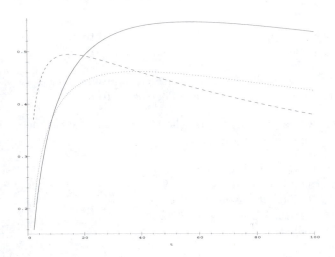

Figure 1.13

If σ is large then the maximum is reached early in the life. Therefore, the lognormal distribution is also used to model situations when the risk of failure is decreasing.

Loglogistic distribution $LL(\theta, \nu)$

$$S(t, \theta, \nu) = \frac{1}{1 + (\frac{t}{\theta})^\nu} \quad (\theta, \nu > 0); \tag{1.13}$$

$$\alpha(t, \theta, \nu) = \frac{\nu}{\theta^\nu}t^{\nu-1}\left(1 + (\frac{t}{\theta})^\nu\right)^{-1};$$

$$p(t, \theta, \nu) = \frac{\nu}{\theta^\nu}t^{\nu-1}\left(1 + (\frac{t}{\theta})^\nu\right)^{-2};$$

$$t_p = \theta(\frac{p}{1-p})^{1/\nu}; \quad 0 < p < 1.$$

For $0 < \nu \le 1$ the mean does not exist. For $\nu > 1$

$$\mathbf{E}(T) = \theta\,\Gamma(1 + 1/\nu)\,\Gamma(1 - 1/\nu).$$

The variance exists for $\nu > 2$:

$$\mathbf{Var}(T) = \theta^2 \{\Gamma(1 + 2/\nu)\,\Gamma(1 - 2/\nu) - \Gamma^2(1 + 1/\nu)\,\Gamma^2(1 - 1/\nu)\}.$$

With $\nu > 1$ the hazard rate increases from 0 to its maximum value and then decreases to 0, i.e. it is $\cap$-shaped (Figure 1.14).

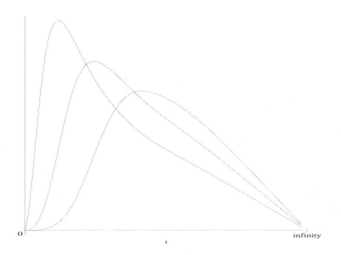

Figure 1.14

Inverse Gaussian distribution $IG(\nu, \theta)$

$$F(t, \theta, \nu) = \Phi\left(\sqrt{\nu}\left(\sqrt{\frac{t}{\theta}} - \sqrt{\frac{\theta}{t}}\right)\right) + e^{2\nu}\,\Phi\left(-\sqrt{\nu}\left(\sqrt{\frac{t}{\theta}} + \sqrt{\frac{\theta}{t}}\right)\right); \quad (1.14)$$

$$p(t, \theta, \nu) = \sqrt{\nu\theta}\,t^{-3/2}\varphi\left(\sqrt{\nu}\left(\sqrt{\frac{t}{\theta}} - \sqrt{\frac{\theta}{t}}\right)\right), \quad (\theta, \nu > 0); \quad t \geq 0;$$

$$\alpha(t, \theta, \nu) = \frac{p(t, \theta, \nu)}{1 - F(t, \theta, \nu)};$$

$$\mathbf{E}(T) = \theta, \quad \mathbf{Var}(T) = \theta^2/\nu.$$

The hazard rate increases from 0 to its maximum value and then decreases to $\nu/2\theta$, i.e. it is $\cap$-shaped (Figure 1.15).

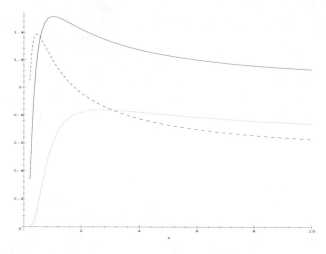

Figure 1.15

Birnbaum and Saunders (1969) distribution $BS(\nu, \theta)$

$$F(t, \theta, \nu) = \Phi\left(\frac{1}{\nu}\left(\sqrt{\frac{t}{\theta}} - \sqrt{\frac{\theta}{t}}\right)\right) \quad (\theta, \nu > 0); \quad t \geq 0;$$

$$p(t, \theta, \nu) = \frac{1}{2\nu t}\left(\sqrt{\frac{t}{\theta}} + \sqrt{\frac{\theta}{t}}\right)\varphi\left(\frac{1}{\nu}\left(\sqrt{\frac{t}{\theta}} - \sqrt{\frac{\theta}{t}}\right)\right);$$

$$t_p = \frac{\theta}{4}\left\{\nu\Phi^{-1}(p) + \sqrt{4 + \{\nu\Phi^{-1}(p)\}^2}\right\}^2; \quad 0 < p < 1;$$

$$\alpha(t, \theta, \nu) = \frac{p(t, \theta, \nu)}{1 - F(t, \theta, \nu)};$$

$$\mathbf{E}(T) = \theta\left(1 + \frac{\nu^2}{2}\right), \quad \mathbf{Var}(T) = (\frac{\theta}{\nu})^2\left(1 + \frac{5}{4}\nu^2\right).$$

The hazard rate increases from 0 to its maximum value and then decreases to $1/2\theta\nu^2$, i.e., it is $\cap$-shaped.

The BS family is very similar to the IG family of distributions.

Generalized Weibull distribution $GW(\theta, \nu, \gamma)$

The generalized Weibull distribution was given by (1.10).

With $\gamma > \nu > 1$ the hazard rate is increasing from 0 to its maximum value

$$c = \frac{\nu}{\gamma\theta}\left(\frac{\gamma(\nu - 1)}{\gamma - \nu}\right)^{\frac{\nu - 1}{\nu}}\left(\frac{\nu(\gamma - 1)}{\gamma - \nu}\right)^{\frac{1 - \gamma}{\gamma}} \tag{1.15}$$

and then decreases to 0, i.e., it is ∩-shaped (Figure 1.16).

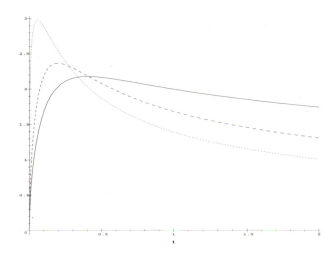

Figure 1.16

Exponentiated Weibull distribution $EW(\theta, \nu, \gamma)$

The exponentiated Weibull distribution was given by (1.11).

With $\gamma < \nu < 1$ the hazard rate is increasing from 0 to its maximum value $c > 0$ and then decreases to 0, i.e., it is ∩-shaped.

Summary: For the values of parameters where the hazard rate is ∩-shaped we have:

$LN(\mu, \sigma)$: the hazard rate increases from 0 to its maximum value and then decreases to 0;

$LL(\theta, \nu)$: the hazard rate increases from 0 to its maximum value $c > 0$ and then decreases to 0;

$IG(\nu, \theta)$: the hazard rate increases from 0 to its maximum value $c > 0$ and then decreases to c_1, $0 < c_1 < c$;

$BS(\nu, \theta)$: the hazard rate increases from 0 to its maximum value $c > 0$ and then decreases to c_1, $0 < c_1 < c$;

$GW(\theta, \nu, \gamma)$: the hazard rate increases from 0 to its maximum value $c > 0$ and then decreases to 0;

$EW(\theta, \nu, \gamma)$: the hazard rate increases from 0 to its maximum value $c > 0$ and then decreases to 0.

1.2.4 ∪-shaped hazard rate

Generalized Weibull distribution $GW(\theta, \nu, \gamma)$

The generalized Weibull distribution was given by (1.10).

If $0 < \gamma < \nu < 1$ then the hazard rate is decreasing from ∞ to its minimum value

$$c = \frac{\nu}{\gamma\theta}\left(\frac{\gamma(1-\nu)}{\nu-\gamma}\right)^{\frac{\nu-1}{\nu}}\left(\frac{\nu(1-\gamma)}{\nu\gamma}\right)^{\frac{1-\gamma}{\gamma}}$$

and then increases to ∞, i.e., it is ∪-shaped (Figure 1.17).

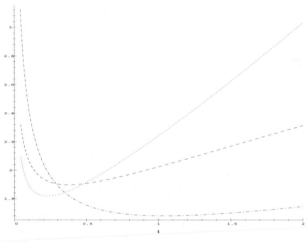

Figure 1.17

Exponentiated Weibull distribution $EW(\theta, \nu, \gamma)$

If $\gamma > \nu > 1$ then the hazard rate is decreasing from ∞ to its minimum value c and then increases to ∞.

Thus, for the values of parameters where the hazard rate is ∪-shaped we have $GW(\theta, \nu, \gamma)$ and $EW(\theta, \nu, \gamma)$: the hazard rate decreases from ∞ to its minimum value $c > 0$ and then increases to ∞.

In this chapter we consider only several basic probability models and notions, which are often used in reliability and survival analysis. There are many other important univariate continuous distributions which are useful in applications. Bagdonavičius and Nikulin (1995), Barlow and Proschan (1975), Bayer (1994), Beichelt and Franken (1983), Bogdanoff and Kozin (1985), Bonneuil (1997), Gertsbakh and Kordonsky (1969), Gertsbakh (1989), Courgeau, D., Lelièvre, E. (1989), Gnedenko and Ushakov (1995), Singpurwalla and Wilson (1999), Voinov and Nikulin (1993), Xie (2000) provide detailed information on a wide range of different probabilistic parametric families and their applications arising in reliability and survival analysis. Statistical analysis of these families can be found in Bain and Engelhardt (1991), Gerville-Réache and Nikulin (2000), Greenwood and Nikulin (1996), Kalbfleisch and Prentice (1980), Lawless (1982), Mann, Schafer and Singpurwalla (1974), Meeker and Escobar (1998), Nelson (1990), Nikulin and Voinov (2000), Stacy (1962), Voinov and Nikulin (1996), and Zacks (1992).

CHAPTER 2

Accelerated life models

2.1 Introduction

Accelerated life models relate the lifetime distribution to the explanatory variable (stress, covariate, regressor). This distribution can be defined by the survival, cumulative distribution, or probability density functions. Nevertheless, the sense of accelerated life models is best seen if they are formulated in terms of the hazard rate function.

Suppose at first that the explanatory variable $x(\cdot)$ is a deterministic time function:

$$x(\cdot) = (x_1(\cdot), ..., x_m(\cdot))^T \ : [0, \infty) \to B \in \mathbf{R}^m.$$

If $x(\cdot)$ is constant in time, we shall write x instead of $x(\cdot)$ in all formulas.

Denote informally by $T_{x(\cdot)}$ the failure time under $x(\cdot)$ and by

$$S_{x(\cdot)}(t) = \mathbf{P}\{T_{x(\cdot)} \geq t\}, \quad F_{x(\cdot)}(t) = \mathbf{P}\{T_{x(\cdot)} < t\}, \quad p_{x(\cdot)}(t) = -S'_{x(\cdot)}(t),$$

the survival, cumulative distribution, and probability density function, respectively.

The hazard rate function under $x(\cdot)$ is

$$\alpha_{x(\cdot)}(t) = \lim_{h \downarrow 0} \frac{1}{h} \mathbf{P}\{T_{x(\cdot)} \in [t, t+h) \mid T_{x(\cdot)} \geq t\} = -\frac{S'_{x(\cdot)}(t)}{S_{x(\cdot)}(t)}.$$

Denote by

$$A_{x(\cdot)}(t) = \int_0^t \alpha_{x(\cdot)}(u)du = -ln\{S_{x(\cdot)}(t)\}$$

the cumulative hazard under $x(\cdot)$.

Each specified accelerated life model relates the hazard rate (or other function) to the explanatory variable in some particular way.

If the explanatory variable is a stochastic process $X(t)$, $t \geq 0$, and $T_{X(\cdot)}$ is the failure time under $X(\cdot)$, then denote by

$$S_{x(\cdot)}(t) = \mathbf{P}\{T_{X(\cdot)} \geq t | X(s) = x(s), 0 \leq s \leq t\},$$

$$\alpha_{x(\cdot)}(t) = -S'_{x(\cdot)}(t)/S_{x(\cdot)}(t), \quad A_{x(\cdot)}(t) = -\ln\{S_{x(\cdot)}(t)\}$$

the conditional survival, hazard rate, and cumulative hazard functions. In this case the definitions of models should be understood in terms of these conditional functions.

To be concise the word stress will be used for explanatory variable in this chapter.

2.2 Generalized Sedyakin's model

2.2.1 Definition of the model

Accelerated life models could be at first formulated for constant explanatory variables. Nevertheless, before formulating them, let us consider a method for generalizing such models to the case of time-varying stresses.

In 1966 N. Sedyakin formulated the *physical principle in reliability* which states that for two identical populations of units functioning under different stresses x_1 and x_2, two moments t_1 and t_2 are equivalent if the probabilities of survival until these moments are equal:

$$\mathbf{P}\{T_{x_1} \geq t_1\} = S_{x_1}(t_1) = S_{x_2}(t_2) = \mathbf{P}\{T_{x_2} \geq t_2\}.$$

If after these equivalent moments the units of both groups are observed under the same stress x_2, i.e. the first population is observed under the step-stress

$$x(\tau) = \begin{cases} x_1, & 0 \leq \tau < t_1, \\ x_2, & \tau \geq t_1, \end{cases}$$

and the second all time under the constant stress x_2, then for all $s > 0$

$$\alpha_{x(\cdot)}(t_1 + s) = \alpha_{x_2}(t_2 + s).$$

Using the idea of Sedyakin, Bagdonavičius (1978) generalized the model to the case of any time-varying stresses by supposing that the hazard rate $\alpha_{x(\cdot)}(t)$ at any moment t is a function of the value of the stress at this moment and of the probability of survival until this moment. It is formalized by the following definition.

Definition 2.1 *The generalized Sedyakin's (GS) model holds on a set of stresses E if there exists on $E \times \mathbf{R}^+$ a positive function g such that for all $x(\cdot) \in E$*

$$\alpha_{x(\cdot)}(t) = g\left(x(t), S_{x(\cdot)}(t)\right). \tag{2.1}$$

Equivalently, the model can be written in the form

$$\alpha_{x(\cdot)}(t) = g_1\left(x(t), A_{x(\cdot)}(t)\right). \tag{2.2}$$

with $g_1(x,s) = g(x, exp\{-s\})$.

On sets of constant stresses the GS model is not a model at all: it always holds. It is seen from the following proposition.

Proposition 2.1. *If the hazard rates $\alpha_x(t) > 0$, $t > 0$ exist on a set of constant stresses E_1 then the GS model holds on E_1.*

Proof. For all $x \in E_1$ we have:

$$\alpha_x(t) = \alpha_x(A_x^{-1}(A_x(t))) = g_1\left(x, A_x(t)\right),$$

with $g_1(x,s) = \alpha_x(A_x^{-1}(s))$.

Thus, the GS model does not give any relations between the hazard rates (or the survival functions) under different constant stresses. This model only shows the influence of stress variability in time on survival and gives the rule of the hazard rate (or survival) function construction under any time-varying stress from the hazard rate (or survival) functions under different constant stresses. It is seen from the following proposition.

Proposition 2.2. *If the GS model holds on a set* $E \supset E_1$ *of stresses* $x(\cdot)$: $\mathbf{R}^+ \to E_1$, *then for all* $x(\cdot) \in E$

$$\alpha_{x(\cdot)}(t) = \alpha_{x_t} \left(A_{x_t}^{-1}(A_{x(\cdot)}(t)) \right), \qquad (2.3)$$

where x_t *is a constant stress equal to the value of the time-varying stress* $x(\cdot)$ *at the moment* t.

Proof. If the GS model holds on a set $E \supset E_1$ then the formula (2.2) implies that for all $x \in E_1$

$$g_1(x, s) = g_1 \left\{ x, A_x(A_x^{-1}(s)) \right\} = \alpha_x(A_x^{-1}(s)).$$

Thus,

$$\alpha_{x(\cdot)}(t) = g_1 \left\{ x(t), A_{x(\cdot)}(t) \right\} = \alpha_{x_t} \left\{ A_{x_t}^{-1}(A_x(\cdot)(t)) \right\}.$$

The fact that the GS model does not give relations between the survival under different constant stresses is a cause of non-applicability of this model for estimation of reliability under the design (usual) stress from accelerated experiments. On the other hand, restrictions of this model when not only the rule (2.3) but also some relations between survival under different constant stressses are assumed, can be considered. These narrower models can be formulated by using models for constant stressses and the rule (2.3). For example, it will be shown later that the well known and mostly used accelerated failure time model for time-varying stresses is a restriction of the GS model when the survival functions under constant stresses differ only in scale.

2.2.2 GS model for step-stresses

The mostly used time-varying stresses in accelerated life testing (ALT) are step-stresses: units are placed on test at an initial low stress and if they do not fail in a predetermined time t_1, the stress is increased. If they do not fail in a predetermined time $t_2 > t_1$, the stress is increased once more, and so on. Thus step-stresses have the form

$$x(u) = \begin{cases} x_1, & 0 \le u < t_1, \\ x_2, & t_1 \le u < t_2, \\ \cdots & \cdots \\ x_m, & t_{m-1} \le u < t_m, \end{cases} \qquad (2.4)$$

where $x_1, \cdots, x_m$ are constant stresses. If $m = 1$, a step-stress is called *simple*.

Sets of step-stresses of the form (2.4) will be denoted by E_m.

Let us consider the meaning of the rule (2.3) for step-stresses.

Let E_1 be a set of constant stresses and E_2 be a set of simple step-stresses of the form

$$x(u) = \begin{cases} x_1, & 0 \le u < t_1, \\ x_2, & u \ge t_1, \end{cases} \tag{2.5}$$

where $x_1, x_2 \in E_1$.

In the GS model the survival function under the simple (and general) step-stress is obtained from the survival functions under constant stresses by the *rule of time-shift*.

Proposition 2.3. *If the GS model holds on E_2 then the survival function and the hazard rate under the stress $x(\cdot) \in E_2$ satisfy the equalities*

$$S_{x(\cdot)}(t) = \begin{cases} S_{x_1}(t), & 0 \le t < t_1, \\ S_{x_2}(t - t_1 + t_1^*), & t \ge t_1, \end{cases} \tag{2.6}$$

and

$$\alpha_{x(\cdot)}(t) = \begin{cases} \alpha_{x_1}(t), & 0 \le t < t_1, \\ \alpha_{x_2}(t - t_1 + t_1^*), & t \ge t_1, \end{cases} \tag{2.7}$$

respectively; the moment t_1^* is determined by the equality $S_{x_1}(t_1) = S_{x_2}(t_1^*)$.

Proof. Set

$$a = A_{x_1}(t_1) = A_{x(\cdot)}(t_1) = A_{x_2}(t_1^*). \tag{2.8}$$

The equality (2.2) implies that the cumulative hazard satisfies the integral equation

$$A_{x(\cdot)}(t) = \int_0^t g\left(x(u), A_{x(\cdot)}(u)\right) du. \tag{2.9}$$

The equalities (2.8)-(2.9) imply that for all $t \ge t_1$

$$A_{x(\cdot)}(t) = a + \int_{t_1}^t g\left(x_2, A_{x(\cdot)}(u)\right) du$$

and

$$A_{x_2}(t - t_1 + t_1^*) = a + \int_{t_1}^t g\left(x_2, A_{x_2}(u - t_1 + t_1^*)\right) du.$$

So for all $t \ge t_1$ the functions $A_{x(\cdot)}(t)$ and $A_{x_2}(t - t_1 + t_1^*)$ satisfy the integral equation

$$h(t) = a + \int_{t_1}^t g\left(x_2, h(u)\right) du$$

with the initial condition $h(t_1) = a$. The solution of this equation is unique, therefore we have

$$A_{x(\cdot)}(t) = A_{x_2}(t - t_1 + t_1^*), \quad \text{for all} \quad t \ge t_1.$$

It implies the equalities (2.6) and (2.7).

Corollary 2.1 *Under conditions of Proposition 2.3 for all $s \ge 0$*

$$\alpha_{x(\cdot)}(t_1 + s) = \alpha_{x_2}(t_1^* + s).$$

It is the model of Sedyakin.

Let us consider a set E_m of more general stepwise stresses of the form (2.4). Set $t_0 = 0$. We shall show that the rule of time-shift holds and for the general step-stress.

Proposition 2.4. *If the GS model holds on E_m then the survival function $S_{x(\cdot)}(t)$ satisfies the equalities:*

$$S_{x(\cdot)}(t) = S_{x_i}(t - t_{i-1} + t^*_{i-1}), \quad \text{if} \quad t \in [t_{i-1}, t_i), \ (i = 1, 2, \ldots, m), \quad (2.10)$$

where t^*_i satisfy the equations

$$S_{x_1}(t_1) = S_{x_2}(t^*_1), \ldots, S_{x_i}(t_i - t_{i-1} + t^*_{i-1}) = S_{x_{i+1}}(t^*_i), \ (i = 1, \ldots, m-1). \quad (2.11)$$

Proof. Proposition 2.3 implies that Proposition 2.4 holds for $m = 2$, i.e. we have

$$A_{x(\cdot)}(t) = A_{x_2}(t - t_1 + t^*_1), \quad \text{for all} \quad t \in [t_1, t_2),$$

where t^*_1 verifies the equality $A_{x_1}(t_1) = A_{x_2}(t^*_1)$.

Suppose that (2.10) holds for $m = j - 1$. Then

$$A_{x(\cdot)}(t) = A_{x_i}(t - t_{i-1} + t^*_{i-1}) \text{ if } t \in [t_{i-1}, t_i), \ (i = 1, \cdots, j-1), \quad (2.12)$$

where

$$A_{x_1}(t_1) = A_{x_2}(t^*_1), \ldots, A_{x_i}(t_i - t_{i-1} + t^*_{i-1}) = A_{x_{i+1}}(t^*_i), (t_0 = 0, i = 1, \ldots, j-2).$$

We shall prove that (2.10) holds for $m = j$. Continuity of the functions $A_{x(\cdot)}(t)$ and $A_{x_2}(t)$, and the equalities (2.12) imply

$$A_{x(\cdot)}(t_{j-1}) = A_{x_{j-1}}(t_{j-1} - t_{j-2} + t^*_{j-2}).$$

So the equation (2.2) implies for all $t \in [t_{j-1}, t_j)$

$$A_{x(\cdot)}(t) = A_{x(\cdot)}(t_{j-1}) + \int_{t_{j-1}}^{t} g\left(x_j, A_{x(\cdot)}(u)\right) du =$$

$$A_{x_{j-1}}(t_{j-1} - t_{j-2} + t^*_{j-2}) + \int_{t_{j-1}}^{t} g\left(x_j, A_{x(\cdot)}(u)\right) du. \quad (2.13)$$

The definition of t^*_{j-1}, given in (2.11), and the equation (2.9) imply for all $t \in [t_1, t_2)$

$$A_{x_j}(t - t_{j-1} + t^*_{j-1}) = A_{x_j}(t^*_{j-1}) + \int_{t^*_{j-1}}^{t - t_{j-1} + t^*_{j-1}} g\left(x_j, A_{x_j}(u)\right) du =$$

$$A_{x_{j-1}}(t_{j-1} - t_{j-2} + t^*_{j-2}) + \int_{t_{j-1}}^{t} g\left(x_j, A_{x_2}(u - t_{j-1} + t^*_{j-1})\right) du. \quad (2.14)$$

The equalities (2.13) and (2.14) imply that the functions $A_{x(\cdot)}(t)$ and $A_{x_j}(t - t_{j-1} + t^*_{j-1})$ satisfy the integral equation

$$h(t) = a + \int_{t_{j-1}}^{t} g\left(x_j, h(u)\right) du \quad \text{for all} \quad t \in [t_{j-1}, t_j)$$

with the initial condition $h(t_{j-1}) = b = A_{x_{j-1}}(t_{j-1} - t_{j-2} + t^*_{j-2})$. The solution of this equation is unique, therefore for all $t \in [t_{j-1}, t_j)$ we have

$$A_{x(\cdot)}(t) = A_{x_j}(t - t_{j-1} + t^*_{j-1}).$$

In the literature on ALT (see Nelson (1990)) the model (2.10) is also called the *basic cumulative exposure model*.

In terms of graphs of the cumulative hazard rate functions $A_{x(\cdot)}(t)$ (thick curve) and $A_{x_i}(t)$ ($m = 3, i = 1, 2, 3$) the result of the proposition 2.2 is illustrated by the Figure 2.1.

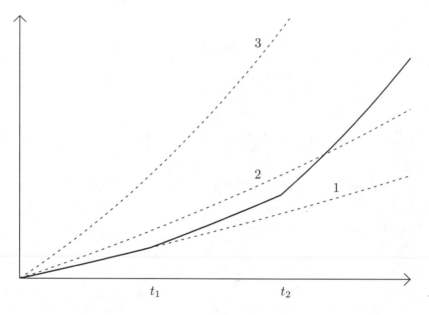

Figure 2.1.

N. Sedyakin called his model the physical principle in reliability meaning that this model is very wide. Nevertheless, this model and its generalization can be not appropriate in situations of periodic and quick change of the stress level or when switch-up's of the stress from one level to the another can imply failures or shorten the life. Generalizations will be considered later.

2.3 Accelerated failure time model

2.3.1 Definition of the model for constant stresses

Suppose that under different constant stresses the survival functions differ only in scale: for any $x \in E_1$

$$S_x(t) = G\{r(x)\,t\}, \tag{2.15}$$

where the survival function G does not depend on x.

Applicability of this model in accelerated life testing was first noted by Pieruschka (1961). It is the most simple and the most used model in FTR data analysis and ALT.

Under the AFT model the distribution of the random variable

$$R = r(x)T_x$$

does not depend on $x \in E_1$ and its survival function is G. Denote by m, σ^2 and t_p the mean, the variance, and the p-quantile of R, respectively.

The AFT model implies

$$\mathbf{E}(T_x) = m/r(x), \quad \mathbf{Var}(T_x) = \sigma^2/r^2(x), \quad t_x(p) = t_p/r(x),$$

where $t_x(p)$ is the p-quantile of T_x.

The coefficient of variation

$$\frac{\mathbf{E}(T_x)}{\sqrt{\mathbf{Var}(T_x)}} = \frac{m}{\sigma}$$

does not depend on x.

If the AFT model holds on E_1 and the survival distribution under any $x \subset E_1$ belongs to any of the classes given in Chapter 1 then the survival distribution under any other $y \in E_1$ also belongs to that class. Only the scale parameter changes.

The survival functions under any $x_1, x_2 \in E_1$ are related in the following way:

$$S_{x_2}(t) = S_{x_1}\{\rho(x_1, x_2)\, t\},$$

where $\rho(x_1, x_2) = r(x_2)/r(x_1)$.

Set $\varepsilon = \ln\{r(x)\} + \ln\{T_x\}$, $a(x) = -\ln\{r(x)\}$. Then

$$\ln\{T_x\} = a(x) + \varepsilon.$$

The distribution of the random variable ε does not depend on x. The last equality implies

$$\mathbf{Var}(\ln T_x) = \mathbf{Var}(\varepsilon).$$

The variance of the random variable $\ln\{T_x\}$ does not depend on x.

2.3.2 Definition of the model for time-varying stresses

The model (2.15) was generalized to the case of time-varying stresses by Bagdonavičius (1978) supposing that the GS model also holds, i.e. the hazard rates under time-varying stresses were obtained from the hazard rates under constant stresses by the rule (2.3). It is seen from the following proposition.

Proposition 2.5. *The GS model with the survival functions (2.15) on E_1 holds on $E \supset E_1$ if and only if there exist on E a positive function r and on $[0, \infty)$ a positive function q such that for all $x(\cdot) \in E$*

$$\alpha_{x(\cdot)}(t) = r\{x(t)\}\, q\{S_{x(\cdot)}(t))\}. \tag{2.16}$$

Proof. Necessity: Denote by x_t the constant stress equal to the value of the time-varying stress $x(\cdot)$ at the moment t. The formula (2.15) implies that

$$\alpha_{x_t}(s) = \alpha\{r(x_t)s\}\, r(x_t) = \alpha\{r(x(t))s\}\, r(x(t)), \quad A_{x_t}^{-1}(s) = \frac{1}{r(x(t))} H(e^{-s}),$$

where $\alpha = -G'/G$, $H = G^{-1}$. The formula (2.3) can be written in the form

$$\alpha_{x(\cdot)}(t) = r\{x(t)\}\, q\{S_{x(\cdot)}(t))\},$$

where $q(p) = \alpha\{H(p)\}$.

Sufficiency: The model (2.16) is the particular case of the GS model. For $x \in E_1$ it implies

$$\alpha_x(t) = r(x)\, q\{S_x(t))\}$$

or

$$-\frac{S_x'(t)}{S_x(t))\, q\{S_x(t))\}} = r(x).$$

Hence

$$S_x(t) = G\{r(x)\, t\},$$

with

$$G = H^{-1}, \quad H(p) = \int_0^p \frac{du}{uq(u)}.$$

So we have a restriction of the GS model with the survival functions (2.15) on E_1. Proposition 2.5 suggests the following model.

Definition 2.2. *The accelerated failure time (AFT) model holds on E if there exists on E a positive function r and on $[0, \infty)$ a positive function q such that for all $x(\cdot) \in E$ the formula (2.16) holds.*

Under the AFT model the hazard rate $\alpha_{x(\cdot)}(t)$ at any moment t is proportional to a function of the stress applied at this moment and to a function of the probability of survival until t under $x(\cdot)$.

Let us find the expression of the survival function under time-varying stresses.

Proposition 2.6. *Suppose that the integral*

$$\int_0^x \frac{dv}{q(v)} \tag{2.17}$$

converges for all $x \geq 0$.

The AFT model holds on a set of stresses E if and only if there exists a survival function G such that for all $x(\cdot) \in E$

$$S_{x(\cdot)}(t) = G\left(\int_0^t r\{x(u)\}\, du\right). \tag{2.18}$$

Proof. The equation (2.16) is equivalent to the integral equation

$$\int_0^{S_{x(\cdot)}(t)} \frac{dv}{vq(v)} = \int_0^t r\{x(u)\}\, du. \tag{2.19}$$

The result follows immediately. The AFT model in the form (2.18) was given by Cox and Oakes (1984).

2.3.3 AFT model for step-stresses

Let us find the form of the survival functions under simple step-stresses.

Proposition 2.7. *If the AFT model holds on E_2 then the survival function under any stress $x(\cdot) \in E_2$ of the form (2.5) verifies the equality*

$$S_{x(\cdot)}(t) = \begin{cases} S_{x_1}(t), & 0 \leq \tau < t_1, \\ S_{x_2}(t - t_1 + t_1^*), & \tau \geq t_1, \end{cases} \tag{2.20}$$

where

$$t_1^* = \frac{r(x_1)}{r(x_2)} t_1. \tag{2.21}$$

Proof. The equality (2.15) implies

$$S_{x_1}(t) = S_{x_2}\left(\frac{r(x_1)}{r(x_2)} t\right), \tag{2.22}$$

therefore the moment t_1^*, given in Proposition 2.3, is defined by (2.21). $\square$

If the AFT model holds on E_m, the equality (2.22) implies that the moment t_i^*, defined by (2.11), has the form

$$t_i^* = \frac{1}{r(x_{i+1})} \sum_{j=1}^{i} r(x_j)(t_j - t_{j-1}). \tag{2.23}$$

Proposition 2.8. *If the AFT model holds on E_m then the survival function $S_{x(\cdot)}(t)$ verifies the equalities:*

$$S_{x(\cdot)}(t) = G\left\{ \sum_{j=1}^{i-1} r(x_j)(t_j - t_{j-1}) + r(x_i)(t - t_{i-1}) \right\} =$$

$$S_{x_i}\left\{ t - t_{i-1} + \frac{1}{r(x_i)} \sum_{j=1}^{i-1} r(x_j)(t_j - t_{j-1}) \right\}, \quad t \in [t_{i-1}, t_i), (i = 1, 2, ..., m). \tag{2.24}$$

Proof. The first equality is implied by the formula (2.18), the second by the formulas (2.10) and (2.22).

2.3.4 Relations between the means and the quantiles

Suppose that $x(\cdot)$ is a time-varying stress. Denote by $t_{x(\cdot)}(p)$ the p-quantile of the random variable $T_{x(\cdot)}$, and by $x_\tau = x(\tau)\mathbf{1}_{\{t \geq 0\}}$ a constant stress equal to the value of time-varying stress $x(\cdot)$ at the moment τ.

Proposition 2.9. *Suppose that the AFT model holds on E and $x(\cdot)$, $x_t \in E$ for all $t \geq 0$. Then*

$$\int_0^{t_{x(\cdot)}(p)} \frac{d\tau}{t_{x_\tau}(p)} = 1. \tag{2.25}$$

If the means $\mathbf{E}(T_{x(\cdot)})$, $\mathbf{E}(T_{x_\tau})$ exist and are positive then

$$\mathbf{E}\left(\int_0^{T_{x(\cdot)}} \frac{d\tau}{\mathbf{E}(T_{x_\tau})}\right) = 1. \tag{2.26}$$

Proof. If the AFT model holds, the equality (2.18) implies that for any $x(\cdot) \in E$ the cumulative distribution function G of the random variable

$$\int_0^{T_{x(\cdot)}} r\{x(s)\}\, ds \tag{2.27}$$

does not depend on $x(\cdot)$. Denote by m the mean of this random variable. For the constant in time stress x_τ holds

$$\int_0^{T_{x_\tau}} r(x_\tau)\, ds = r\{x_\tau\} T_{x_\tau} = r\{x(\tau)\} T_{x_\tau}.$$

Taking the expectations of both sides we get

$$m = r\{x(\tau)\}\, \mathbf{E}(T_{x_\tau}). \tag{2.28}$$

The equalities (2.27) and (2.28) imply

$$m = \mathbf{E}\left(\int_0^{T_{x(\cdot)}} r\{x(\tau)\}\, d\tau\right) = \mathbf{E}\left(\int_0^{T_{x(\cdot)}} \frac{m}{\mathbf{E}(T_{x_\tau})}\, d\tau\right)$$

$$= m\, \mathbf{E}\left(\int_0^{T_{x(\cdot)}} \frac{d\tau}{\mathbf{E}(T_{x_\tau})}\right),$$

and the equality (2.26) is obtained. The model (2.26) is the *model of Miner* (1945).

Denote by $t(p)$ the p-quantile of the random variable (2.27).

If τ is fixed, the equality (2.15) implies

$$t(p) = r\{x(\tau)\}\, t_{x_\tau}(p). \tag{2.29}$$

We have

$$p = \mathbf{P}\{T_{x(\cdot)} \leq t_{x(\cdot)}(p)\} = \mathbf{P}\left\{\int_0^{T_{x(\cdot)}} r\{x(\tau)\}\, d\tau \leq \int_0^{t_{x(\cdot)}(p)} r\{x(\tau)\}\, d\tau\right\}.$$

It implies

$$t(p) = \int_0^{t_{x(\cdot)}(p)} r\{x(\tau)\} d\tau.$$

Using the last equality and the equality (2.29), the equality (2.25) is obtained.

Corollary 2.2. *For the stress of the form (2.4) the formula (2.26) implies*

$$\sum_{k=1}^{m} \frac{\mathbf{E}(T_k)}{\mathbf{E}(T_{x_k})} = 1, \tag{2.30}$$

where

$$T_k = \begin{cases} 0, & T_{x(\cdot)} < t_{k-1}, \\ T_{x(\cdot)} - t_{k-1}, & t_{k-1} \leq T_{x(\cdot)} < t_k, \\ t_k - t_{k-1}, & T_{x(\cdot)} \geq t_k, \end{cases}$$

is the life in the interval $[t_{k-1}, t_k)$ for the unit tested under the stress $x(\cdot)$.

The formula (2.25) implies that for $t_{x(\cdot)}(p) \in [t_{k-1}, t_k)$ the following equality holds:

$$\sum_{i=1}^{k-1} \frac{t_i - t_{i-1}}{t_{x_i}(p)} + \frac{t_{x(\cdot)}(p) - t_{k-1}}{t_{x_k}(p)} = 1. \tag{2.31}$$

The model (2.31) is the model of Peshes-Stepanova (see, Kartashov, 1979). So all of the models (2.18),(2.24),(2.25), (2.30), (2.31) are implied by the AFT model and illustrate various properties of this model.

In the case $m = 2$, the formula (2.30) can be written in the form

$$\frac{\mathbf{E}(T_1)}{\mathbf{E}(T_{x_1})} + \frac{\mathbf{E}(T_2)}{\mathbf{E}(T_{x_2})} = 1, \tag{2.32}$$

and the formula (2.31) can be written in the form

$$\frac{t_1}{t_{x_1}(p)} + \frac{t_{x(\cdot)}(p) - t_1}{t_{x_2}(p)} = 1. \tag{2.33}$$

So

$$\mathbf{E}(T_{x_1}) = \frac{\mathbf{E}(T_1)}{1 - \frac{\mathbf{E}(T_2)}{\mathbf{E}(T_{x_2})}}, \tag{2.34}$$

and

$$t_{x_1}(p) = \frac{t_1}{1 - \frac{t_{x(\cdot)}(p) - t_1}{t_{x_2}(p)}}, \quad \text{if } t_{x(\cdot)}(p) \geq t_1. \tag{2.35}$$

Thus, if the AFT model holds on E_2 then $\mathbf{E}(T_1)$, $\mathbf{E}(T_2)$ and $\mathbf{E}(T_{x_2})$ determine $\mathbf{E}(T_{x_1})$, and $t_{x(\cdot)}(p)$ and $t_{x_2}(p)$ determine $t_{x_1}(p)$.

2.4 Proportional hazards model

2.4.1 Definition of the model for constant stresses

In survival analysis the most widely used model describing the influence of covariates on the lifetime distribution is the *proportional hazards* (PH) or Cox model, introduced by D.Cox (1972).

Suppose that under different constant stresses $x \in E_1$ the hazard rates are proportional to a baseline hazard rate:

$$\alpha_x(t) = r(x)\,\alpha_0(t). \qquad (2.36)$$

For $x \in E_1$ the survival functions have the form

$$S_x(t) = S_0^{r(x)}(t) = \exp\{-r(x)A_0(t)\}, \qquad (2.37)$$

where

$$S_0(t) = \exp\left\{-\int_0^t \alpha_0(u)du\right\}, \quad A_0(t) = \int_0^t \alpha_0(u)du = -\ln S_0(t).$$

For any $x_0 \in E_1$ the PH model implies

$$\alpha_x(t) = \rho(x_0, x)\,\alpha_{x_0}(t), \quad S_x(t) = S_{x_0}^{\rho(x_0,x)}(t), \qquad (2.38)$$

where

$$\rho(x_0, x) = r(x)/r(x_0). \qquad (2.39)$$

2.4.2 Definition of the model for time-varying stresses

In the statistical literature the following formal generalization of the PH model to the case of time-varying stresses is used.

Definition 2.3. *The proportional hazards (PH) model holds on a set of stresses E if for all $x(\cdot) \in E$*

$$\alpha_{x(\cdot)}(t) = r\{x(t)\}\,\alpha_0(t), \qquad (2.40)$$

This definition implies that

$$A_{x(\cdot)}(t) = \int_0^t r\{x(u)\}dA_0(u). \qquad (2.41)$$

In terms of survival functions the PH model is written :

$$S_{x(\cdot)}(t) = \exp\left\{-\int_0^t r\{x(u)\}dA_0(u)\right\}. \qquad (2.42)$$

The model (2.40) is not natural when units are aging under constant stresses. Indeed, denote by x_t constant in time stress equal to the value of time-varying stress $x(\cdot)$ at the moment t. Then

$$\alpha_{x_t}(t) = r\{x(t)\}\alpha_0(t),$$

which implies

$$\alpha_{x(\cdot)}(t) = \alpha_{x_t}(t). \qquad (2.43)$$

For any t the hazard rate under the time-varying stress $x(\cdot)$ at the moment t does not depend on the values of the stress $x(\cdot)$ before the moment t but only on the value of it at this moment. It is not natural when the hazard rates are not constant under constant stresses, i.e. when times-to-failure are not exponential under constant stresses.

Nevertheless, the PH model with time-varying stresses is very useful because the PH model with constant or time-varying stresses and time-varying coefficients (see Sections 2.8.2 and 7.3.5) can be written as the usual PH model with constant coefficients and time-varying "explanatory variables".

Note also that the PH model can be considered as a conditional model when heterogeneity of units is observed. For example, if the AFT model with the exponential distribution under constant stresses holds on a set $E = E^{(1)} \times E^{(2)}$ of stresses of the form $(x^{(1)}(\cdot), x^{(2)}(\cdot))$:

$$\alpha_{x(\cdot)}(t) = r_1\{x_1(t)\}\, r_2\{x_2(t)\},$$

then conditionally, given a fixed time-varying component $x^{(2)}(\cdot)$, the PH model holds on the set $E^{(1)}$:

$$\alpha_{x^{(1)}(\cdot)}(t) = r_1\{x_1(t)\}\, \alpha_0(t),$$

where $\alpha_0(t) = r_2\{x_2(t)\}$. Note that even for constant $x^{(1)}$ the conditional distribution of $T_{x^{(1)}}$ given $x^{(2)}(\cdot)$ is not necessarily exponential.

A natural generalization of the PH model for time-varying stresses is a restriction of the GS model coinciding with the PH model on sets of constant stresses.

Proposition 2.10. *The GS model with the survival functions (2.37) on E_1 holds on $E \supset E_1$ iff for all $x(\cdot) \in E$*

$$\alpha_{x(\cdot)}(t) = r\{x(t)\}\, \alpha_0\left\{ A_0^{-1}\left(\frac{A_{x(\cdot)}(t)}{r\{x(t)\}} \right) \right\}. \tag{2.44}$$

Proof. Necessity. The formula (2.36) implies that

$$\alpha_{x_t}(s) = r(x_t)\,\alpha_0(s) = r\{x(t)\}\,\alpha_0(s), \quad A_{x_t}^{-1}(s) = A_0^{-1}\left\{ \frac{s}{r(x(t))} \right\}.$$

The formula (2.3) can be written in the form (2.43)

Sufficiency. The model (2.44) is the particular case of the GS model. For $x \in E_1$ it implies

$$\alpha_x(t) = r(x)\,\alpha_0\left\{ A_0^{-1}\left(\frac{A_x(t)}{r(x)} \right) \right\}$$

or

$$\frac{\partial}{\partial t} A_0^{-1}\left(\frac{A_x(t)}{r(x)} \right) = 1.$$

Hence

$$A_x(t) = r(x)A_0(t), \quad \alpha_x(t) = r(x)\,\alpha_0(t).$$

So we have a restriction of the GS model with the survival functions (2.37) on E_1. $\square$

Proposition 2.10 suggests the following model.

Definition 2.4. *The modified proportional hazards (MPH) model holds on*

E if there exists on E a positive function r and on $(0,\infty)$ a positive function α_0 such that for all $x(\cdot) \in E$ the equality (2.44) with

$$A_0(t) = \int_0^t \alpha_0(u)du$$

holds.

Let us consider the PH and MPH models for step-stresses.

2.4.3 PH and MPH models for simple step-stresses

Let us consider the expressions of the hazard rate, cumulative hazard, and survival functions under the PH model for simple step-stresses $x(\cdot) \in E_2$.
   The definition of the PH model implies the following result.

Proposition 2.11. *If the PH model holds on E_2 then for any $x(\cdot) \in E_2$*

$$\alpha_{x(\cdot)}(t) = \begin{cases} \alpha_{x_1}(t), & 0 \leq t < t_1, \\ \alpha_{x_2}(t), & t \geq t_1 \end{cases} = \begin{cases} r(x_1)\,\alpha_0(t), & 0 \leq t < t_1, \\ r(x_2)\,\alpha_0(t), & t \geq t_1, \end{cases}$$

$$S_{x(\cdot)}(t) = \begin{cases} S_{x_1}(t)^{r(x_1)}, & 0 \leq t < t_1, \\ S_{x_1}(t_1)\frac{S_{x_2}(t)}{S_{x_2}(t_1)}, & t \geq t_1, \end{cases} =$$

$$= \begin{cases} S_0(t)^{r(x_1)}, & 0 \leq t < t_1, \\ S_0(t_1)^{r(x_1)}\left(\frac{S_0(t)}{S_0(t_1)}\right)^{r(x_2)}, & t \geq t_1. \end{cases} \tag{2.45}$$

The proposition implies for any $x_0 \in E_1$

$$\alpha_{x(\cdot)}(t) = \begin{cases} \rho(x_0,x_1)\,\alpha_{x_0}(t), & 0 \leq t < t_1, \\ \rho(x_0,x_2)\,\alpha_{x_0}(t), & t \geq t_1. \end{cases}$$

$$S_{x(\cdot)}(t) = \begin{cases} S_{x_0}^{\rho(x_0,x_1)}(t), & 0 \leq t < t_1, \\ S_{x_0}^{\rho(x_0,x_1)}(t_1)\left(\frac{S_{x_0}(t)}{S_{x_0}(t_1)}\right)^{\rho(x_0,x_2)}, & t \geq t_1. \end{cases} \tag{2.46}$$

Taking $x_0 = x_1$ we have

$$\alpha_{x(\cdot)}(t) = \begin{cases} \alpha_{x_1}(t), & 0 \leq t < t_1, \\ \rho(x_1,x_2)\,\alpha_{x_1}(t), & t \geq t_1. \end{cases} \tag{2.47}$$

$$S_{x(\cdot)}(t) = \begin{cases} S_{x_1}(t), & 0 \leq t < t_1, \\ S_{x_1}(t_1)\left(\frac{S_{x_1}(t)}{S_{x_1}(t_1)}\right)^{\rho(x_1,x_2)}, & t \geq t_1. \end{cases}$$

The PH model for simple step-stresses of the form (2.47) is called the *tampered failure rate* (TFR) model (Bhattacharyya & Stoejoeti (1989)). The remark concerning applicability of the PH hazards model for time-varying stresses holds and for TFR model.
   Let us consider the MPH model.

Proposition 2.12. *If the MPH model holds on a set of simple step-stresses E_2 then the survival function under stress $x(\cdot) \in E_2$ is*

$$S_{x(\cdot)}(t) = \begin{cases} S_{x_1}(t), & 0 \leq t < t_1, \\ S_{x_2}(t - t_1 + t_1^*), & t \geq t_1, \end{cases} \qquad (2.48)$$

where

$$t_1^* = S_{x_1}^{-1}\left((S_{x_1}(t_1))^{\rho(x_2,x_1)} \right) = S_{x_2}^{-1}\left((S_{x_2}(t_1))^{\rho(x_2,x_1)} \right). \qquad (2.49)$$

Proof. The equality (2.37) implies that

$$S_{x_2}(t) = (S_{x_1}(t))^{\rho(x_1,x_2)}.$$

Thus the moment t_1^* in Proposition 2.3, is defined by (2.49).

2.4.4 PH and MPH models for general step-stresses

If $x \in E_m$ is a general step-stress, then the PH model can be written in the following forms: for any $t \in [t_{i-1}, t_i)$

$$\alpha_{x(\cdot)}(t) = r(x_i)\,\alpha_0(t), \ (t_0 = 0, i = 1, ..., m), \qquad (2.50)$$

$$S_{x(\cdot)}(t) = \prod_{j=1}^{i-1} \left(\frac{S_0(t_j)}{S_0(t_{j-1})} \right)^{r(x_j)} \left(\frac{S_0(t)}{S_0(t_{i-1})} \right)^{r(x_i)}. \qquad (2.51)$$

For any $x_0 \in E_1$ and $t \in [t_{i-1}, t_i)$

$$\alpha_{x(\cdot)}(t) = \rho(x_0, x_i)\,\alpha_{x_0}(t),$$

$$S_{x(\cdot)}(t) = \prod_{j=1}^{i-1} \left(\frac{S_{x_0}(t_j)}{S_{x_0}(t_{j-1})} \right)^{\rho(x_0,x_j)} \left(\frac{S_{x_0}(t)}{S_{x_0}(t_{i-1})} \right)^{\rho(x_0,x_i)}. \qquad (2.52)$$

Proposition 2.13. *If the MPH model holds on a set of step-stresses E_m then the survival function under stress $x(\cdot) \in E_m$ is: for $t \in [t_{i-1}, t_i)$*

$$S_{x(\cdot)}(t) = S_{x_i}(t - t_{i-1} + t_{i-1}^*) = S_0^{r(x_i)}(t - t_{i-1} + t_{i-1}^*) \ (i = 1, \cdots, m), \quad (2.53)$$

where t_1^* is defined by (2.48), and

$$t_i^* = S_{x_i}^{-1}\left((S_{x_i}(t_i - t_{i-1} + t_{i-1}^*))^{\rho(x_{i+1},x_i)} \right). \qquad (2.54)$$

Proof. The result of the proposition is implied by the formulas (2.10), (2.11), and (2.37).

For any $x_0 \in E_1$, not necessary equal to x_i,

$$S_{x(\cdot)}(t) = S_{x_0}^{\rho(x_0,x_i)}(t - t_{i-1} + t_{i-1}^*), \ t \in [t_{i-1}, t_i) \ (i = 1, \cdots, m), \qquad (2.55)$$

where $t_1^* = S_{x_0}^{-1}\left((S_{x_0}(t_1))^{\rho(x_2,x_1)} \right)$ and

$$t_i^* = S_{x_0}^{-1}\left((S_{x_0}(t_i - t_{i-1} + t_{i-1}^*))^{\rho(x_{i+1},x_i)} \right). \qquad (2.56)$$

2.4.5 Relations between the PH and the AFT models

When does the PH model coincide with the AFT model? The answer is given in the following two propositions. The first states that both models coincide on sets of constant stresses if and only if the lifetime distributions are Weibull. The second states that on wider sets (when a simple step-stress is added) both models coincide if and only if the lifetime distributions under constant stresses are exponential.

We suppose that if the AFT (or PH) model holds on E_1 then the set $r(E_1)$ has an interior point.

Proposition 2.14. *The PH and AFT models are equivalent on E_1 if and only if the failure-time distribution is Weibull for all $x \in E_1$*

Proof. Sufficiency. If the failure-time distribution is Weibull and the PH model holds on E_1 then for all $x \in E_1$

$$S_x(t) = e^{-(\frac{t}{\theta(x)})^{\alpha(x)}} = S_0(t)^{r(x)},$$

Taking twice the logarithm of both sides, we obtain that for all $t > 0$

$$\alpha(x)(\ln t - \ln \theta(x)) = \ln r(x) + \ln(-\ln S_0(t)).$$

The function $\ln(-\ln S_0(t))$ does not depend on x, hence $\alpha(x) = \alpha = const$ for all $x \in E_1$, which implies

$$S_x(t) = e^{-(\frac{t}{\theta(x)})^{\alpha}},$$

i.e. the AFT model holds on E_1.

If the failure-time distribution is Weibull and the AFT model holds then

$$S_x(t) = e^{-(\frac{t}{\theta(x)})^{\alpha}},$$

$$A_x(t) = (\frac{t}{\theta(x)})^{\alpha}, \quad \alpha_x(t) = r(x)\,\alpha_0(t),$$

where $r(x) = \{\theta(x)\}^{-\alpha}$ and $\alpha_0(t) = t^{\alpha}$. So the PH model holds.

Necessity. Suppose that both the PH and AFT models hold on E_1. It means that functions S_0, S_1, r and ρ exist such that for all $x \in E_1$

$$S_1(\rho(x)t) = S_0(t)^{r(x)}.$$

Taking twice the logarithm of both sides, we obtain for all $t > 0$

$$\ln\{-\ln S_1(\rho(x)t)\} = \ln r(x) + \ln(-\ln S_0(t)). \qquad (2.57)$$

Set

$$g_1(v) = \ln(-\ln S_1(e^v)), \quad g_0(v) = \ln(-\ln S_0(e^v)),$$

$$\alpha(x) = \ln \rho(x), \quad \beta(x) = \ln r(x).$$

The equality (2.57) can be written in the following way: for all $u \in \mathbf{R}$, $x \in E_1$

$$g_1(u + \alpha(x)) = \beta(x) + g_0(u).$$

The set $r(E_1)$ has an interior point, i.e. contains an interval, so the set $\rho(E_1)$ also has an interior point. Take $x_1, x_2, x_3 \in E_1$ such that

$$\rho(x_2)/\rho(x_1) \neq \rho(x_3)/\rho(x_2).$$

Then for all $i, j = 1, 2, 3$

$$g_1(u + \alpha(x_i)) - g_1(u + \alpha(x_j)) = \beta(x_i) - \beta(x_j).$$

Setting

$$k_1 = \alpha(x_2) - \alpha(x_1), \quad k_2 = \alpha(x_3) - \alpha(x_2),$$
$$l_1 = \beta(x_2) - \beta(x_1), \quad l_2 = \beta(x_3) - \beta(x_2),$$

the last equality is written in the form: for all $v \in \mathbf{R}$

$$g_1(v + k_i) = g_1(v) + l_i \ (i = 1, 2, \ k_1 \neq k_2).$$

It implies that

$$g_1(v) = av + b, \ S_1(t) = \exp\{-e^b t^a\}$$

and consequently

$$S_x(t) = \exp\{-e^b(\rho(x)t)^a\}.$$

So the lifetime distribution is Weibull for all $x \in E_1$.

Suppose that E_1 is the set of constant stresses defined in Proposition 2.14, $x_1, x_2 \in E_1$ are two fixed constant stresses and a step-stress $x_s(\cdot)$ has the form

$$x_s(\tau) = \begin{cases} x_1, & 0 \leq \tau < s, \\ x_2, & \tau \geq s, \end{cases} \tag{2.58}$$

where s is a fixed positive number.

Proposition 2.15. *Suppose that a set E includes E_1 and $x_s(\cdot)$ for some $s > 0$. The AFT and PH models are equivalent on E if and only if the time to-failure is exponential for all $x \in E_1$.*

Proof. Necessity. Suppose that both the PH and the AFT models hold on E. Proposition 2.14 implies for all $x \in E_1$

$$S_x(t) = \exp\left\{-\left(\frac{t}{\theta(x)}\right)^\alpha\right\}. \tag{2.59}$$

Set $\theta_i = \theta(x_i)$, $i = 1, 2$. Then

$$S_{x_i}(t) = \exp\left\{-\left(\frac{t}{\theta_i}\right)^\alpha\right\}, \quad \alpha_{x_i}(t) = \frac{\alpha}{\theta_i^\alpha} t^{\alpha-1}. \tag{2.60}$$

The PH model implies

$$\alpha_{x_s(\cdot)}(t) = \begin{cases} \alpha_{x_1}(t), & 0 < t < s, \\ \alpha_{x_2}(t), & t \geq s, \end{cases}$$

and for all $t > s$

$$S_{x_s(\cdot)}(t) = \exp\{-\int_0^t \alpha_{x_s(\cdot)}(u)du\} = \exp\{-\int_0^s \alpha_{x_1}(u)du - \int_s^t \alpha_{x_2}(u)du\} =$$

$$\exp\left\{-\left(\frac{s}{\theta_1}\right)^\alpha - \left(\frac{t}{\theta_2}\right)^\alpha + \left(\frac{s}{\theta_2}\right)^\alpha\right\}. \tag{2.61}$$

The AFT model implies for all $t > s$

$$S_{x_s(\cdot)}(t) = \exp\left\{-\left(\frac{s}{\theta_1} + \frac{t-s}{\theta_2}\right)^\alpha\right\} \tag{2.62}$$

The equalities (2.61) and (2.62) imply that for all $t > s$

$$\left(\frac{s}{\theta_1}\right)^\alpha + \left(\frac{t}{\theta_2}\right)^\alpha - \left(\frac{s}{\theta_2}\right)^\alpha = -\left(\frac{s}{\theta_1} + \frac{t-s}{\theta_2}\right)^\alpha. \tag{2.63}$$

If $\alpha = 1$, this equality is verified. Suppose that $\alpha \neq 1$. For all $t > s$ put

$$g(t) = \left(\frac{s}{\theta_1}\right)^\alpha + \left(\frac{t}{\theta_2}\right)^\alpha - \left(\frac{s}{\theta_2}\right)^\alpha - \left(\frac{s}{\theta_1} + \frac{t-s}{\theta_2}\right)^\alpha. \tag{2.64}$$

The derivative of $g(t)$ is

$$g'(t) = \frac{\alpha}{\theta_2^\alpha}t^{\alpha-1}\left(1 - \left(\frac{\theta_2 - \theta_1}{\theta_1}\frac{s}{t} + 1\right)^{\alpha-1}\right) \neq 0 \tag{2.65}$$

and for all $t > s$ has the same sign for fixed $\theta_1 \neq \theta_2$ and $\alpha \neq 1$. So the function g is increasing or decreasing but not constant in t which contradicts the equality (2.63). The assumption $\alpha \neq 1$ was false. So $\alpha = 1$, and the equality (2.59) implies that the lifetime distribution under any $x \in E_1$ is exponential:

$$S_x(t) = \exp\{-\frac{t}{\theta(x)}\}, \quad t \geq 0.$$

Sufficiency. Suppose that the PH model holds on E and the failure-time is exponential for all $x \in E_1$. The formula (2.41) implies that for all $x \in E_1$

$$S_x(t) = \exp\{-r(x)A_0(t)\}.$$

Exponentiality of the times-to-failure under $x \in E_1$ and the last formula imply that $A_0(t) = ct$. The constant c can be included in $r(x)$, so we have $A_0(t) = t$. The formula (2.35) implies that

$$S_{x(\cdot)}(t) = \exp\left\{-\int_0^t r\{x(u)\}du\right\},$$

i.e. the AFT model holds on E.

Suppose that the AFT model holds on E and the failure-time is exponential for all $x \in E_1$, i.e. $\alpha_x(t) = r(x)$. The formula (2.16) implies $q(u) \equiv 1$ and consequently

$$\alpha_{x(\cdot)}(t) = r(x(t)),$$

i.e. the PH model with $\alpha_0(t) = 1$ holds on E.

2.4.6 Relations between the GS and the PH models

The GS model is more general then the AFT model. When is the PH model also a GS model? It is given in the following proposition.

Proposition 2.16. *Suppose that the PH model holds on the set E including E_1 and all stresses of the form (2.31) with $s < \delta$, where δ is any positive number.*

The PH model is a GS model on E if and only if the time to-failure is exponential for all $x \in E_1$ and the PH model holds.

Proof. The PH model implies for all $s < \delta$

$$\alpha_{x_s(\cdot)}(t) = \alpha_{x_2}(t), \ t > s.$$

If the GS model also holds on E, then for all $s < \delta$

$$\alpha_{x_s(\cdot)}(t) = \alpha_{x_2}(t - s + \varphi(s)), \ t > s,$$

where

$$\varphi(s) = A_{x_2}^{-1}(A_{x_1}(s)) \tag{2.66}$$

is an increasing function. It implies that if both the GS and PH models hold on E then for all $s_1 < \delta$ and $s_2 < \delta$

$$\alpha_{x_2}(t - s_1 + \varphi(s_1)) = \alpha_{x_2}(t - s_2 + \varphi(s_2)), \ t > max(s_1, s_2).$$

Any function $\alpha_{x_2}(t) \equiv const$ verifies this. Assume that the function $\alpha_{x_2}(t)$ is not constant. Then

$$\varphi(s) - s = c = \quad const \quad \text{for all } s > 0,$$

because otherwise the function $\alpha_{x_2}(t)$ has two or more different periods. Note that $c \neq 0$, because

$$A_{x_2}(\varphi(s)) = A_{x_1}(s) \neq A_{x_2}(s).$$

The equalities

$$\lim_{s \to 0} A_{x_2}(\varphi(s)) = \lim_{s \to 0} A_{x_1}(s) = 0$$

and the monotonicity of $\varphi(s)$ imply that $\lim_{s \to 0} \varphi(s) = 0$. So exists $\delta_0 \in (0, \delta)$ such that

$$\mid \varphi(s) - s \mid < \mid c \mid, \quad \text{if } 0 < s < \delta_0.$$

It contradicts the implication that $\varphi(s) - s = c$ for any $s > 0$. It means that the assumption that $\alpha_{x_2}(t)$ is not constant was false. So $\alpha_{x_2}(t) = \alpha = const$ which implies

$$S_{x_2}(t) = e^{-\alpha t}$$

and the PH model leads for all $x \in E_1$ to

$$S_x(t) = S_0(t)^{r(x)} = e^{-r(x)t}, \tag{2.67}$$

i.e., the failure time distribution is exponential for all $x \in E_1$.

Suppose that for all $x \in E_1$ the failure-time distribution is exponential and the PH model holds. The proof of Proposition 2.15 implies that the AFT model and consequently the GS model also holds on E.

2.5 Generalized proportional hazards models

2.5.1 Introduction

The AFT, PH, and MPH models are rather restrictive.

Under the PH and MPH models lifetime distributions under constant stresses are from the narrow class (2.3) of distributions: the ratio of the hazard rates under any two different constant stresses is constant over time.

Under the AFT model the stress changes (locally, if the stress is not constant) only the scale.

We shall consider generalizations of these two models in two directions. Generalized proportional hazards models allow the ratios of hazard rates under constant stresses to be not only constant but also increasing, decreasing, and even have the cross-effects. Changing shape and scale models allow not only scale but also shape change.

2.5.2 Definitions of the generalized proportional hazards models

Let us consider models which include AFT and PH models as particular cases. This generalization was given by Bagdonavičius and Nikulin (1994), supposing that the hazard rate at any moment t is proportional not only to a function of the stress applied at this moment and to a baseline rate, but also to a function of the probability of survival until t (or, equivalently, to the cumulative hazard at t). This is formalized by the following definition.

Definition 2.5. *The first generalized proportional hazards* (GPH1) *model holds on* E *if for all* $x(\cdot) \in E$

$$\alpha_{x(\cdot)}(t) = r\{x(t)\} \, q\{A_{x(\cdot)}(t)\} \, \alpha_0(t). \qquad (2.68)$$

The particular cases of the GPH1 model are the PH model ($q(u) \equiv 1$) and the AFT model ($\alpha_0(t) \equiv \alpha_0 = const$).

Similar generalization of the GS and PH models is the following model.

Definition 2.6. *The second generalized proportional hazards* (GPH2) *model holds on* E *if for all* $x(\cdot) \in E$

$$\alpha_{x(\cdot)}(t) = u\{x(t), A_{x(\cdot)}(t)\} \, \alpha_0(t). \qquad (2.69)$$

The particular cases of the GPH2 model are the GS model ($\alpha_0(t) \equiv \alpha_0 = const$) and GPH1 model ($u(x, s) = r(x) \, q(s)$)

Models of different levels of generality can be obtained by completely specifying q, parametrizing q, or considering q as unknown.

Proposition 2.17. *The GPH1 model (2.68) holds on* E *if and only if sur-*

vival functions G and S_0 exist such that for all $x(\cdot) \in E$

$$S_{x(\cdot)}(t) = G\left\{\int_0^t r(x(\tau))dH(S_0(\tau))\right\}; \qquad (2.70)$$

here $H = G^{-1}$ is the inverse to G function.

Proof. Suppose that the GPH1 model (2.68) holds on E. Define the function $H(u)$ by the formula

$$H(u) = \int_0^{-\ln u} \frac{dv}{q(v)}.$$

Then

$$\frac{\partial H(S_{x(\cdot)}(t))}{\partial t} = \frac{1}{q\{A_{x(\cdot)}(t)\}} \alpha_{x(\cdot)}(t) = r\{x(t)\}\alpha_0(t).$$

So

$$S_{x(\cdot)}(t) = G\left\{\int_0^t r(x(\tau))dA_0(t)\right\},$$

where

$$A_0(t) = \int_0^t \alpha_0(u)du.$$

Set $S_0(u) = G(A_0(u))$. Then (2.70) holds.

Vice versa, if the equality (2.70) holds for all $x(\cdot) \in E$ then

$$\alpha_{x(\cdot)}(t) = \alpha(H(S_{x(\cdot)}(t)))r\{x(t)\}H'\{S_0(t)\}S_0'(t),$$

where $\alpha = -G'/G$. Set

$$q(u) = \alpha(H(e^{-u})), \quad \alpha_0(t) = -H'\{S_0(t)\}S_0'(t).$$

Then for all $x(\cdot) \in E$

$$\alpha_{x(\cdot)}(t) = r\{x(t)\}\,q\{A_{x(\cdot)}(t)\}\,\alpha_0(t).$$

So the GPH1 holds.

Set $A_0(u) = H(S_0(u))$. In terms of survival functions the GPH1 model is written

$$S_{x(\cdot)}(t) = G\left(\int_0^t r\{x(u)\}dA_0(u)\right), \qquad (2.71)$$

where

$$A_0(u) = \int_0^t \alpha_0(u)du.$$

The formula (2.68) implies for the GPH1 model and constant stresses $x \in E_1$

$$\alpha_x(t) = r\{x\}\,q\{A_x(t)\}\,\alpha_0(t), \quad S_x(t) = G\{r(x)H(S_0(t))\}. \qquad (2.72)$$

For any $x_1, x_2 \in E_1$

$$S_{x_2}(t) = G\{\rho(x_1, x_2)\,H(S_{x_1}(t))\}, \qquad (2.73)$$

where $\rho(x_1, x_2) = r(x_2)/r(x_1)$.

Under the GPH2 model

$$\alpha_x(t) = u\{x, A_x(t)\}\, \alpha_0(t). \tag{2.74}$$

2.5.3 Interpretation of the GPH and AFT models in terms of resource usage

Models of accelerated life can be formulated using the notion of the resource introduced by Bagdonavičius and Nikulin (1995, 1999a, 2001b).

Let Ω be a population of units and suppose that the failure-time of units under stress $x(\cdot)$ is defined by a non-negative absolutely continuous random variable $T_{x(\cdot)} = T_{x(\cdot)}(\omega), \omega \in \Omega$, with the survival function $S_{x(\cdot)}(t)$ and the cumulative distribution function $F_{x(\cdot)}(t)$. The moment of failure of a concrete item $\omega_0 \in \Omega$ is given by a nonnegative number $T_{x(\cdot)}(\omega_0)$.

The proportion $F_{x(\cdot)}(t)$ of units from Ω which fail until the moment t under the stress $x(\cdot)$ is also called the *uniform resource of population used until the moment t*. The same population of units Ω, observed under different stresses $x_1(\cdot)$ and $x_2(\cdot)$ use different resources until the same moment t if $F_{x_1(\cdot)}(t) \neq F_{x_2(\cdot)}(t)$. In sense of equality of used resource the moments t_1 and t_2 are equivalent if $F_{x_1(\cdot)}(t_1) = F_{x_2(\cdot)}(t_2)$.

The random variable

$$R^U = F_{x(\cdot)}(T_{x(\cdot)}) = 1 - S_{x(\cdot)}(T_{x(\cdot)})$$

is called the *uniform resource*.

The distribution of the random variable R^U does not depend on $x(\cdot)$ and is uniform on $[0, 1)$. The uniform resource of any concrete item $\omega_0 \in \Omega$ is $R^U(\omega_0)$. It shows the proportion of the population Ω which fails until the moment of the unit's ω_0 failure $T_{x(\cdot)}(\omega_0)$.

For any $x(\cdot)$ there exists one-to-one application between the set of values of the r.v. $T_{x(\cdot)}$ and the set of values of the r.v. R^U.

The considered definition of the resource is not unique. Take any decreasing and continuous function $H : (0, 1] \to \mathbf{R}$ such that the inverse $G = H^{-1}$ of H is a survival function. Then exists one-to-one application between the set of values of the r.v. $T_{x(\cdot)}$ and the set of values of the r.v. $R^G = H(S_{x(\cdot)}(T_{x(\cdot)}))$, too. In the case of the uniform resource $H(p) = 1 - p, p \in (0, 1]$. The distribution of the random variable R^G doesn't depend on $x(\cdot)$ and the survival function of R^G is G.

The random variable R^G is called the *G-resource* and the number

$$f^G_{x(\cdot)}(t) = H(S_{x(\cdot)}(t)),$$

is called the *G-resource used until the moment t*.

Accelerated life models can be formulated specifying the way of resource usage.

Note that all definitions of accelerated life models were formulated in terms of *exponential resource usage*, when $G(t) = e^{-t}, t \geq 0$ because the exponential resource usage rate is nothing but the hazard rate and the used resource is the cumulative hazard rate.

Let $\alpha_{x(\cdot)}(t)$ and $A_{x(\cdot)}(t)$ be the hazard rate and the cumulative hazard rate under $x(\cdot)$. The exponential resource is obtained by taking $G(t) = e^{-t}$, $t \geq 0$ and $H(p) = G^{-1}(p) = -\ln p$, so it is the random variable

$$R = A_{x(\cdot)}(T_{x(\cdot)})$$

with standard exponential distribution.

For any t the number $A_{x(\cdot)}(t) \in [0, \infty)$ is the *exponential resource used until the moment t* under stress $x(\cdot)$. The rate of exponential resource usage is the hazard rate $\alpha_{x(\cdot)}(t)$.

The GPH1 and AFT models can be formulated in terms of other resources than exponential.

Let us consider at first one particular resource. Suppose that x_0 is a fixed (for example, usual) stress and $G = S_{x_0}$. For any $x(\cdot) \in E \supset E_1$ set

$$f_{x(\cdot)}(t) = S_{x_0}^{-1}(S_{x(\cdot)}(t)).$$

Then the moment t under any stress $x(\cdot) \in E$ is equivalent to the moment $f_{x(\cdot)}(t)$ under the usual stress x_0. The survival function of the resource R is S_{x_0}.

Under the AFT model:

$$S_{x(\cdot)}(t) = S_{x_0}\left(\int_0^t r\{x(u)\}du\right),$$

the S_{x_0}-resource usage rate is

$$\frac{\partial}{\partial t}f_{x(\cdot)}(t) = r\{x(t)\}.$$

If $x \in E_1$ is constant then

$$\frac{\partial}{\partial t}f_x(t) = r(x),$$

and the resource usage rate is constant in time.

Let us consider now any survival function G, not necessarily equal to S_{x_0}.

Definition 2.7. *The generalized multiplicative (GM) model with the resource survival function G holds on E if there exist a positive function r and a survival function S_0 such that for all $x(\cdot) \in E$*

$$\frac{\partial f_{x(\cdot)}^G(t)}{\partial t} = r\{x(t)\}\frac{\partial f_0^G(t)}{\partial t},$$

where $f_0^G(t) = H(S_0(t))$. The definition implies that for all $x(\cdot) \in E$

$$S_{x(\cdot)}(t) = G\left\{\int_0^t r(x(\tau))\,dH(S_0(\tau))\right\}.$$

Proposition 2.17 implies that the GPH1 model holds on E if and only if there exists a survival function G such that the GM model holds on E.

Thus, the GPH1 (or, equivalently, the GM) model means that the G-resource usage rate at the moment t is proportional to a baseline rate with the proportionality constant depending on the value of stress applied at this moment.

2.5.4 Characterization of the GPH1 model with constant stresses

Consider the choice of the function G in the GPH1 model. It will be shown that if G is chosen from the class of survival functions of the form

$$G(t; \theta, \nu, \gamma) = G_0 \{(t/\theta)^\nu, \gamma\},$$

then for any values of θ, ν the same GPH1 model for constant stresses is defined. So we can take $\theta = \nu = 1$. It shows that the number of unknown parameters in the GPH1 models is the same (if γ is absent) as in the PH model or one complementary parameter is added (if γ is one-dimensional). The last case is the most interesting because it gives the possibility to formulate flexible models with increasing or decreasing ratios of hazards.

In what follows we skip the parameter γ in all expressions.

Let the function G be continuous and strictly decreasing on $[0, \infty[$, and set

$$G_1(u) = G((u/\theta)^\nu).$$

Let $E_1 = [x_0, x_1] \subset \mathbf{R}$ be an interval of constant in time one-dimensional stresses, $\{S_x, \ x \in [x_0, x_1]\}$ be a class of continuous survival functions, such that $S_x(t) > S_y(t)$ for all $x, y \in E_1$, $x < y$, $t > 0$,

$$H = G^{-1} :]0, 1] \to [0, \infty] \text{ and } H_1 = G_1^{-1}$$

be the inverse functions of G and G_1, respectively. If the GPH1 model with the resource survival function G holds on E_1, then the equality (2.73) implies that

$$H(S_x(t)) = \lambda(x)H(S_{x_0}(t)), \ t > 0, \ x \in [x_0, x_1], \tag{2.75}$$

where $\lambda(x) = \rho(x_0, x)$. Then

$$H_1(S_x(t)) = \lambda^{1/\nu}(x)H_1(S_{x_0}(t)), \ t > 0, \ x \in [x_0, x_1]. \tag{2.76}$$

The inverse result also takes place:

Proposition 2.24. *Assume that the function G is continuous and strictly decreasing on $[0, \infty[$ and the equality (2.75) holds. Then the equality (2.76) also holds if and only if*

$$G_1(u) = G((u/\theta)^\nu), \ u \in [0, \infty),$$

for some positive constants θ and ν.

Proof.

1) It was just shown that if the GPH1 model holds for the survival function G and $G_1(t) = G((t/\theta)^\nu)$ then the GPH1 model holds for the survival function G_1.

2) Suppose that the GPH1 model holds for the survival functions G and G_1, i.e. the equalities (2.75) and (2.76) hold. Introduce a function $D : [0, \infty[\to$

$[0, \infty[$ such that $D(u) = H_1(G(u))$, $u \in [0, \infty[$. Then $H_1(p) = D(H(p))$, $p \in {]0, 1]}$, and the relation (2.76) can be rewritten as follows:

$$D(H(S_x(t))) = \lambda^{1/\nu}(x) D(H(S_{x_0}(t))), \ t > 0, \ x \in [x_0, x_1].$$

Using (2.75) we obtain that

$$D(\lambda(x) H(S_x(t))) = \lambda^{1/\nu}(x) D(H(S_{x_0}(t))), \ t > 0, \ x \in [x_0, x_1]$$

with the initial conditions $D(0) = 0$ and $\lim_{u \to \infty} D(u) = \infty$. Setting $y = H(S_{x_0}(t))$ we obtain that

$$D(\lambda(x) y) = \lambda^{1/\nu}(x) D(y), \ y \in [0, \infty[, \ x \in [x_0, x_1],$$

or for $v = \ln y$

$$Q(\ln \lambda(x) + v) = \frac{1}{\nu} \ln(\lambda(x)) + Q(v), \ v \in \mathbf{R}, \ x \in [x_0, x_1],$$

where $Q(v) = \ln(D(e^v)))$. This equality leads to

$$Q(v) = av + b, \quad a = \frac{1}{\nu}.$$

It implies that $D(y) = \theta y^a$, where $\theta = e^b$. Consequently,

$$G(y) = G_1(D(y)) = G_1(\theta y^a) \quad \text{and} \quad G_1(u) = G((u/\theta)^\nu), \ u \in [0, \infty[.$$

This proposition implies that, in particular, the PH model is a sub-model of the GM (or GPH1) model when G is not only the standard exponential but when it is any exponential or two-parameter Weibull survival function. So sub-models of the GPH1 model can be obtained by fixing classes of resource distributions.

2.5.5 Relations with the frailty models

Let us consider relations between the GPH1 models and the *frailty models with covariates*.

The hazard rate can be influenced not only by the observable stress $x(\cdot)$ but also by a non-observable positive random covariate Z, called the *frailty variable*, see Hougaard (1995). Suppose that for all $x(\cdot) \in E$

$$\alpha_{x(\cdot)}(t | Z = z) = z \, r(x(t)) \, \alpha_0(t).$$

Then

$$S_{x(\cdot)}(t | Z = z) = exp\{-z \int_0^t r(x(\tau)) \, dA_0(\tau)\}$$

and

$$S_{x(\cdot)}(t) = \mathbf{E} \, exp\{-Z \int_0^t r(x(\tau)) \, dA_0(\tau)\} = G\{\int_0^t r(x(\tau)) dA_0(\tau)\},$$

where $G(s) = \mathbf{E}e^{-sZ}$. If we set $S_0(t) = G(A_0(t))$, then for all $x(\cdot) \in E$

$$S_{x(\cdot)}(t) = G\{\int_0^t r(x(\tau))\, dH(S_0(\tau))\},$$

where $H = G^{-1}$. We obtained that the frailty model defined by a frailty variable Z, the GM model with the survival function of the resource $G(s) = \mathbf{E}e^{-sZ}$, and the GPH1 model (cf. Proposition 2.17) with the function $q(u) = -e^u G'(H(e^{-u}))$ give the same survival function under any stress $x(\cdot) \in E$.

2.5.6 Relations with the linear transformation models

Under constant stresses the GPH1 model is related with the *linear transormation* (LT), Dabrowska & Doksum (1988), Cheng, Wei, Ying (1995).

Consider the set E_1 of constant in time stresses and let T_x denote the time-to-failure under the explanatory variable $x \in E_1$. The LT model holds on E_1 if for all $x \in E_1$

$$h(T_x) = -\beta^T x + \varepsilon,$$

where $h : [0, \infty) \to [0, \infty)$ is a strictly increasing function, and ε is a random error with distribution function Q. Under this model

$$S_x(t) = G\{e^{\beta^T x + h(t)}\} = G\{e^{\beta^T x} H(S_0(t))\},$$

where $G(t) = 1 - Q(\ln t)$, $S_0(t) = G\{e^{h(t)}\}$. Therefore, in the case of *constant in time* stresses, the frailty model defined by the frailty variable Z, the GM model with the survival function of the resource $G(s) = Ee^{-sZ}$, the GPH1 model with the function $q(u) = -e^u G'(H(e^{-u}))$, and the LT model with the distribution function $Q(x) = 1 - G(\ln x)$ of the random error ε give the same expression of survival functions.

2.5.7 The main classes of the GPH models

Particular classes of the GPH models are important for survival analysis and accelerated life testing. The numerous examples of real data show that taking two constant in time covariates, say x_1 and x_2, the ratio $\alpha_{x_2}(t)/\alpha_{x_1}(t)$ (which is constant under the PH model), can be increasing or decreasing in time and even a cross-effect of hazard rates can be observed.

Such data can be modeled by sub-models of the GPH1 or more general GPH2 model. Let us consider some of them.

GPH models with monotone ratios of hazard rates

Let us consider the choice of the function q in the GPH1 model

$$\alpha_{x(\cdot)}(t) = r\{x(t)\}\, q\{A_{x(\cdot)}(t)\}\, \alpha_0(t).$$

The purpose is to obtain models with monotone ratios of the hazard rates under constant stresses.

Suppose that $q(0) = 1$. Otherwise one can consider the functions $q_1(u) = q(u)/q(0)$ and $r_1(x) = q(0)r(x)$. We shall not consider complicated models with more then one unknown parameter in the function q. The function q being positive, the natural choice is taking one of the following functions (having power and exponential rates):

$$q(u) = (1+u)^\gamma, \quad e^{\gamma u}, \quad (1+\gamma u)^{-1} \tag{2.77}$$

where $\gamma \in \mathbb{R}$ is an unknown scalar parameter.

1) GPH_{GW} model

The most satisfying model is obtained by choosing the first from the functions (2.77) (with reparametrization $-\gamma + 1$ instead of γ):

$$q(u) = (1+u)^{-\gamma+1} \quad (\gamma > 0). \tag{2.78}$$

We have the model:

$$\alpha_{x(\cdot)}(t) = r\{x(t)\}(1 + A_{x(\cdot)}(t))^{-\gamma+1}\alpha_0(t). \tag{2.79}$$

The particular case of this model with $\gamma = 1$ is the PH model.

Under this model the ratios of the hazard rates under constant stresses may be increasing or decreasing in various rates. Indeed, take $x_1, x_2 \in E_1$ and set

$$c_0 = r(x_2)/r(x_1).$$

Suppose that $c_0 > 1$. Then

$$\alpha_{x_2}(t)/\alpha_{x_1}(t) = c_0 \left\{ \frac{1 + \gamma\, r(x_2)\, A_0(t)}{1 + \gamma\, r(x_1)\, A_0(t)} \right\}^{-1+\frac{1}{\gamma}} \rightarrow c_0^{\frac{1}{\gamma}} \quad \text{as } t \rightarrow \infty;$$

here

$$A_0(t) = \int_0^t \alpha_0(u)du.$$

The ratio $\alpha_{x_2}(t)/\alpha_{x_1}(t)$ has the following properties:

a) if $0 < \gamma < 1$, then the ratio $\alpha_{x_2}(t)/\alpha_{x_1}(t)$ increases from the value $c_0 > 1$ to the value

$$c_\infty = c_0^{\frac{1}{\gamma}} \in (c_0, \infty).$$

b) if $\gamma = 1$ (PH model), the ratio $\alpha_{x_2}(t)/\alpha_{x_1}(t)$ is constant in time.

c) if $\gamma > 1$, then the ratio $\alpha_{x_2}(t)/\alpha_{x_1}(t)$ decreases from the value $c_0 > 1$ to the value $c_\infty \in (1, c_0)$.

So this model can be used in the case when the hazard rates under different constant explanatory variables approach one another and in the case when they are going away one from another.

In Proposition 2.17 we obtained that the survival function of the resource, defined in 2.5.3, is $G = H^{-1}$ with

$$H(u) = \int_0^{-\ln u} \frac{dv}{q(v)}. \tag{2.80}$$

It implies that under the model (2.79) the survival function of the resource has the form

$$G(t) = \exp\left\{1 - (1 + \gamma t)^{\frac{1}{\gamma}}\right\}. \tag{2.81}$$

Proposition 2.24 implies that under constant stresses the model (2.79) is obtained by taking the resource with any survival function from the family of the generalized Weibull distribution:

$$G(t) = \exp\left\{1 - \left(1 + (\frac{t}{\theta})^\nu\right)^{\frac{1}{\gamma}}\right\}. \tag{2.82}$$

So the model (2.79) will be called the *GPH model with the generalized Weibull distribution of the resource*, the GPH_{GW} model.

If the failure-time $T_{x(\cdot)}$ distributions have finite supports $[0, sp_{x(\cdot)}), sp_{x(\cdot)} < \infty$ then the GPH_{GW} model can be modified, taking $\gamma < 0$ in (2.79). Finite supports are very possible in ALT: failures of units at different accelerated stresses are concentrated in intervals with different finite right limits.

If $\gamma < 0$ then the survival function of the resource is

$$G(t) = \exp\left\{1 - (1 + \gamma t)^{1/\gamma}\right\} \mathbf{1}_{[0, -1/\gamma)}(t), \tag{2.83}$$

and the right limit of the support of $\alpha_{x(\cdot)}(t)$ verifies the equality

$$\int_0^{sp_{x(\cdot)}} r\{x(u)\} dH(S_0(u)) = -1/\gamma, \tag{2.84}$$

where $S_0 = \exp\{-A_0\}$.

Similarly as in the case of $\gamma > 0$ the ratio of hazard rates $\alpha_{x_2}(t)/\alpha_{x_1}(t)$ can be increasing or decreasing in various rates on $[0, sp_{x_2(\cdot)})$. On $[sp_{x_2(\cdot)}, sp_{x_1(\cdot)})$ the hazard rates are: $\alpha_{x_2}(t) = 0$ and $\alpha_{x_1}(t) > 0$.

Take notice that for the GPH_{GW} model is a generalization of the *positive stable frailty model* (PSFM) (see Hougaard (1986), Bar-Lev and Enis (1986), Nikulin (1991)) *with explanatory variables*.

Indeed, suppose that $\gamma > 1$ and the frailty variable Z follows the positive stable distribution with the density

$$p_Z(z) = -\frac{1}{\pi z} exp\{-\alpha z + 1\} \sum_{k=1}^\infty \frac{(-1)^k}{k!} \sin(\pi\alpha k) \frac{\Gamma(\alpha k + 1)}{z^{\alpha k}}, \quad z > 0,$$

where $\alpha = 1/\gamma$ is a stable index, $0 < \alpha < 1$. The survival function of the resource

$$G(s) = \mathbf{E}e^{-sZ} = \exp\{1 - (1 + \gamma t)^{1/\gamma}\}.$$

Then

$$q(u) = (1 + u)^{-\gamma + 1}, \quad \gamma > 1.$$

We have the positive stable frailty model with explanatory variables.

2) GPH_{GLL} model

Taking the second from the functions (2.57), we have the model:

$$\alpha_{x(\cdot)}(t) = r(x(t)) e^{\gamma A_{x(\cdot)}(t)} \alpha_0(t). \qquad (2.85)$$

If $\gamma = 0$, it becomes the usual PH model. For $\gamma < 0$

$$\alpha_{x_2}(t)/\alpha_{x_1}(t) = c_0 \left\{ \frac{1 - \gamma r(x_1) A_0(t)}{1 - \gamma r(x_2) A_0(t)} \right\} \to 1 \quad \text{as} \quad t \to \infty;$$

here $c_0 = r(x_2)/r(x_1)$.

The ratio $\alpha_{x_2}(t)/\alpha_{x_1}(t)$ *decreases* from the value $c_0 > 1$ to 1.

This model is not so interesting as the GPH_{GW} model because it gives less possibilities of choice: under it the hazard rates approach one another and meet at infinity. Under the GPH_{GW} model the hazard rates may approach (but not meet) or go away one from another.

The formula (2.80) implies that for $\gamma < 0$ the survival function of the resource has the form

$$G(t) = (1 - \gamma t)^{\frac{1}{\gamma}}. \qquad (2.86)$$

Proposition 2.24 implies that under constant stresses the model (2.85) is obtained by taking the resource with any survival function from the family of the generalized loglogistic distributions:

$$G(t) = (1 - (t/\theta)^\nu)^{1/\gamma}. \qquad (2.87)$$

So the model (2.85) shall be called the *GPH model with the generalized loglogistic distribution of the resource*, the GPH_{GLL} model.

If the failure-time $T_{x(\cdot)}$ distributions have finite supports $[0, sp_{x(\cdot)}), sp_{x(\cdot)} < \infty$ then $\gamma < 0$ in (2.85). In this case the survival function of the resource is

$$G(t) = (1 - \gamma t)^{1/\gamma} \mathbf{1}_{[0,1/\gamma)}(t), \qquad (2.88)$$

and the right limit of the support of $\alpha_{x(\cdot)}(t)$ verifies the equality

$$\int_0^{sp_{x(\cdot)}} r\{x(u)\} dS_0^\gamma(u)) = -1. \qquad (2.89)$$

Take notice that the GPH_{GLL} model is a generalization of the *gamma frailty model* (GFM) (see Vaupel et al. (1979)) *with explanatory variables*.

Indeed, suppose that the frailty variable Z follows a gamma distribution with the scale parameter $\theta > 0$, the shape parameter $k > 0$, and the density

$$p_Z(z) = \frac{z^{k-1}}{\theta^k \Gamma(k)} e^{-z/\theta}, \quad z > 0.$$

The survival function of the resource

$$G(s) = \mathbf{E} e^{-sZ} = (1 + \theta t)^{-k}.$$

Set $\gamma = -1/k < 0$. Then

$$q(u) = -\frac{1}{\lambda\gamma}e^{\gamma u}, \, \gamma < 0.$$

The proportionality constant can be included in α_0 and $q(u)$ can be written in the form

$$q(u) = e^{\gamma u}, \, \gamma < 0.$$

We have the gamma frailty model with explanatory variables.

For constant in time stresses $x_1, x_2 \in E_1$ and $\gamma < 0$ we obtain the *generalized proportional odds-rate* (GPOR) *model* (Dabrowska & Doksum (1988)):

$$\frac{S_{x_2}^{-\gamma}(t) - 1}{S_{x_2}^{-\gamma}(t)} = \frac{r(x_2)}{r(x_1)} \frac{S_{x_1}^{-\gamma}(t) - 1}{S_{x_1}^{-\gamma}(t)}.$$

Let us consider the frailty model with the density $p_Z(z)$ which is the inverse Laplace transformation of the survival function

$$G(t) = (1 - \gamma t)^{1/\gamma} \, \mathbf{1}_{[0,1/\gamma)}(t).$$

Then

$$q(u) = e^{\gamma u}, \, \gamma > 0.$$

The survival function of the resource is

$$G(t) = (1 - \gamma t)^{1/\gamma}, \quad \gamma < 0.$$

For $\gamma > 0$ the support of G is $[0, 1/\gamma)$

3) Inverse gaussian frailty model with explanatory variables.

Let us consider the GPH1 model with parametrization

$$q(u) = \frac{1}{1 + \gamma u}, \quad \gamma > 0.$$

We have the model

$$\alpha_{x(\cdot)}(t) = r(x(t)) \frac{\alpha_0(t)}{1 + \gamma A_{x(\cdot)}(t)}.$$

Take notice that this model is the *inverse gaussian frailty model with covariates*.

Indeed, suppose that the frailty variable Z has the inverse gaussian distribution with the density

$$p_Z(z) = \left(\frac{\sigma}{\pi}\right)^{1/2} e^{\sqrt{4\sigma\theta}} z^{-3/2} e^{-\theta z - \frac{\sigma}{z}}, \quad z > 0.$$

Then

$$q(u) = \frac{2\sigma}{u + (4\sigma\theta)^{1/2}}.$$

Including the proportionality constant $\sqrt{\theta/\sigma}$ in α_0 and taking $\gamma = (4\sigma\theta)^{-1/2}$, we obtain $q(u) = (1 + \gamma u)^{-1}$.

Under this model

$$\alpha_{x_2}(t)/\alpha_{x_1}(t) = c_0 \frac{(1 + \gamma\, r(x_1)\, A_0(t))^{1/2} - 1}{(1 + \gamma\, r(x_2)\, A_0(t))^{1/2} - 1},$$

where $c_0 = r(x_2)/r(x_1)$. Suppose that $c_0 > 1$. The ratio $\alpha_{x_2}(t)/\alpha_{x_1}(t)$ increases from the value 1 to $\sqrt{c_0} > 1$. So the hazard rates are equal at the beginning of functionning and are going away one from another when t increases. So under the inverse gaussian frailty model the hazard rates are supposed to meet at zero. So it is not so interesting as the GPH_{GW} model.

The survival function of the resource has the form

$$G(t) = \exp\left\{\frac{1}{\gamma}(1 - \sqrt{1 + 2\gamma t}\,)\right\}.$$

Proposition 2.24 implies that under constant stresses the model (2.79) is obtained by taking the resource with any survival function from the family:

$$G(t) = \exp\left\{\frac{1}{\gamma}\left(1 - \sqrt{1 + (\frac{t}{\theta})^{\nu}}\right)\right\}.$$

GPH model with cross-effects of hazard rates

In some situations cross-effects of hazard rates may be observed. For example, if the burn-in period exists and the stress influences mainly only units with particular defects then the hazard rate under x_2 may be greater than that under x_1 at the beginning of life but to the end of the burn-in under x_2 the burn-in under x_1 is not yet finished and the hazard rate under x_2 may be smaller than that under x_1.

To obtain a cross-effect of hazard rates consider the following sub-models of GPH2:

1) First model with cross-effects Bagdonavicius and Nikulin (2000d)

$$\alpha_{x(\cdot)}(t) = r(x(t))(1 + A_{x(\cdot)}(t))^{\gamma^T x(t) + 1} \alpha_0(t). \tag{2.90}$$

Suppose that $c_0 = r(x_2)/r(x_1) > 1, (x_1, x_2 \in E_1)$, and $\gamma^T x_2 < \gamma^T x_1 < 0$. Then

$$\alpha_{x_2}(t)/\alpha_{x_1}(t) = c_0 \frac{(1 - \gamma^T x_2\, r(x_2) A_0(t))^{-1 - \frac{1}{\gamma^T x_2}}}{(1 - \gamma^T x_1\, r(x_1) A_0(t))^{-1 - \frac{1}{\gamma^T x_1}}},$$

and

$$\alpha_{x_2}(0)/\alpha_{x_1}(0) = c_0 > 1, \quad \lim_{t \to \infty} \alpha_{x_2}(t)/\alpha_{x_1}(t) = 0.$$

So we have a cross-effect of the hazard rates.

We shall denote this model by CRE1.

2) Second model with cross-effects (Devarajan and Ebrahimi (1998))

$$A_{x(\cdot)}(t) = r(x(t))(A_0(t))^{e^{\gamma^T x(t)}}.$$

The ratio

$$\alpha_{x_2}(t)/\alpha_{x_1}(t) = c(x_1, x_2)(A_0(t))^{e^{\gamma^T x_2} - e^{\gamma^T x_1}},$$

is monotone and takes values from 0 to ∞. So we have a cross-effect of the hazard rates.

We shall denote this model by CRE2.

GPH1 models with specified G.

Let us consider GM models with G specified. These models are alternative to the PH model, so rather restrictive, as the PH model is.

1) If the distribution of the resource is loglogistic, i.e.

$$G(t) = \frac{1}{1+t}\mathbf{1}_{\{t\geq 0\}}, \tag{2.91}$$

then $q(t) = e^{-t}$ and the GM model can be formulated in the following way:

$$\frac{\alpha_{x(\cdot)}(t)}{S_{x(\cdot)}(t)} = r\{x(t)\}\frac{\alpha_0(t)}{S_0(t)}. \tag{2.92}$$

If stresses are constant in time then we obtain the model

$$\frac{1}{S_x(t)} - 1 = r(x)\,(\frac{1}{S_0(t)} - 1). \tag{2.93}$$

It is the analogue of the logistic regression model which is used for analysis of dichotomous data when the probability of "success" in dependence of some factors is analyzed. The obtained model is near to the PH model when t is small.

2) If the resource is lognormal, then

$$G(t) = 1 - \Phi(\log t), \quad t > 0,$$

where Φ is the distribution function of the standard normal law. If covariates are constant in time then in terms of survival functions the GM model can be written as follows:

$$\Phi^{-1}\left(S_x(t)\right) = \log\left(r(x)\right) + \Phi^{-1}\left(S_0(t)\right). \tag{2.94}$$

It is the *generalized probit model* (see Dabrowska & Doksum, 1988).

2.5.8 Modification of the GPH1 models when the time-shift rule holds

The GPH1 models do not verify the rule (2.3). As for the PH model, a natural situation when such models can be true when some non-observable stress influences the reliability of units. For example, suppose that on a set $E = E^{(1)} \times E^{(2)}$ of stresses of the form $(x^{(1)}(\cdot), x^{(2)}(\cdot))$ the AFT model holds:

$$\alpha_{x(\cdot)}(t) = r_1\{x_1(t)\}\, r_2\{x_2(t)\}q\{S_{x(\cdot)}(t)\}.$$

Then conditionally, given a fixed time-varying component $x^{(2)}(\cdot)$, the GPH1 model holds on the set $E^{(1)}$:

$$\alpha_{x^{(1)}(\cdot)}(t) = r_1\{x_1(t)\}\, q\{S_{x^{(1)}(\cdot)}(t)\}\, \alpha_0(t),$$

where $\alpha_0(t) = r_2\{x_2(t)\}$.

If the rule (2.3) is suggested by the data, the restriction of the GS model coinciding with the GPH1 model on sets of constant stresses can be considered.

Proposition 2.18. *The GS model with the survival functions (2.72) on E_1 holds on $E \supset E_1$ iff for all $x(\cdot) \in E$*

$$\alpha_{x(\cdot)}(t) = r\{x(t)\}\, \alpha\left\{A^{-1}\left(A_{x(\cdot)}(t)\right)\right\}\, \alpha_0\left\{A_0^{-1}\left(\frac{A^{-1}(A_{x(\cdot)}(t))}{r\{x(t)\}}\right)\right\}, \quad (2.95)$$

where $A = -\ln G$, $\alpha = A'$.

The proof is analogous to the proof of Proposition 2.17.

Proposition 2.18 suggests the following model.

Definition 2.8. *The first modified generalized proportional hazards (MGPH1) model holds on E if there exist a positive on E function r and hazard rates α_0 and α such that for all $x(\cdot) \in E$ the equality (2.95) with*

$$A(t) = \int_0^t \alpha(u)du, \quad A_0(t) = \int_0^t \alpha_0(u)du,$$

holds.

2.5.9 Relations between the survival functions under constant and non-constant stresses.

As in the case of the AFT model, consider some useful relations between survival functions under constant and time-varying stresses.

Proposition 2.19. *Suppose that $x(\cdot)$, $x(\tau)$, $x_0 \in E$ for all $\tau \geq 0$. If the GM model holds on E, then*

$$S_{x(\cdot)}(t) = G\left(\int_0^t \frac{H(S_{x(\tau)}(t))}{H(S_{x_0}(t))}\, dH(S_{x_0}(\tau))\right) \quad (2.96)$$

$$= G\left(\int_0^t H(S_{x(\tau)}(\tau))\, d\log H(S_{x_0}(\tau))\right).$$

Proof.

The equality (2.70) implies that there exists the functional $r_1 : E \to [0, \infty)$ such that

$$S_{x(\cdot)}(t) = G\left\{\int_0^t r_1[x(\tau)]dH(S_{x_0}(\tau))\right\}. \quad (2.97)$$

So for all fixed τ, $\tau \leq t$

$$r_1\{x(\tau)\} = H(S_{x(\tau)}(t))/H(S_{x_0}(t)). \quad (2.98)$$

Putting $r_1\{x(\tau)\}$ in the equality (2.97), the first of the equalities (2.96) is obtained. Putting $t = \tau$ in (2.98) and the obtained expression of $r_1\{x(\tau)\}$ in (2.97), the second of equalities (2.96) is obtained.

Let us consider the PH and MPH models for step-stresses.

2.5.10 GPH1 and MGPH1 models for simple step-stresses

Propositon 2.20. *If the GPH1 model holds on E_2, $x_0 \in E_1$, then for any*
$x(\cdot) \in E_2$

$$S_{x(\cdot)}(t) = \begin{cases} G\{r(x_1)A_0(t)\}, & 0 \le t < t_1, \\ G\{r(x_1)A_0(t) + r(x_2)(A_0(t) - A_0(t_1))\}, & t \ge t_1. \end{cases}$$

$$= \begin{cases} S_{x_1}(t), & 0 \le t < t_1, \\ G\{H(S_{x_1}(t)) + \rho(x_1, x_2)(H(S_{x_1}(t)) - S_{x_1}(t_1)))\}, & t \ge t_1. \end{cases}$$

$$= \begin{cases} G\{\rho(x_0, x_1)(H(S_{x_0}(t)), & 0 \le t < t_1, \\ G\{\rho(x_0, x_1)H(S_{x_0}(t_1)) + \rho(x_0, x_2)(H(S_{x_0}(t)) - H(S_{x_0}(t_1)))\}, & t \ge t_1. \end{cases}$$

(2.99)

where $\rho(x_1, x_2) = r(x_2)/r(x_1)$.

Let us consider the MGPH1 model.

Proposition 2.21. *If the MGPH1 model holds on a set E_2 of simple step-stresses then for any $x(\cdot) \in E_2$, $x_0 \in E_1$*

$$S_{x(\cdot)}(t) = \begin{cases} S_{x_1}(t), & 0 \le t < t_1, \\ S_{x_2}(t - t_1 + t_1^*), & t \ge t_1. \end{cases} \qquad (2.100)$$

where

$$t_1^* = S_{x_0}^{-1}\{G(\rho(x_2, x_1)H(S_{x_0}(t_1)))\}$$

If x_2 is design stress, one can take $x_0 = x_2$ and we have the model

$$S_{x(\cdot)}(t) = \begin{cases} G\{\rho(x_2, x_1)H(S_{x_0}(t))\}, & 0 \le t < t_1, \\ S_{x_0}(t - t_1 + t_1^*), & t \ge t_1. \end{cases}$$

where

$$t_1^* = S_{x_0}^{-1}\{G(\rho(x_0, x_1)H(S_{x_0}(t_1)))\}$$

Thus, the form of the survival function under stress $x(\cdot)$ is determined by the survival function under stress x_0.

2.5.11 General step-stresses

Proposition 2.22. *If the GPH1 model holds on a set of general step-stresses E_m, then for any $x \in E_m$ of the form (2.4) and for any $t \in [t_{i-1}, t_i)$ $(i > 1)$*

$$S_{x(\cdot)}(t) = G\left\{\sum_{j=1}^{i-1} r(x_j)(A_0(t_j) - A_0(t_{j-1}) + r(x_i)(A_0(t) - A_0(t_{i-1}))\right\}$$

$$= G\left\{H(S_{x_i}(t)) - H(S_{x_i}(t_{i-1}))) + \sum_{j=1}^{i-1} \rho(x_i, x_j)(H(S_{x_i}(t_j)) - H(S_{x_i}(t_{j-1})))\right\}$$

$$= G\left\{\rho(x_0, x_i)H(S_{x_0}(t)) + \sum_{j=1}^{i-1} \rho(x_0, x_j)(H(S_{x_0}(t_i)) - H(S_{x_0}(t_{i-1}))) \right\}.$$

(2.101)

Let us consider the MGPH1 model.

Proposition 2.23. *If the MGPH1 model holds on a set of step-stresses of the form (2.4) then for any $t \in [t_{i-1}, t_i)$*

$$S_{x(\cdot)}(t) = S_{x_i}(t - t_{i-1} + t_{i-1}^*),$$

(2.102)

where

$$t_1^* = S_{x_0}^{-1}\left\{G(\rho(x_2, x_1)H(S_{x_0}(t_1)))\right\},$$

$$t_i^* = S_{x_0}^{-1}\left\{G(\rho(x_{i+1}, x_i)H(S_{x_0}(t_i - t_{i-1} + t_{i-1}^*)))\right\} \quad (i = 2, \cdots, m).$$

2.6 GAH and GAMH models

Definition 2.9. *The generalized additive hazards (GAH) model holds on E (Bagdonavičius and Nikulin (1995)), if there exist a function a on E and a survival function S_0 such that for all $x(\cdot) \in E$*

$$\frac{\partial f_{x(\cdot)}^G(t)}{\partial t} = \frac{\partial f_0^G(t)}{\partial t} + a(x(t))$$

(2.103)

with the initial conditions $f_0^G(0) = f_{x(\cdot)}^G(0) = 0$; here $f_0^G(t) = H(S_0(t))$.

So stress influences additively the rate of resource usage. The last equation implies that

$$S_{x(\cdot)}(t) = G\left(H(S_0(t)) + \int_0^t a(x(\tau))d\tau\right).$$

(2.104)

In terms of exponential resource usage the GAH model can be written in the form

$$\alpha_{x(\cdot)}(t) = q\{A_{x(\cdot)}(t)\}(\alpha_0(t) + a(x(t))).$$

The particular case of the GAH model is the *additive hazards model* (AH) (Aalen (1980)):

$$\alpha_{x(\cdot)}(t) = \alpha_0(t) + a(x(t)).$$

(2.105)

Both the GPH1 and the GAH models can be included into the following model.

Definition 2.10 *The generalized additive-multiplicative hazards (GAMH) model (Bagdonavicius and Nikulin (1997a)) holds on E if there exist functions a and r (positive) on E and a survival function S_0 such that for all $x(\cdot) \in E$*

$$\frac{\partial f_{x(\cdot)}^G(t)}{\partial t} = r\{x(t)\}\frac{\partial f_0^G(t)}{\partial t} + a(x(t))$$

(2.106)

with the initial conditions $f_0^G(0) = f_{x(\cdot)}^G(0) = 0$; here $f_0^G(t) = H(S_0(t))$.

So stress influences the rate of resource usage as multiplicatively as additively. The last equation implies that

$$S_{x(\cdot)}(t) = G\left(\int_0^t r\{x(\tau)\}dH(S_0(\tau)) + \int_0^t a(x(\tau))d\tau\right). \qquad (2.107)$$

In terms of exponential resource usage the GAM model can be written in the form:

$$\alpha_{x(\cdot)}(t) = q\{A_{x(\cdot)}(t)\}(r\{x(t)\}\alpha_0(t) + a(x(t))).$$

In the particular case of the exponential resource we obtain the additive-multiplicative hazards (AMH) model (see Lin and Ying (1996)).

$$\alpha_{x(\cdot)}(t) = r\{x(t)\}\alpha_0(t) + a(x(t)). \qquad (2.108)$$

The functions a and q are parametrized as the function $\ln r$ in the GM models and the function q in the GPH models, respectively.

2.7 Changing shape and scale models

2.7.1 Definition of the model for constant stresses

Natural generalization of the AFT model (see Mann et al (1974)) is obtained by supposing that different constant stresses $x \in E_1$ influence not only the scale but also the shape of survival distribution: there exist positive functions on E_1 $\theta(x)$ and $\nu(x)$ such that for any $x \in E_1$

$$S_x(t) = S_{x_0}\left\{\left(\frac{t}{\theta(x)}\right)^{\nu(x)}\right\}; \qquad (2.109)$$

here x_0 is fixed stress, for example, design (usual) stress.

This model has the following interpretation. For any $x(\cdot) \in E \supset E_1$ set

$$f_{x(\cdot)}(t) = S_{x_0}^{-1}(S_{x(\cdot)}(t)),$$

which is the S_{x_0}-resource used until the moment t.

Under the model (2.109) the resource usage rate under $x \in E_1$ is

$$\frac{\partial}{\partial t}f_x(t) = r(x)\,t^{\nu(x)-1},$$

where $r(x) = \nu(x)/\theta(x)^{\nu(x)}$.

The model (2.108) means that resource usage rate under stress x is increasing, if $\nu(x) > 1$, decreasing if $0 < \nu(x) < 1$, and constant if $\nu(x) = 1$.

In the case $\nu(x) = 1$ we have the AFT model. So under thr AFT model resource usage rate is constant in time.

Let us consider generalizations of the model (2.108) to the case of time varying stresses.

2.7.2 Definition of the model for time-varying stresses

Definition 2.11. *The changing shape and scale (CHSS) model (Bagdon-*

avičius and Nikulin (2000d)) holds on E if there exist positive functions on E
r and ν such that for all $x(\cdot) \in E$

$$\frac{\partial f_{x(\cdot)}(t)}{\partial t} = r\{x(t)\}\, t^{\nu(x(t))-1}. \tag{2.110}$$

This equality implies that

$$S_{x(\cdot)}(t) = S_{x_0}\left(\int_0^t r\{x(\tau)\}\tau^{\nu(x(\tau))-1}d\tau\right). \tag{2.111}$$

Variation of stress changes locally not only the scale but also the shape of distribution.

In terms of the hazard rate the model can be written in the form:

$$\alpha_{x(\cdot)}(t) = r\{x(t)\}\, q(A_{x(\cdot)}(t))\, t^{\nu(x(t))-1}. \tag{2.112}$$

This model is not in the class of the GS models because the hazard rate $\alpha_{x(\cdot)}(t)$ depends not only on $x(t)$ and $A_{x(\cdot)}(t)$ but also on t.

If Sedyakin's rule is suggested by the data, restriction of the GS model coinciding with the model (2.100) on E_1 can be considered.

Proposition 2.25. The GS model with the survival function (2.108) on E_1 holds on $E \supset E_1$ if and only if for any $x(\cdot) \in E$

$$\alpha_{x(\cdot)}(t) = \{r(x(t))\}^{1/\nu(x(t))}\alpha_{x_0}\left\{A_{x_0}^{-1}(A_{x(\cdot)}(t))\right\}\left\{\nu(x(t))A_{x_0}^{-1}(A_{x(\cdot)}(t))\right\}^{1-1/\nu(x_t)}. \tag{2.113}$$

The proof is analogous to the proof of Proposition 2.10.

The proposition implies the following model.

Definition 2.12. *The modified changing scale and shape model holds on E if for any $x(\cdot) \in E$ the equality (2.113) holds.*

In the particular case when $\nu(x) = 1$ for any x, we have the MPH model.

2.7.3 CHSS and MCHSS models for simple step-stresses

Proposition 2.26. *If the CHSS model holds on E_2 then for any $x(\cdot) \in E_2$, $x_0 \in E_1 \subset E_2$*

$$S_{x(\cdot)}(t) = \begin{cases} S_{x_0}\left\{\left(\frac{t}{\theta(x_1)}\right)^{\nu(x_1)}\right\}, & 0 \le t < t_1, \\[2ex] S_{x_0}\left\{\left(\frac{t_1}{\theta(x_1)}\right)^{\nu(x_1)} + \left(\frac{t}{\theta(x_2)}\right)^{\nu(x_2)} - \left(\frac{t_1}{\theta(x_2)}\right)^{\nu(x_2)}\right\}, & t \ge t_1, \end{cases} \tag{2.114}$$

where $\theta(x) = \left(\frac{\nu(x)}{r(x)}\right)^{1/\nu(x)}$.

In the particular case when $x_0 = x_2$ we have $\nu(x_2) = \theta(x_2) = 1$ and

$$S_{x(\cdot)}(t) = \begin{cases} S_{x_0}\left\{\left(\frac{t}{\theta(x_1)}\right)^{\nu(x_1)}\right\}, & 0 \leq t < t_1, \\ S_{x_0}\left\{\left(\frac{t_1}{\theta(x_1)}\right)^{\nu(x_1)} + t - t_1\right\}, & t \geq t_1. \end{cases} \quad (2.115)$$

Proposition 2.27. *If the MCHSS model holds on E_2 then for any $x(\cdot) \in E_2$*

$$S_{x(\cdot)}(t) = \begin{cases} S_{x_1}(t), & 0 \leq t < t_1, \\ S_{x_2}(t - t_1 + t_1^*), & t \geq t_1, \end{cases} \quad (2.116)$$

where

$$t_1^* = \theta(x_2)(t_1/\theta(x_1))^{\nu(x_1)/\nu(x_2)}.$$

2.7.4 CHSS and MCHSS models for general step-stresses

Proposition 2.28. *If the CHSS model holds on E_m then for any $x(\cdot) \in E_m$ and $t \in [t_{i-1}, t_i)$*

$$S_{x(\cdot)}(t) =$$

$$G\left\{\sum_{j=1}^{i-1}\left(\left(\frac{t_j}{\theta(x_j)}\right)^{\nu(x_j)} - \left(\frac{t_{j-1}}{\theta(x_j)}\right)^{\nu(x_j)}\right) + \left(\frac{t}{\theta(x_i)}\right)^{\nu(x_i)} - \left(\frac{t_{i-1}}{\theta(x_i)}\right)^{\nu(x_i)}\right\} =$$

$$S_{x_i}\left\{\theta(x_i)\left(\sum_{j=1}^{i-1}\left(\left(\frac{t_j}{\theta(x_j)}\right)^{\nu(x_j)} - \left(\frac{t_{j-1}}{\theta(x_j)}\right)^{\nu(x_j)}\right) + \right.\right.$$

$$\left.\left.\left(\frac{t}{\theta(x_i)}\right)^{\nu(x_i)} - \left(\frac{t_{i-1}}{\theta(x_i)}\right)^{\nu(x_i)}\right)^{1/\nu(x_i)}\right\}.$$

Proposition 2.29. *If the MCHSS model holds on E_m then for any $x(\cdot) \in E_2$*

$$S_{x(\cdot)}(t) = S_{x_i}(t - t_{i-1} + t_{i-1}^*), \ t \in [t_{i-1}, t_i),$$

where

$$t_1^* = \theta(x_2)(t_1/\theta(x_1))^{\nu(x_1)/\nu(x_2)}, \quad t_i^* = \theta(x_{i+1})\{(t_i - t_{i-1} + t_{i-1}^*)/\theta(x_i)\}^{\nu(x_i)/\nu(x_{i+1})}.$$

2.8 Generalizations

Schaebe and Viertl (1995) considered an axiomatic approach to model building.

Proposition 2.30. (Schaebe and Viertl (1995)). *Suppose that there exists a functional*

$$a : E \times E \times [0, \infty) \to [0, \infty)$$

such that for any $x_1(\cdot), x_2(\cdot) \in E$ it is differentiable and increasing in t,

$$a(x_1(\cdot), x_2(\cdot), 0) = 0$$

and

$$T_{x_2(\cdot)} \sim a(x_1(\cdot), x_2(\cdot), T_{x_1(\cdot)}),$$

where $\sim$ denotes equality in distribution.

For any differentiable on $[0, \infty)$ c.d.f. F exists a functional $b : E \times [0, \infty) \to [0, \infty)$ such that for all $x(\cdot) \in E$

$$F_{x(\cdot)}(t) = F\left(\int_0^t b(x(\cdot), u)du\right). \tag{2.117}$$

Proof. Fix $x_0(\cdot) \in E$ and for all $x(\cdot) \in E$ put

$$a_0(x(\cdot), t) = F^{-1}(F_{x_0(\cdot)}(a(x(\cdot), x_0(\cdot), t))).$$

The distribution of the random variable $R = a_0(x(\cdot), T_{x(\cdot)})$ does not depend on $x(\cdot)$ and its c.d.f. is F. Put

$$b(x(\cdot), t) = \frac{\partial}{\partial t} a_0(x(\cdot), t).$$

Then

$$a_0(x(\cdot), t) = \int_0^t b(x(u), u)du,$$

which implies

$$F_{x(\cdot)}(t) = \mathbf{P}\{T_{x(\cdot)} < t\} = \mathbf{P}\{R < a_0(x(\cdot), t)\} = F\left(\int_0^t b(x(u), u)du\right).$$

$\square$

Remark 2.8. Set $G(t) = 1 - F(t)$, $S_{x(\cdot)}(t) = 1 - F_{x(\cdot)}(t)$, $H = G^{-1}$, $f_{x(\cdot)}^G(t) = H(S_{x(\cdot)}(t))$. The equality (2.117) implies that

$$\frac{\partial}{\partial t} f_{x(\cdot)}^G(t) = b(x(\cdot), t). \tag{2.118}$$

This model means that the rate of the G-resource usage is a functional of stress and the time.

The above considered models are sub-models of this general model:

1) If $b(x(\cdot), t) = r(x(t))$, we have the AFT model.

2) If $b(x(\cdot), t) = r(x(t))\alpha_0(t)$, we have the GM (or, equivalently, GPH1) model.

3) If $b(x(\cdot), t) = r(x(t))\alpha_0(t)$ and the resource is exponential, i.e. $G(t) = e^{-t}, t \geq 0$, we have the PH model.

4) If $b(x(\cdot), t) = r(x(t))t^{\nu(x(t))-1}$, we have the CHSH model.

2.8.1 AFT model with time dependent regression coefficients

The AFT model usually is parametrized (see Section 5) in the following form:

$$S_{x(\cdot)}(t) = G\left\{\int_0^t e^{-\beta^T x(u)}du\right\}, \tag{2.119}$$

where $\beta = (\beta_0, \cdots, \beta_m)^T$ is a vector of unknown parameters. At any moment t the resource usage rate

$$\frac{\partial}{\partial t} f_{x(\cdot)}^G(t) = e^{-\beta^T x(t)}$$

depends only on the value of the explanatory variable at the moment t; here

$$x(t) = (x_0(t), x_1(t), \cdots, x_m(t))^T, \quad x_0(t) \equiv 1.$$

Flexible models can be obtained by supposing that the coefficients β are time-dependent, i.e. taking

$$\frac{\partial}{\partial t} f_{x(\cdot)}^G(t) = e^{-\beta^T(t)x(t)} = e^{-\sum_{i=0}^m \beta_i(t)x_i(t)}.$$

If the function $\beta_i(\cdot)$ is increasing or decreasing in time then the effect of ith component of the explanatory variable is increasing or decreasing in time.

So we have the model

$$S_{x(\cdot)}(t) = G \left\{ \int_0^t e^{-\beta^T(u)x(u)} du \right\}. \tag{2.120}$$

It is the AFT model with time-dependent regression coefficients.

We shall consider the coefficients $\beta_i(t)$ in the form

$$\beta_i(t) = \beta_i + \gamma_i g_i(t), \quad (i = 1, 2, ..., m),$$

where $g_i(t)$ are some specified deterministic functions or realizations of predictable processes. In such a case the AFT model with time dependent coefficients and constant or time dependent explanatory variables can be written in the usual form (2.119) with different interpretation of the explanatory variables. Indeed, set

$$\theta = (\theta_0, \theta_1, \cdots, \theta_{2m})^T = (\beta_0, \beta_1, \cdots, \beta_m, \gamma_1, \cdots, \gamma_m)^T,$$

$$z(\cdot) = (z_0(\cdot), z_1(\cdot), \cdots, z_{2m}(\cdot))^T =$$

$$(1, x_1(\cdot), \cdots, x_m(\cdot), x_1(\cdot)g_1(\cdot), \cdots, x_m(\cdot)g_m(\cdot))^T. \tag{2.121}$$

Then

$$\beta^T(u)x(u) = \beta_0 + \sum_{i=1}^m (\beta_i + \gamma_i g_i(t))x_i(t) = \theta^T z(u).$$

So the AFT model with the time dependent regression coefficients can be written in the form

$$S_{x(\cdot)} = G \left\{ \int_0^t e^{-\theta^T z(u)} du \right\}. \tag{2.122}$$

We have the AFT model where the unknown parameters and the explanatory variables are defined by (2.121).

2.8.2 PH model with time dependent regression coefficients

The PH model usually is parametrized (see Section 7) in the following form:

$$\alpha_{x(\cdot)}(t) = e^{-\beta^T x(t)} \alpha_0(t).$$

Similarly, as in the case of the AFT model, the PH model with time-dependent regression coefficients has the form:

$$\alpha_{x(\cdot)}(t) = e^{-\beta^T(t)x(t)} \alpha_0(t). \qquad (2.123)$$

As in the case of the AFT model with time-dependent coefficients, the model (2.123) with $\beta_i(t) = \beta_i + \gamma_i g_i(t)$ can be written in the form of the usual PH model

$$\alpha_{x(\cdot)}(t) = e^{-\theta^T z(t)} \alpha_0(t),$$

where θ and z are defined by (2.121).

2.8.3 Partly parametric additive risk and Aalen's models

McKeague and Sasieni (1994) give the following generalization of the additive hazards model (2.105) with constant explanatory variables:

$$\alpha_x(t) = x_1^T \alpha(t) + \beta^T x_2,$$

where x_1 and x_2 are q and p dimensional components of the explanatory variable x. Here $\alpha(t) = (\alpha_1, \cdots, \alpha_q)$ and $\beta = (\beta_1, \cdots, \beta_p)^T$ are unknown. It generalizes also the Aalen's (1980) model

$$\alpha_x(t) = x^T \alpha(t).$$

2.9 Models including switch-up and cycling effects

Considering the GS model, it was noted that this (and also AFT) model may not be appropriate when stress is periodic due to a quick change of its values. The greater the number of stress cycles, the shorter the life of units. So the effect of cycling must be included in the model. The GS model can be not verified when switch-up's of stress can imply failures of units or influence their reliability in the future. We shall follow here Bagdonavičius and Nikulin (2001d). Suppose that a periodic stress is differentiable. Then the number of cycles in the interval $[0, t]$ is

$$n(t) = \int_0^t |\, d\mathbf{1}\{x'(u) > 0\} \,| \,.$$

Generalizing the GPH1 (or GM) model we suppose that the G-resource used until the moment t has the form.

$$f_{x(\cdot)}^G(t) = \int_0^t r_1\{x(u)\} dH(S_0(u)) + \int_0^t r_2\{x(u)\} d \,|\, \mathbf{1}\{x'(u) > 0\} \,| \quad (2.124)$$

The second term includes the effect of cycling on resource usage. In terms of survival functions

$$S_{x(\cdot)}(t) = G\left\{ \int_0^t r_1\{x(u)\}dH(S_0(u)) + \int_0^t r_2\{x(u)\} \mid d\mathbf{1}\{x'(u) > 0\} \right\}.$$

$$(2.125)$$

If amplitude is constant, $r_2\{x(u)\} = c$ can be considered.

The AFT model is generalized by the model

$$S_{x(\cdot)}(t) = G\left\{ \int_0^t r_1\{x(u)\}du + \int_0^t r_2\{x(u)\} \mid d\mathbf{1}\{x'(u) > 0\} \mid \right\}. \qquad (2.126)$$

The GS and AFT models are not appropriate if $x(\cdot)$ is a step-stress with many switch ons and switch offs which shorten the life of units.

An alternative to the GS model under step-stresses can be obtained by taking into account the influence of switch-ups of stresses on reliability of units. Switch-ups can imply failures of units. Suppose that an item is observed under stress (2.4) and after the switch-off at the moment t_i from the stress x_i to the stress x_{i+1} the survival function has a jump:

$$S_{x(\cdot)}(t_i) = S_{x(\cdot)}(t_i-)\,\delta_i;$$

here δ_i is the probability for an item not to fail because of the switch-off at the moment t_i. In this case the GS model for step-stresses can be modified as follows:

$$S_{x(\cdot)}(t) = S_{x_i}(t - t_{i-1} + t^{**}_{i-1}), \qquad (2.127)$$

where

$$t^{**}_1 = S^{-1}_{x_2}\{S_{x_1}(t_1)\,\delta_1\}, \quad t^{**}_i = S^{-1}_{x_{i+1}}\{S_{x_i}(t_i - t_{i-1} + t^{**}_{i-1})\,\delta_i\}. \qquad (2.128)$$

Thus the time shift is modified by the jumps.

In this case the following model can be considered:

$$f^G_{x(\cdot)}(t) = \int_0^t r_1\{x(u)\}dH(S_0(u)) + \int_0^t r_2\{x(u)\}\mathbf{1}(\Delta x(u) > 0)\frac{\mid dx(u)\mid}{\mid \Delta x(u)\mid}$$

$$+ \int_0^t r_3\{x(u)\}\mathbf{1}(\Delta x(u) < 0)\frac{\mid dx(u)\mid}{\mid \Delta x(u)\mid}. \qquad (2.129)$$

The second and the third terms include the effect of switch-ons and switch-offs (or *vice versa*), respectively, on resource usage. If the step-stress has two values, the functions r_2 and r_3 can be constants.

2.10 Heredity hypothesis

Suppose that a process of production is *unstable*, i.e. the reliability of units produced in non-intersecting time intervals $I_1 = (t_0, t_1], \cdots, I_m(t_{m-1}, t_m]$ is different: under the same stress conditions the survival functions of units, produced in the intervals I_i and I_j $(i \neq j)$, are different. Here we shall follow Bagdonavičius and Nikulin (1997f). Suppose that a failure time $T_x^{(i)}$ of units produced in the interval I_i and functioning under a constant stress $x \in E_1$ is a non-negative random variable with the reliability function $S_x^{(i)}(t) = \mathbf{P}\{T_x^{(i)} > t\}$.

The models of accelerated life are some hypotheses about the influence of the applied stress on the reliability. For many models the reliability characteristics under the usual stress $x^{(0)}$ often can be written via the reliability characteristics under the accelerated stress $x^{(1)}$ and some function $\rho^{(i)}(x^{(0)}, x^{(1)})$ of the stresses $x^{(0)}$ and $x^{(1)}$.

AFT model:

$$S_{x^{(0)}}^{(i)}(t) = S_{x^{(1)}}(t/\rho^{(i)}(x^{(0)}, x^{(1)})), \qquad (2.130)$$

where

$$\rho^{(i)}(x_0, x_1) = r^{(i)}(x_1)/r^{(i)}(x_0) > 1.$$

GPH1 model:

$$S_{x^{(0)}}^{(i)}(t) = G\left\{H \circ S_{x^{(1)}}(t)/\rho^{(i)}(x^{(0)}, x^{(1)})\right\}, \qquad (2.131)$$

GAH model:

$$S_{x^{(0)}}^{(i)}(t) = G\left\{H \circ S_{x^{(1)}}(t) - b^{(i)}(x^{(0)}, x^{(1)})t\right\}, \qquad (2.132)$$

where

$$b^{(i)}(x^{(0)}, x^{(1)}) = a^{(i)}(x^{(1)}) - a^{(i)}(x^{(0)}) > 0.$$

Definition 2.13. *If the process of production is unstable, the model AFT(or GPH1, GAH) holds for units produced in each of the intervals I_i, and $\rho^{(i)}(x_1, x_2) = \rho(x_1, x_2)$ for all i (the models AFT or GM) or $b^{(i)}(x_1, x_2) = b(x_1, x_2)$ (the model GA) are the same for groups of units produced in different time intervals, then the heredity hypothesis holds.*

If the heredity hypothesis is satisfied on E and sufficiently large usual and accelerated data are accumulated during a long period of observation then good estimators of the functions $\rho(x^{(0)}, x^{(1)})$ (or $b(x^{(0)}, x^{(1)})$) can be obtained. The reliability of newly produced units under the usual stress x_0 can be estimated from accelerated life data under the stress $x^{(1)} > x^{(0)}$, using the estimators $\hat{\rho}(x^{(0)}, x^{(1)})$ or $\hat{b}(x^{(0)}, x^{(1)})$ and without using the experiment under the normal stress. The formulated hypothesis is called the heredity hypothesis as it has some associations with the "heredity principle" of Kartashov and Perrote (1968) and is motivated by it.

The heredity principle is formulated as follows. Suppose units of ith group are observed under the stress x and are characterized by some multivariate technical parameter $W_i(x)$. The parameter W_i is good if

$$c \leq W_i(x) \leq d, \quad c, d \in R^k.$$

The parameter W_i is some function of the interior physical parameters ν_i of units:

$$W_i(x) = f_i(\nu_i(x)).$$

It is supposed that there exists a function ϕ such that

$$\nu_2(x) = \phi(\nu_1(x), x).$$

The heredity principle states that the distribution of the random vectors $\nu_i(x)$ can change going from one group of units to another but the functions f_i and ϕ are invariant.

Suppose that x_1 is a usual stress and $x_2 > x_1$ an accelerated stress.

If one of the models AFT, GM, or GA and the heredity principle hold, then sufficiently large data can be cumulative during a long period of observation and good estimators of the functions $\rho(x_1, x_2)$ or $b(x_1, x_2)$ can be obtained. The reliability of newly produced units under the usual stress x_1 can be estimated from accelerated life data obtained under the accelerated stress x_2, using the estimators $\hat{\rho}(x_1, x_2)$ or $\hat{b}(x_1, x_2)$.

2.11 Summary

Models relating the lifetime distribution to possibly time dependent explanatory variables were considered in this chapter. As a rule they were defined in terms of the hazard function. We give here a short survey of them.

1) **Accelerated failure time (AFT) model:**

$$\alpha_{x(\cdot)}(t) = r\{x(t)\}q\{S_{x(\cdot)}(t)\} \quad \sim \quad S_{x(\cdot)}(t) = G\left\{\int_0^t r\{x(u)\}du\right\}.$$

The model is parametric if the function G is from a specified parametric family of distributions (such families are given in Chapter 1) and the function r is parametrized (see Section 5.1):

$$r(x) = e^{\beta^T \varphi(x)}, \qquad (2.133)$$

where $\beta = (\beta_0, \cdots, \beta_m)^T$ is a vector of unknown parameters,

$$\varphi(x) = (\varphi_0(x), \cdots, \varphi_m(x))^T, \quad \varphi_0(x) \equiv 1,$$

being a vector of specified real functions on the set of values of the explanatory variable $x(\cdot) = (x_0(\cdot), \cdots, x_m(\cdot))$, $x_0 \equiv 1$.

The model is semiparametric if one of the functions G or r is completely unknown and other is parametrized.

The model is nonparametric if both functions G or r are completely unknown. Estimation in this case is possible only under special plans of experiments.

2) **Proportional hazards (PH) model:**

$$\alpha_{x(\cdot)}(t) = r\{x(t)\}\alpha(t).$$

The model is parametric if the function r is parametrized (as a rule in the form (2.130), see Section 7.2) and α is from a specified parametric class of hazard functions. The model is semiparametric if one of the functions r and α is completely unknown and other is parametrized.

3) **Additive hazards (AH) model**:

$$\alpha_{x(\cdot)}(t) = \alpha_0(t) + a(x(t)).$$

The function a generally is parametrized in the form $a(x) = \gamma^T x$.

These models being narrow (explanatory variables influence locally only the scale of the survival distribution for the AFT model, ratios or differences of hazard rates are constant over time for the PH or AH model, respectively), the last two being not very natural for aging units, a number of alternative or wider models may be considered.

The natural generalizations of the AFT model are the following models.

4) **Changing shape and scale (CHSS) model**:

$$\alpha_{x(\cdot)}(t) = r\{x(t)\}t^{\nu\{x(t)\}-1}q\{A_{x(\cdot)}(t)\}$$

$$\sim \quad S_{x(\cdot)}(t) = G\left\{\int_0^t r\{x(u)\}u^{\nu\{x(u)\}-1}du\right\}.$$

Under this model, explanatory variables influence locally not only the scale but also the shape of survival distribution.

The model may be considered as parametric or semiparametric. Generally the function r is parametrized in the form (2.130) and the function ν in the form $\nu(x) = \exp\{\gamma^T x\}$.

5) **AFT model with time-dependent regression coefficients:**

$$S_{x(\cdot)}(t) = G\left\{\int_0^t e^{-\beta^T(u)x(u)}du\right\}. \tag{2.134}$$

with coefficients $\beta_i(t)$ in the form

$$\beta_i(t) = \beta_i + \gamma_i g_i(t), \quad (i = 1, 2, ..., m),$$

where $g_i(t)$ are some specified deterministic functions or realizations of predictable processes. It can be written in the form of the usual AFT model

$$S_{x(\cdot)} = G\left\{\int_0^t e^{-\theta^T z(u)}du\right\}, \tag{2.135}$$

where

$$\theta = (\theta_0, \theta_1, \cdots, \theta_{2m})^T = (\beta_0, \beta_1, \cdots, \beta_m, \gamma_1, \cdots, \gamma_m)^T,$$

$$z(\cdot) = (z_0(\cdot), z_1(\cdot), \cdots, z_{2m}(\cdot))^T =$$

$$(1, x_1(\cdot), \cdots, x_m(\cdot), x_1(\cdot)g_1(\cdot), \cdots, x_m(\cdot)g_m(\cdot))^T.$$

Even if the explanatory variable x is constant in time, the explanatory variable $z(\cdot)$ is time-dependent. So statistical analysis of this model can be done using methods of statistical estimation for the usual AFT model with constant regression coefficients and time-dependent explanatory variables.

The following class of models generalizes in one or another sense both the AFT and PH models.

6) First generalized proportional hazards (GPH1) model:

$$\alpha_{x(\cdot)}(t) = r\{x(t)\}q\{A_{x(\cdot)}(t)\}\alpha(t).$$

It coincides with the AFT model if $\alpha(t) \equiv \text{const}$ or with the PH model if $q(u) \equiv \text{const}$.

The model is parametric if the functions r and q are parametrized and α is from a specified parametric family of hazard rates. We consider here semiparametric models with r parametrized and q a specified parameter free or parametrized function, α being completely unknown. The function r is parametrized in the form (2.130). The possible parametrizations of the function q are $q(u,\gamma) = (1+u)^{\gamma}, e^{\gamma u}, (1+\gamma u)^{-1}$. Under these parametrizations the situations with approaching or going away hazard rates under various constant explanatory variables may be modeled. Take notice that only one complementary parameter γ is included in these models with respect to the PH model.

Taking $q(u) = e^{-u}$, we obtain the analogue of the logistic regression model, taking

$$q(u) = \varphi(v)e^{u-v}, \quad v = \Phi^{-1}(1 - e^{-u}),$$

the generalized probit model is obtained. The last two models are alternatives to the PH model.

7) Second generalized proportional hazards (GPH2) model:

$$\alpha_{x(\cdot)}(t) = u\{x(t), A_{x(\cdot)}(t)\}\alpha(t).$$

It includes the GPH1 model as the particular case. The model CRE1 with $u(x,s) = r(x)(1+s)^{\gamma^T x+1}$ and the model CRE2 with $A_x(t) = r(x)A(t)^{e^{\gamma^T x}}$ are in the class of GPH2 models and situations with intersecting hazard rates may be modeled using them.

8) PH model with time-dependent regression coefficients

$$\alpha_{x(\cdot)}(t) = e^{-\beta^T(t)x(t)}\alpha_0(t),$$

with coefficients $\beta_i(t)$ in the form

$$\beta_i(t) = \beta_i + \gamma_i g_i(t), \quad (i = 1, 2, ..., m),$$

where $g_i(t)$ are specified deterministic functions or realizations of predictable processes. It can be written in the form of the usual PH model

$$\alpha_{x(\cdot)}(t) = e^{\theta^T z(t)}\alpha_0(t),$$

where

$$\theta = (\theta_0, \theta_1, \cdots, \theta_{2m})^T = (\beta_0, \beta_1, \cdots, \beta_m, \gamma_1, \cdots, \gamma_m)^T,$$

$$z(\cdot) = (z_0(\cdot), z_1(\cdot), \cdots, z_{2m}(\cdot))^T =$$

$$(1, x_1(\cdot), \cdots, x_m(\cdot), x_1(\cdot)g_1(\cdot), \cdots, x_m(\cdot)g_m(\cdot))^T.$$

Even if the explanatory variable x is constant in time, the explanatory variable $z(\cdot)$ is time-dependent. Statistical analysis of this model can be done using methods of statistical estimation for the usual PH model with constant regression coefficients and time-dependent explanatory variables.

The following model includes both the PH and AH models.

9) **Additive-multiplicative hazards (AMH) model:**

$$\alpha_{x(\cdot)}(t) = r\{x(t)\}\alpha_0(t) + a(x(t)).$$

As the GPH1 model generalizes the PH model, the following model generalizes the AH model.

10) **Generalized additive hazards (GAH) model:**

$$\alpha_{x(\cdot)}(t) = q\{A_{x(\cdot)}(t)\}(\alpha_0(t) + a(x(t))).$$

The GAH model is the AH model when $q(u) \equiv 1$. The function q is parametrized as in the case of GPH1 models.

11) **Aalen's additive risk (AAR) model:**

$$\alpha_x(t) = x^T \alpha(t).$$

This model allows the influence of each explanatory variable to vary separately over time.

12) **Partly parametric additive risk (PPAR) model**

$$\alpha_x(t) = x_1^T \alpha(t) + \beta^T x_2,$$

where x_1 and x_2 are q and p dimensional components of the explanatory variable x, $\alpha(t) = (\alpha_1(t), \cdots, \alpha_q(t))^T$, $\beta = (\beta_1, \cdots, \beta_p)^T$ are unknown. It includes AAR and AH models as particular cases.

13) **Generalized additive-multiplicative hazards (GAMH) model:**

$$\alpha_{x(\cdot)}(t) = q\{A_{x(\cdot)}(t)\}(r\{x(t)\}\alpha_0(t) + a(x(t))).$$

If $q(u) \equiv 1$, then GAMH model is the AMH model.

A natural property of the models with time-varying explanatory variables is given by Sedyakin's rule which states that the hazard rate at any moment t depends only on the value of the explanatory variable at this moment and the probability to survive up to this moment.

14) **Generalized Sedyakin's (GS) model**

$$\alpha_{x(\cdot)}(t) = g\left(x(t), S_{x(\cdot)}(t)\right).$$

This model is too wide for AFT data analysis but it is useful for construction of narrower models. Note that from the above considered models only the AFT model verifies this rule. We discussed how to modify models to verify the rule. Such models are rather complicated for statistical analysis in the case of general time-varying stresses. Nevertheless, if the explanatory variables are step functions, then they are simple.

The GS model (and also AFT model) may not be appropriate when stress is periodic with quick change of its values. The greater the number of stress cycles, the shorter the life of units.

The GS and AFT' models are also not appropriate if $x(\cdot)$ is a step-stress with many switch ons and switch offs which shorten the life of units.

Models including the effect of cycling and the influence of switch-ups of stresses on reliability were discussed in Section 2.9.

Most of the considered models may be used for analysis of data collected during different time periods of unstable production process. It may be done if the models have invariants which do not change going from one group of produced units to another (see Section 2.10).

Good sources of information and further references on accelerated life models are Cox and Oakes (1984), Derringer (1982), Finkelstein (1999), Gertsbakh and Kordonskiy (1969), Hirose (1997a,b), Hsieh (2000), Iuculano and Zanini (1986), Lin and Ying (1995), LuValle (2000), Mazzuchi and Soyer (1992), Meeker and Escobar (1998), Miner (1945), Meeter and Meeker (1994), Nelson (1980, 1990), Nelson and Meeker (1991), Rukhin and Hsieh (1987), Schaebe and Viertl (1995), Schaebe (1998), Schmoyer (1991), Sedyakin (1966), Shaked and Singpurwalla (1983), Sethuraman and Singpurwalla (1982), Singpurwalla (1971, 1987), Singpurwalla , Castellino and Goldschen (1975), Tibshirani and Ciampi (1983), Viertl (1988), Viertl and Gurker (1995, 1998), Viertl and Spencer (1991).

CHAPTER 3

Accelerated degradation models

3.1 Introduction

Failures of highly-reliable units are rare and failure time data can be very scarce. Two ways of obtaining additional information about reliability of units can be used. One way is to use higher levels of experimental factors or stresses to increase the number of failures, and, hence, to obtain reliability information quickly. Another way is to measure some parameters characterizing degradation (aging) of the product in time. Both methods can be combined: degradation and failure time data can be obtained at higher levels of stress, (see, for example, Meeker, Escobar and Lu (1998), Singpurwalla (1995)). Analysis of such data is possible if accelerated degradation models relating degradation and failure times to the accelerating factors (stresses) are well chosen.

Degradation models with the explanatory variables may also be used to estimate reliability when the environment is dynamic (see, Singpurwalla (1995)). The explanatory variables may be uncontrollable by an experimenter in such a case. For example, tire wear rate and failure times depend on quality of roads, temperature, and other factors.

Accelerated degradation models can be used when optimal values of explanatory variables are needed to maximize the reliability of the product, are needed. For example, degradation of light emitting diodes is characterized by their decreasing luminosity, and the rate of degradation depends on such factors as type of silver, epoxy coating, epoxy lens material, initial curing temperature, and curing duration, (see, Hamada (1995), Chiao and Hamada (1996)).

Modeling accelerated degradation one must keep in mind that an unit may be treated as failed when its degradation reaches a critical level (non-traumatic failure) or when a traumatic event occurs. The probability of the traumatic event may depend on the degradation level and on the explanatory variables. For example, a puncture of a tire is more probable if thickness of the tire protector (degradation measure) is smaller and the load (the explanatory variable) is heavier.

Thus, in the most general situations an accelerated degradation model must include:

1. The stochastic process describing changing of the degradation level in time;

2. Dependence of degradation process parameters on the explanatory variables;

3. The stochastic process characterizing traumatic events;

4. Dependence of this process on degradation and the explanatory variables.

3.2 Degradation models

Suppose that an increasing stochastic process $Z(t)$ describes the degradation level of an item. Sometimes the functional form of the mean degradation $m(t) = \mathbf{E}(Z(t))$ is known, sometimes it is not. For example, the mean tire protector wear rate $m'(t)$ gradually decreases in time and after some short period becomes practically constant. It can be modelled by $m'(t) = \gamma_0 + \gamma_1 e^{-\gamma_2 t}$. In this case the mean degradation

$$m(t) = \gamma_0 t + \frac{\gamma_1}{\gamma_2}(1 - e^{-\gamma_2 t}), \quad \gamma_i > 0. \tag{3.1}$$

If the initial accelerated wear period is absent, then $m(t) = \gamma_0 t$ is a linear function.

The mean relative luminosity of light-emitting diodes (see Mitsuo (1991)) is a nonlinear function $d(t) = (1 + \gamma_0 t^{\gamma_1})^{-1}$ and the mean degradation can be determined as

$$m(t) = \gamma_0 \, t^{\gamma_1}. \tag{3.2}$$

In such a case $m(0) = 0$, and the mean degradation is increasing. There exist a one-to-one application between the values of $m(t)$ and $d(t)$.

In many situations (see Carey and Koenig (1991)) degradation approaches a saturation point where deterioration ends (for example, where oxidation ceases). The mean degradation can be defined by the function

$$m(t) = \gamma_0(1 - e^{-\gamma_1 t^{\gamma_2}}), \quad \gamma_i > 0. \tag{3.3}$$

Usually one of the three general shapes for the mean degradation curve $m = m(t)$ is observed: linear, convex, or concave (see Dowling (1993), Lu and Meeker (1993), Boulanger and Escobar (1994), Meeker and Escobar (1998), Tseng, Hamada and Chiao (1994), Suzuki, Maki and Yokogawa (1993)). This shape can be suggested by data analysis or by knowlege of physics of the degradation process. See, also, Chang (1992), Fukuda (1991), Nelson (1990), Klinger (1992), Pieper and Tiedge (1983), Yu and Tseng (1999), Yanasigava (1997), etc.

In what follows we suppose that $m(t)$ is increasing and continuously differentiable on $[0, \infty]$ and $m(0) = 0$.

Many uncontrollable factors imply variability of individual degradation curves. Thus, the degradation curve of a concrete item ω is a trajectory $Z(t, \omega), t \geq 0$ of a stochastic process Z.

Let us consider several classes of stochastic processes used to model the degradation.

We assumed that the degradation process is non-negative and non-decreasing. So we do not consider such well known stochastic processes as Wiener process and it's generalizations because for such models it is possible that degradation is decreasing in any interval.

We do not also consider models (they are numerous) defined by stochastic processes with numerically untractable finite-dimensional distributions.

3.2.1 General degradation path model

The degradation process is modeled by a random process

$$Z = d(t, \theta), \quad \theta = (\theta_1, \theta_2), \quad \theta_1 \in \mathbf{R}^k, \theta_2 \in \mathbf{R}^l,$$

where θ_1 is a possibly multidimentional random variable and θ_2 is a non-random parameter.

The form of the function $d(t, \theta)$ can be suggested by the form of the mean degradation function or by individual degradation curves. For example, the linear mean degradation model suggests the *linear degradation path model*:

$$Z = \theta t,$$

where θ is a positive random variable.

If tires are functioning at the same stress conditions, their wear is well modeled by such a model.

The mean degradation model (3.2) suggests the degradation path model

$$Z = \theta_1 t^{\theta_2}.$$

If individual degradation curves do not intersect, the parameter θ_1 should be non-random and the parameter θ_2 random, or vice versa, the parameter θ_1 should be random and the parameter θ_2 non-random. If these curves intersect, both θ_1 and θ_2 should be random.

3.2.2 Gamma-process

Stochastic process is a gamma process with the shape parameter $\nu(t)$ and the scale parameter σ, denoted by $Z(t) \in G(\nu(t), 1/\sigma)$, if

a) $Z(0) = 0$;

b) $Z(t)$ has independent increments, i.e. for any $0 < t_1 < \cdots < t_m$ the random variables $Z(t_1), Z(t_2) - Z(t_1), \cdots, Z(t_m) - Z(t_{m-1})$ are independent;

c) the distribution of $Z(t) - Z(s)$ is gamma with the density

$$p_{Z(t)-Z(s)}(x) = \frac{1}{\Gamma(\nu(t) - \nu(s))} \, x^{\nu(t)-\nu(s)-1} \, \sigma^{-(\nu(t)-\nu(s))} \, e^{-x/\sigma}, \quad x \geq 0.$$

The gamma process is non-decreasing and its increments $\Delta Z(t) = Z(t + \Delta t) - Z(t)$ are from the same family of gamma distributions.

The mean of the process $Z(t)$ is $\sigma \nu(t)$. Thus, the degradation process with the non-decreasing mean $m(t)$ is modeled by the gamma process $Z(t) \in G(m(t)/\sigma, 1/\sigma)$. In such a case

$$\mathbf{E}(Z(t)) = m(t), \quad \mathbf{Var}(Z(t)) = \sigma m(t), \quad \mathbf{Cov}(Z(s), Z(t)) = \sigma m(s \wedge t).$$

If the mean degradation is linear, $m(t) = at$, then the gamma process is stationary, i.e. the distribution of $Z(t) - Z(s)$ is a function of $t - s$. For detailed treatment of gamma processes in degradation models see Cinlar (1980), Gaver (1963), Singpurwalla (1995), Singpurwalla and Youngren (1998).

3.2.3 Shock processes

Assume that degradation results from shocks, each of them leading to an increment of degradation.

Let $T_n, (n \geq 1)$ be the time of the nth shock and X_n the nth increment of the degradation level. Denote by $N(t)$ the number of shocks in the interval $[0, t]$. Set $X_0 = 0$.

The degradation process is given by

$$Z(t) = \sum_{n=1}^{\infty} \mathbf{1}\{T_n \leq t\} X_n = \sum_{n=0}^{N(t)} X_n. \qquad (3.4)$$

Suppose that T_n are the moments of transition of the doubly stochastic Poisson (DSP) process. DSP process is a Poisson process with an intensity function that is also random.

Suppose that this random intensity has the form

$$\lambda(t) = Y \eta(t), \qquad (3.5)$$

where $\eta(t)$ is a deterministic function and Y is a nonnegative random variable with finite expectation. So the distribution of the number of shocks up to time t is defined by

$$\mathbf{P}\{N(t) = k\} = \mathbf{E}\left\{ \frac{(Y\eta(t))^k}{k!} \exp\{-Y\eta(t)\} \right\}. \qquad (3.6)$$

If Y is non-random, N is a non-homogenous Poisson process, in particular, when $Y\eta(t) = \lambda t$, N is a homogenous Poisson process.

Assume that $X_1, , X_2, \cdots$ are conditionally independent given $\{T_n\}$ and assume that the c.d.f. and the probability density functions of X_n given $\{T_n\}$ are G and g, respectively.

For any random vector $(Z_1, Z_2), Z_1 \in \mathbf{R}^k, Z_2 \in \mathbf{R}^l$ such that $\mathbf{P}\{Z_2 = z_2\} > 0$, set

$$p_{Z_1, Z_2 = z_2}(z_1) = \frac{\partial}{\partial z_1} \mathbf{P}\{Z_1 \leq z_1, Z_2 = z_2\}.$$

Proposition 3.1. *The distribution of the random vector*

$$(Z(t_1), Z(t_2) - Z(t_1), \cdots, Z(t_m) - Z(t_{m-1}))$$

is given by:

$$\mathbf{P}\{Z(t_1) = 0, Z(t_2) - Z(t_1) = 0, \cdots, Z(t_m) - Z(t_{m-1}) = 0\} = \mathbf{P}\{N(t_m) = 0\}; \qquad (3.7)$$

for any $1 \le k_1 < \cdots < k_s \le m$ and $u_{k_1}, \cdots, u_{k_s} > 0$

$$p_{Z(t_{k_1})-Z(t_{k_1-1}),\cdots,Z(t_{k_s})-Z(t_{k_s-1}),Z(t_l)-Z(t_{l-1})=0,l\neq k_1,\cdots,k_s}(u_{k_1},\cdots,u_{k_s}) =$$

$$\sum_{i_1=1}^{\infty} \cdots \sum_{i_s=1}^{\infty} g_{i_1}(u_{k_1}) \cdots g_{i_s}(u_{k_s}) \mathbf{P}\{N(t_{k_1}) - N(t_{k_1-1}) = i_1, \cdots,$$

$$N(t_{k_s}) - N(t_{k_s-1}) = i_s, N(t_l) - N(t_{l-1}) = 0, l \neq k_1, \cdots k_s\};$$

for any $u_1, \cdots, u_m > 0$

$$p_{Z(t_1),Z(t_2)-Z(t_1),\cdots,Z(t_m)-Z(t_{m-1})}(u_1, u_2, \cdots, u_m) =$$

$$\sum_{i_1=1}^{\infty} \cdots \sum_{i_m=1}^{\infty} g_{i_1}(u_1) \cdots g_{i_m}(u_m)$$

$$\mathbf{P}\{N(t_1) = i_1, N(t_2) - N(t_1) = i_2, \cdots, N(t_m) - N(t_{m-1}) = i_m\},$$

where g_i is the convolution of i densities g,

$$\mathbf{P}\{N(t_1) = i_1, N(t_2) - N(t_1) = i_2, \cdots, N(t_m) - N(t_{m-1}) = i_m\} =$$

$$\frac{\eta(t_1)^{i_1}}{i_1!} \cdots \frac{\{\eta(t_m) - \eta(t_{m-1})\}^{i_m}}{i_m!} \mathbf{E}\{Y^{i_1+\cdots+i_m} e^{-Y\eta(t_m)}\}. \tag{3.8}$$

Proof. Let us prove (3.8). Denote by G_i the c.d.f. of the sum of i i.i.d. random variables with the c.d.f. G. For any $u_{k_1}, \cdots, u_{k_s} > 0$ we have

$$\mathbf{P}\{u_{k_1} < Z_{k_1} - Z_{k_1-1} < u_{k_1} + h_{k_1}, \cdots, u_{k_1} < Z_{k_s} - Z_{k_s-1} < u_{k_s} + h_{k_s},$$

$$Z(t_l) - Z(t_{l-1}) = 0, l \neq k_1, \cdots, k_s\} =$$

$$\sum_{i_1=1}^{\infty} \cdots \sum_{i_s=1}^{\infty} \mathbf{P}\{u_{k_1} < \sum_{j=N_{k_1-1}+1}^{N_{k_1}} X_j < u_{k_1} + h_{k_1}, \cdots, u_{k_s} <$$

$$\sum_{j=N_{k_s-1}+1}^{N_{k_s}} X_j < u_{k_s} + h_{k_s} \mid N_{k_1} - N_{k_1-1} = i_1,$$

$$\cdots, N_{k_s} - N_{k_s-1} = i_s, N(t_l) - N(t_{l-1}) = 0, l \neq k_1, \cdots, k_s\}\times$$

$$\mathbf{P}\{N_{k_1} - N_{k_1-1} = i_1, \cdots, N_{k_s} - N_{k_s-1} = i_s, N(t_l) - N(t_{l-1}) = 0, l \neq k_1, \cdots, k_s\}$$

$$= \sum_{i_1=1}^{\infty} \cdots \sum_{i_s=1}^{\infty} \{G_{i_1}(u_{k_1} + h_{k_1}) - G_{i_1}(u_{k_1})\} \cdots \{G_{i_s}(u_{k_s} + h_{k_s}) - G_{i_s}(u_{k_s})\}\times$$

$$\mathbf{P}\{N_{k_1} - N_{k_1-1} = i_1, \cdots, N_{k_s} - N_{k_s-1} = i_s, N(t_l) - N(t_{l-1}) = 0, l \neq k_1, \cdots, k_s\}.$$

Dividing by $h_{k_1} \cdots h_{k_s}$ and going to the limit when $h_{k_1}, \cdots, h_{k_s} \to 0$, we obtain (3.8).

Examples.

1) If the distribution of Y is gamma:

$$p_Y(y) = \frac{c^b}{\Gamma(b)} y^{b-1} e^{-cy}, y \ge 0,$$

then

$$\mathbf{E}\{Y^k e^{-Y\eta(t_m)}\} = \frac{c^b \Gamma(k+b)}{\Gamma(b)(c+\eta_m)^{k+b}}.$$

2) If the distribution of Y is inverse Gaussian:

$$p_Y(y) = \mathbf{1}\{y \geq 0\}\sqrt{\frac{\beta}{2\pi y^3}}\exp\{-\frac{1}{2}\frac{\beta(y-\mu)^2}{\mu^2 y}\},$$

then

$$\mathbf{E}\{Y^k e^{-Y\eta(t_m)}\}$$

$$= \exp\{-\frac{\beta}{\mu}(\sqrt{1+2\eta(t_m)\mu^2/\beta}-1)\}\left(\frac{\mu}{\sqrt{1+2\eta(t_m)\mu^2/\beta}}\right)^k \times$$

$$\sum_{j=0}^{k-1}\frac{(k-1+j)!}{(k-1-j)!j!}\left(\frac{\mu}{2\beta\sqrt{1+2\eta(t_m)\mu^2/\beta}}\right)^j \quad (k>0),$$

$$\mathbf{E}\{e^{-Y\eta(t_m)}\} = \int_0^\infty e^{-Y\eta(t_m)}\sqrt{\frac{\beta}{2\pi}}y^{-\frac{3}{2}}\exp\{-\frac{1}{2}\frac{\beta}{\mu^2}\frac{(y-\mu)^2}{y}\}.$$

A wide survey on shock processes is given in Wendt (1999). See also Kahle and Wendt (2000), Wilson (2000). From here we shall follow the paper of Bagdonavičius and Nikulin (2001).

3.3 Modeling the influence of explanatory variables

Suppose that the degradation process is observed under a possibly multi-dimensional and time-dependent explanatory variable (stress, explanatory variable, regressor) $x(\cdot) = (x_0(\cdot), \ldots, x_s(\cdot))^T$, consisting of fixed first coordinate $x_0(t) \equiv 1$ and of s one-dimensional stresses.

We assume in what follows that the deterministic or stochastic process $x(\cdot)$ is bounded right continuous with finite left hand limits.

Denote informally by $Z_{x(\cdot)}(t)$ the degradation level under stress $x(\cdot)$ at the moment t.

We suppose that the process $Z_{x(\cdot)}$ can be transformed via a time-transformation to a process

$$Z(t), \quad t \geq 0,$$

which does not depend on $x(\cdot)$.

The moment of a non-traumatic failure under the explanatory variable $x(\cdot)$ is

$$T_{x(\cdot)} = \sup\{t : Z_{x(\cdot)}(t) < z_0\},$$

i.e. it is the moment when the degradation reaches a critical level z_0 under the explanatory variable $x(\cdot)$.

Let

$$S_{x(\cdot)}(t) = \mathbf{P}\{T_{x(\cdot)} > t \mid x(s), 0 \leq s \leq t\} = \mathbf{P}\{Z_{x(\cdot)}(t) < z_0 \mid x(s), 0 \leq s \leq t\}$$

be the survival function of the random variable $T_{x(\cdot)}$, and $x_0(\cdot)$ be fixed (for example, usual) stress. Set

$$f_{x(\cdot)}(t) = S_{x_0(\cdot)}^{-1}\big(S_{x(\cdot)}(t)\big).$$

Then for all $x(\cdot)$

$$S_{x_0(\cdot)}\big(f_{x(\cdot)}(t)\big) = S_{x(\cdot)}(t).$$

In terms of the probability of survival, the moment t under the explanatory variable $x(\cdot)$ is equivalent to the moment $f_{x(\cdot)}(t)$ under the explanatory variable $x_0(\cdot)$.

Thus, it is natural to assume that the distribution of degradation process, observed under stress $x(\cdot)$, at the moment t is the same as the distribution of degradation process, observed under stress $x_0(\cdot)$, at the moment $f_{x(\cdot)}(t)$:

$$Z_{x(\cdot)}(t) = Z_{x_0(\cdot)}\big(f_{x(\cdot)}(t)\big).$$

This model can be practically used if a concrete form of the functional $f_{x(\cdot)}(t)$ is assumed, i.e. an accelerated life model relating failure time to stress is given.

The simplest model is the AFT model which in terms of the functional $f_{x(\cdot)}(t)$ is formulated as follows

$$\frac{\partial\, f_{x(\cdot)}(t)}{\partial t} = r\{x(t)\},$$

with initial condition $f_{x(\cdot)}(0) = 0$; here r is a positive function on $\mathbf{R}^{s+1}$.

This model implies that

$$f_{x(\cdot)}(t) = \int_0^t r(x(\tau))d\tau.$$

The function r is parametrized as follows:

$$r\{x(t)\} = e^{\beta^T y(t)},$$

where $\beta = \big(\beta_0, \ldots, \beta_s\big)^T$ is the vector of unknown parameters,

$$y(t) = \phi(x(t)) \quad \text{and} \quad \phi : R^{s+1} \to R^{s+1},$$

where ϕ is a specified function. Possible forms of the function ϕ are discussed in Chapter 2. We write x instead of y even when it is not so.

Thus, the AFT model implies that the moment t under the explanatory variable $x(\cdot)$ is equivalent to the moment

$$\int\limits_0^t e^{\beta^T x(s)}\, ds$$

under the explanatory variable $x^0(\cdot)$. This implies

Degradation model with explanatory variables:

$$Z_{x(\cdot)}(t) = Z\left(\int\limits_0^t e^{\beta^T x(s)}\, ds\right).$$ (3.9)

Using models different from the AFT model for time-transformation, more complicated degradation models with explanatory variables can be considered.

3.4 Modeling the traumatic event process

Suppose that under explanatory variable $x(\cdot)$ the degradation process is defined by the model (3.11).

Let $C_{x(\cdot)}$ be the moment of the traumatic failure under $x(\cdot)$. Assume that $C_{x(\cdot)}$ is the first transition of a (possibly non-stationary) Poisson process $(N(t), t \geq 0)$ with intensity $\lambda(Z_{x(\cdot)}(t), x(t))$ at the moment t which depends on a value $Z_{x(\cdot)}(t)$ of the degradation and the value $x(t)$ of the explanatory variable at this moment. This means that for any fixed t the conditional distribution (given trajectories of $x(s)$, $Z_{x(\cdot)}(s)$, $0 \leq s \leq t$ which we denote here also $x(s)$, $Z_{x(\cdot)}(s)$, $0 \leq s \leq t$) of the random variable $N(t)$ is the Poisson distribution with the mean $\int_0^t \lambda(Z_{x(\cdot)}(s), x(s))ds$. In particular

$$\mathbf{P}\{C_{x(\cdot)} > t \mid x(s), Z_{x(\cdot)}(s), 0 \leq s \leq t\} =$$

$$\mathbf{P}\{N(t) = 0 \mid x(s), Z_{x(\cdot)}(s), 0 \leq s \leq t\} = \exp\{-\int_0^t \lambda(Z_{x(\cdot)}(s), x(s))ds\}.$$
(3.10)

To understand the sense of the function $\lambda(z, x)$, consider the conditional probability of the traumatic event in a small interval $(t, t+\Delta]$, given that until the moment t the traumatic event did not occur yet and given $x(s), Z_{x(\cdot)}(s), 0 \leq s \leq t$. This probability is

$$\mathbf{P}\{t < C_{x(\cdot)} \leq t + \Delta \mid C_{x(\cdot)} > t,\, x(s), Z_{x(\cdot)}(s), 0 \leq s \leq t\} =$$

$$1 - \exp\{-\int_t^{t+\Delta} \lambda(Z_{x(\cdot)}(s), x(s))ds\} \approx \lambda\{Z_{x(\cdot)}(t), x(t)\}\Delta.$$

Thus if $\lambda\{Z_{x(\cdot)}(\cdot), x(\cdot)\}$ is right-continuous at the point t, then $\lambda\{Z_{x(\cdot)}(t), x(t)\}$ is proportional to the conditional probability of the traumatic event in a small interval given that at time t the traumatic event has not yet occurred, the value of the explanatory variable was $x(t)$, and the level of degradation was $Z_{x(\cdot)}(t)$.

Call $\lambda(z, x)$ the traumatic event intensity (or killing rate). In what follows we suppose that λ is continuous on $[0, \infty) \times \mathbf{R}^s$. In such a case the conditional survival function (3.12) is continuous and the conditional distribution of $C_{x(\cdot)}$ is absolutely continuous with the density

$$p_{C_{x(\cdot)} \mid x(s), Z_{x(\cdot)}(s), 0 \leq s \leq t}(t)$$

$$= \lambda\{Z_{x(\cdot)}(t), x(t)\} \exp\{- \int_0^t \lambda(Z_{x(\cdot)}(s), x(s)) ds\}. \qquad (3.11)$$

Let $U_{x(\cdot)}$ be the time to failure (traumatic or non-traumatic) of items observed under stress $x(\cdot)$ and $T_{x(\cdot)} = \sup\{t : Z_{x(\cdot)}(t) < z_0\}$. Then $U_{x(\cdot)} = \min(T_{x(\cdot)}, C_{x(\cdot)})$.

Denote by

$$Q_{x(\cdot)}(t) = \mathbf{P}\{C_{x(\cdot)} > t \mid x(s), 0 \le s \le t\} \qquad (3.12)$$

the survival function of the traumatic event time under explanatory variable $x(\cdot)$. If $x(\cdot)$ influences only degradation but not the intensity of traumatic events then the function Q does not depend on $x(\cdot)$.

Let

$$G_{x(\cdot)}(t) = \mathbf{P}\{U_{x(\cdot)} > t \mid x(s), 0 \le s \le t\} \qquad (3.13)$$

be the survival function of the time to any kind of failure under the explanatory variable $x(\cdot)$, and

$$S_{x(\cdot)}(t) = \mathbf{P}\{T_{x(\cdot)} > t \mid x(s), 0 \le s \le t\} \qquad (3.14)$$

be the survival function of the time to non-traumatic failure under the explanatory variable $x(\cdot)$.

Proposition 3.2. *The survival functions $G_{x(\cdot)}(t)$ and $Q_{x(\cdot)}(t)$ have the form:*

$$G_{x(\cdot)}(t) = \mathbf{E}\left\{\exp\{- \int_0^t \lambda\left(Z_{x(\cdot)}(s), x(s)\right) ds\}\mathbf{1}_{\{Z_{x(\cdot)}(t)<z_0\}} \mid x(s), 0 \le s \le t\right\}, \qquad (3.15)$$

and

$$Q_{x(\cdot)}(t) = \mathbf{E}\left\{\exp\{- \int_0^t \lambda\left(Z_{x(\cdot)}(s), x(s)\right) ds\} \mid x(s), 0 \le s \le t\right\}. \qquad (3.16)$$

Proof. Using the properties of conditional expectations we have:

$$G_{x(\cdot)}(t) = \mathbf{P}\{T_{x(\cdot)} > t, C_{x(\cdot)} > t \mid x(s), 0 \le s \le t\}$$

$$= \mathbf{P}\{Z_{x(\cdot)}(t) < z_0, C_{x(\cdot)} > t \mid x(s), 0 \le s \le t\} =$$

$$\mathbf{E}\left\{\mathbf{1}_{\{Z_{x(\cdot)}(t)<z_0, C_{x(\cdot)}>t\}} \mid x(s), 0 \le s \le t\right\}$$

$$= \mathbf{E}\left\{\mathbf{E}(\mathbf{1}_{\{Z_{x(\cdot)}(t)<z_0, C_{x(\cdot)}>t\}} \mid x(s), Z_{x(\cdot)}(s), 0 \le s \le t) \mid x(s), 0 \le s \le t\right\}$$

$$= \mathbf{E}\left\{\mathbf{1}_{\{Z_{x(\cdot)}(t)<z_0\}}\mathbf{E}(\mathbf{1}_{\{C_{x(\cdot)}>t\}} \mid x(s), Z_{x(\cdot)}(s), 0 \le s \le t) \mid x(s), 0 \le s \le t\right\}$$

$$= \mathbf{E}\left\{\mathbf{1}_{\{Z_{x(\cdot)}(t)<z_0\}}\mathbf{P}\{C_{x(\cdot)} > t \mid x(s), Z_{x(\cdot)}(s), 0 \le s \le t\} \mid x(s), 0 \le s \le t\right\}$$

$$= \mathbf{E}\left\{\exp\{- \int_0^t \lambda\left(Z_{x(\cdot)}(s), x(s)\right) ds\}\mathbf{1}_{\{Z_{x(\cdot)}(t)<z_0\}} \mid x(s), 0 \le s \le t\right\}.$$

In the last step we used the formula (3.12). The formula (3.18) is proved similarly.

The intensity λ can be of different forms:

1. λ does not depend on the degradation and explanatory variables:

$$\lambda(z, x) = \alpha_0. \tag{3.17}$$

Then the equalities (3.14)-(3.15) imply

$$Q(t) = e^{-\alpha_0 t}, \quad G_{x(\cdot)}(t) = Q(t)\, S_{x(\cdot)}(t). \tag{3.20}$$

2. λ does not depend on degradation but depends on explanatory variables via the AFT model:

$$\lambda(z, x) = e^{\beta^{*T} x}, \tag{3.18}$$

where $\beta^* = (\beta_0^*, \cdots, \beta_s^*)^T$. We do not multiply the exponential by a constant because, as noted at the beginning of Section 3, the first coordinate in x is unity. Then

$$Q_{x(\cdot)}(t) = \exp\left\{ -\int_0^t e^{\beta^{*T} x(s)} ds \right\}, \quad G_{x(\cdot)}(t) = Q_{x(\cdot)}(t)\, S_{x(\cdot)}(t). \tag{3.19}$$

3. λ depends linearly on degradation and on explanatory variables via the degradation:

$$\lambda(z, x) = \alpha_0 + \alpha_1 z. \tag{3.20}$$

Then

$$\mathbf{P}\{C_{x(\cdot)} > t \mid x(s), Z_{x(\cdot)}(s), 0 \le s \le t\} = \exp\{-\alpha_0 t - \alpha_1 \int_0^t Z_{x(\cdot)}(s) ds\}, \tag{3.21}$$

$$Q_{x(\cdot)}(t) = \mathbf{E}\left\{ \exp\{-\alpha_0 t - \alpha_1 \int_0^t Z_{x(\cdot)}(s) ds\} \mid x(s), 0 \le s \le t \right\}, \tag{3.22}$$

$$G_{x(\cdot)}(t) = \mathbf{E}\left\{ \exp\{-\alpha_0 t - \alpha_1 \int_0^t Z_{x(\cdot)}(s) ds\} \mathbf{1}_{\{Z_{x(\cdot)}(t) < z_0\}} \mid x(s), 0 \le s \le t \right\}, \tag{3.23}$$

$$p_{C_{x(\cdot)} \mid x(s), Z_{x(\cdot)}(s), 0 \le s \le t}(t)$$
$$= \{\alpha_0 + \alpha_1 Z_{x(\cdot)}(t)\} \exp\{-\alpha_0 t - \alpha_1 \int_0^t Z_{x(\cdot)}(s) ds\}. \tag{3.24}$$

4. λ depends linearly on degradation and via the degradation and the AFT model on explanatory variables:

$$\lambda(z, x) = e^{\beta^{*T} x}(1 + \alpha z). \tag{3.25}$$

Then

$$\mathbf{P}\{C_{x(\cdot)} > t \mid x(s), Z_{x(\cdot)}(s), 0 \le s \le t\} = \exp\{-\int_0^t e^{\beta^{*T} x(s)}(1 + \alpha Z_{x(\cdot)}(s)) ds\},$$

$$Q_{x(\cdot)}(t) = \mathbf{E}\left\{ \exp\{-\int_0^t e^{\beta^{*T} x(s)}(1 + \alpha Z_{x(\cdot)}(s)) ds\} \mid x(s), 0 \le s \le t \right\}, \tag{3.26}$$

$$G_{x(\cdot)}(t) =$$

$$\mathbf{E}\left\{\exp\{-\int_0^t e^{\beta^{*T}x(s)}(1+\alpha Z_{x(\cdot)}(s))ds\}\mathbf{1}_{\{Z_{x(\cdot)}(t)<z_0\}} \mid x(s), 0 \le s \le t\right\}.$$
$$(3.27)$$

5. λ depends by power rule on degradation and via the degradation and the AFT model on explanatory variables:

$$\lambda(z, x) = e^{\beta^{*T}x}(1 + \alpha_1 z^{\alpha_2}).\qquad(3.28)$$

Then

$$\mathbf{P}\{C_{x(\cdot)} > t \mid x(s), Z_{x(\cdot)}(s), 0 \le s \le t\} =$$

$$\exp\{-\int_0^t e^{\beta^{*T}x(s)}(1 + \alpha_1 Z_{x(\cdot)}^{\alpha_2}(s))ds\},$$

$$Q_{x(\cdot)}(t) = \mathbf{E}\left\{\exp\{-\int_0^t e^{\beta^{*T}x(s)}(1 + \alpha_1 Z_{x(\cdot)}^{\alpha_2}(s))ds\} \mid x(s), 0 \le s \le t\right\},$$
$$(3.29)$$

$$G_{x(\cdot)}(t) =$$

$$\mathbf{E}\left\{\exp\{-\int_0^t e^{\beta^{*T}x(s)}(1 + \alpha_1 Z_{x(\cdot)}^{\alpha_2}(s))ds\}\mathbf{1}_{\{Z_{x(\cdot)}(t)<z_0\}} \mid x(s), 0 \le s \le t\right\}.$$
$$(3.29)$$

Remark 3.1. *The process generating traumatic events needs not to be a Poisson process. In this case the model*

$$G_{x(\cdot)}(t) =$$

$$\mathbf{E}\left\{G\{\int_0^t \lambda\left(Z_{x(\cdot)}(s), x(s)\right)ds\}\mathbf{1}\{T_{x(\cdot)} > t\} \mid x(s), 0 \le s \le t\right\},$$

where G is a non-exponential survival function, can be considered.

At the end we note that Cox (1999), Doksum and Hoyland (1992), Doksum and Normand (1995), Lawless, Hu and Cao (1995), Lehmann (2000), Lu (1995), Whitmore (1995), Whitmore and Schenkelberg (1997), Whitmore, Crowder and Lawless (1998) model degradation by a Wiener diffusion process.

CHAPTER 4

Maximum likelihood estimation for FTR data

4.1 Censored failure time data

Typically failure-time data are right censored. This means that failure time T is known if it does not exceed a value C, called censoring time. Otherwise it is only known that failure time T is greater than C.

Left censoring means that failure time T is known if it is greater or equal to C, also called censoring time. Otherwise it is only known that failure time T is smaller than C.

Left censoring is fairly rare in analysis of reliability data with explanatory variables and is not considered here.

Right censoring mechanisms can be various:

1) If n units are tested a prespecified time t then censoring is called *Type I censoring*. For all units censoring time $C = t$.

2) If a life test is terminated after a specified number r, $r < n$, of failures occurs then censoring is called *Type II censoring*. For all units censoring time C is the moment of the rth failure.

3) If units are put on test at different time points $t_1, \cdots, t_n$, and the data are to be analyzed at a fixed time point t, $t > \max t_i$, then censoring time for the ith unit $C_i = t - t_i$ is non-random. Such censoring is called *progressive right censoring*.

4) If the failure times $T_1, \cdots, T_n$ and the censoring times $C_1, \cdots, C_n$ are mutually independent random variables then censoring will be called *independent right censoring*. For example, if several failure modes are possible and interest is focused on one particular failure mode then failure of any other mode can be considered as random censoring time.

Type I censoring is a particular case of progressive right censoring. Both are particular cases of independent right censoring.

Suppose that data are right censored, T_i and C_i are failure and censoring times. Set

$$X_i = T_i \wedge C_i, \quad \delta_i = \mathbf{1}_{\{T_i \leq C_i\}} \quad (i = 1, \cdots, n),$$

where $a \wedge b = min(a, b)$, $\mathbf{1}_A$ is the indicator of the event A.

Usually right censored data are presented in the following form:

$$(X_1, \delta_1), \cdots, (X_n, \delta_n). \tag{4.1}$$

If $\delta_i = 1$ then it is known that a failure occurs at the moment $T_i = X_i$. If

$\delta_i = 0$ then it is known that the failure occurs after the moment X_i, i.e. the unit is censored at the moment $C_i = X_i$.

There is another way to describe right censored data. Denote by

$$N_i(t) = \mathbf{1}_{\{X_i \le t, \delta_i = 1\}} = \mathbf{1}_{\{T_i \le t, T_i \le C_i\}} \qquad (4.2)$$

the number of failures of the ith unit in the interval $[0, t]$. It is equal to 1 if failure is *observed* in this interval. Otherwise it is equal to 0.

Set

$$Y_i(t) = \mathbf{1}_{\{X_i \ge t\}}.$$

It is equal to 1 when the ith unit is "at risk" (i.e. it is not censored and not failed) just prior the moment t.

Note that $N(t) = \sum_{i=1}^n N_i(t)$ is the number of observed failures of all units in the interval $[0, t]$ and $Y(t) = \sum_{i=1}^n Y_i(t)$ is the number of units at risk just prior the moment t.

The stochastic processes N, N_i are examples of *counting processes* (see Appendix).

The data can be presented in the form

$$(N_1(t), Y_1(t), t \ge 0), \cdots, (N_n(t), Y_n(t), t \ge 0). \qquad (4.3)$$

The two ways of data presentation are equivalent. Indeed, if (X_i, δ_i) are given then $(N_i(t), Y_i(t)), t \ge 0$ can be found using their definition. *Vice versa*, the moment X_i is the moment of the jump of $Y_i(t)$ from 1 to 0. If $N_i(t)$ has a jump at X_i then $X_i = T_i$ and $\delta_i = 1$. If $N_i(t) = 0$ for any $t \ge 0$ then $X_i = C_i$ and $\delta_i = 0$.

One very important advantage of data presentation in the form (4.3) is the following. The processes N_i and Y_i show dynamics of failure and censoring mechanism over time. If the values of $\{N_i(s), Y_i(s), 0 \le s \le t, i = 1, \cdots, n\}$ are known, the *history* of failures and censorings up to the moment t is known. The data (4.3) gives all history of failures and censorings during the experiment. The notion of the history is formalized by the notion of the *filtration* (see Appendix, Huber (2000), Pons and Huber (2000)).

If the history up to the moment t is known then the values of N_i and Y_i (and of N and Y) at the moment t are known, i.e. the stochastic processes N_i, Y_i, N, Y are *adapted* (see Appendix for the formal definitions).

4.2 Parametric likelihood function for right censored FTR data

Suppose that n units are tested. The ith unit is tested under possibly time-varying explanatory variable $x^{(i)}(\cdot)$. Suppose that distributions of all n units under test are absolutely continuous with the survival functions $S_i(t, \theta)$, the probability densities $p_i(t, \theta)$, and the hazard rates $\alpha_i(t, \theta)$, specified by a common possibly multidimensional parameter $\theta \in \Theta \subset \mathbf{R}^s$. The distributions of units tested under different explanatory variables or stresses generally are different.

Suppose that the data

$$(X_1, \delta_1), \cdots, (X_n, \delta_n).$$

are right censored.

Denote by G_i the survival function of the censoring time C_i. We suppose that the function G_i does not depend on θ.

Suppose that $T_1, \cdots, T_n$ and $C_1, \cdots, C_n$ are mutually independent.

Having a concrete realization of the data, an estimator of the parameter θ should be found by maximizing the probability of such realization with respect to θ. Let us formalize this.

Suppose that for $i = i_1, \cdots, i_k$ the realizations t_i of the failure times T_i are observed and for $i \neq i_1, \cdots, i_k$ the realizations c_i of the censoring times C_i are known. Then it is also known that $C_i \geq t_i$ for $i = i_1, \cdots, i_k$ and $T_i > c_i$ for $i \neq i_1, \cdots, i_k$.

The probability of such concrete realization is zero because the failure times are absolutely continuous. Therefore it is evident that under independent right censoring the function to be maximized with respect to θ is

$$\lim_{h_1, \cdots, h_n \downarrow 0} \frac{1}{h_1, \cdots, h_n} \mathbf{P}\{t_i \leq T_i \leq t_i + h_i, T_i \leq C_i, i = i_1, \cdots, i_k,$$

$$c_i \leq C_i \leq c_i + h_i, C_i < T_i, i \neq i_1, \cdots, i_k\}$$

$$= \prod_{i=i_1, \cdots, i_k} p_i(t_i, \theta) \, G_i(t_i) \prod_{i \neq i_1, \cdots, i_k} g_i(c_i) \, S_i(c_i, \theta), \qquad (4.4)$$

when the censoring times are also absolutely continuous, the survival functions S_i and G_i are differentiable at the points t_i and c_i, respectively, $g_i = -G'$.

The members with G_i and g_i do not contain θ, so they can be rejected. Replacing t_i by X_i when $\delta_i = 1$ and c_i by X_i when $\delta_i = 0$, the likelihood function is obtained:

$$L(\theta) = \prod_{i=1}^{n} p_i^{\delta_i}(X_i, \theta) \, S_i^{1-\delta_i}(X_i, \theta) = \prod_{i=1}^{n} \alpha_i^{\delta_i}(X_i, \theta) \, S_i(X_i, \theta). \qquad (4.5)$$

If C_i are constants (Type I and progressive right censoring) then the function to be maximized is

$$\lim_{h_{i_1}, \cdots, h_{i_k} \downarrow 0} \frac{1}{h_{i_1}, \cdots, h_{i_k}} \mathbf{P}\{t_i \leq T_i \leq t_i + h_i, T_i < C_i, i = i_1, \cdots, i_k,$$

$$C_i < T_i, i \neq i_1, \cdots, i_k\} = \prod_{i=i_1, \cdots, i_k} p_i(t_i, \theta) \prod_{i \neq i_1, \cdots, i_k} S_i(c_i, \theta). \qquad (4.6)$$

The functions G_i do not depend on θ, so the likelihood function has the same form (4.5).

Type II censoring is rarely used in real applications because the time of testing is not known before the rth failure occurs. In accelerated life testing it is sometimes used for each of several groups of units tested under *the same stress* when the failure time distribution is exponential for each value of the stress. In such a case exact confidence intervals for reliability characteristics can be obtained even for small samples.

So in the case of Type II censoring we assume that $p_i = p, S_i = S, \alpha_i = \alpha$.

Under Type II censoring a realization $t_1 \leq \cdots \leq t_r$ of the first r order statistics $T_{1n} \leq \cdots \leq T_{rn}$ is observed. Then the function to be maximized is

$$\lim_{h_1,\cdots,h_r \downarrow 0} \frac{1}{h_1, \cdots, h_r} \mathbf{P}\{t_1 \leq T_{1n} \leq t_1 + h_1, \cdots, t_r \leq T_{rn} \leq t_r + h_r\} =$$

$$\frac{n!}{(n-r)!} \prod_{i=1}^{r} p(t_i, \theta)\, S^{n-r}(t_r, \theta). \tag{4.7}$$

The constant $n!/(n-r)!$ does not depend on θ and can be rejected. The likelihood function is

$$L(\theta) = \prod_{i=1}^{r} p(T_{in}, \theta)\, S^{n-r}(T_{rn}, \theta) = \prod_{i=1}^{n} p^{\delta_i}(X_i, \theta)\, S^{1-\delta_i}(X_i, \theta), \tag{4.8}$$

The last equality is implied by the following: if $T_j = T_{in}$ $(i = 1, \cdots, r)$ then $X_j = T_j \wedge T_{rn} = T_{in}$ and $\delta_j = \mathbf{1}_{\{T_j \leq T_{rn}\}} = 1$; if $T_j > T_{rn}$ then $X_j = T_j \wedge T_{rn} = T_{rn}$ and $\delta_j = \mathbf{1}_{\{T_j \leq T_{rn}\}} = 0$.

The likelihhood function (4.8) has the form (4.5) as in the case of other types of censorings.

The *maximum likelihood (ML) estimator* $\hat{\theta}$ of the parameter θ maximizes the likelihood function (4.5). See also Hjort (1992).

4.3 Score function

Let us consider now the likelihood function (4.5). The logarithm of it is

$$\ln L(\theta) = \sum_{i=1}^{n} \delta_i \log\{\alpha_i(X_i, \theta)\} + \sum_{i=1}^{n} \log\{S_i(X_i, \theta)\}. \tag{4.9}$$

It is maximized in the same point as the likelihood function.

If the functions $\alpha_i(u, \theta)$ are sufficiently smooth then the ML estimator verifies the equation:

$$U(\hat{\theta}) = 0,$$

where U is the *score function*:

$$U(\theta) = \frac{\partial}{\partial \theta} \ln L(\theta) = \left(\frac{\partial}{\partial \theta_1} \ln L(\theta), \cdots, \frac{\partial}{\partial \theta_s} \ln L(\theta) \right)^T. \tag{4.10}$$

Set

$$U_j(\theta) = \frac{\partial}{\partial \theta_j} \ln L(\theta). \tag{4.11}$$

The score function is the vector $U(\theta) = (U_1(\theta), \cdots, U_s(\theta))^T$.

The equality (4.9) implies

$$U_j(\theta) = \sum_{i=1}^{n} \delta_i \frac{\partial}{\partial \theta_j} \log\{\alpha_i(X_i, \theta)\} + \sum_{i=1}^{n} \frac{\partial}{\partial \theta_j} \log\{S_i(X_i, \theta)\}. \qquad (4.12)$$

It was mentioned that the data (4.1) can be given in the form (4.3). Let us write $\ln L(\theta)$ and $U(\theta)$ in terms of the processes N_i and Y_i.

The trajectories of N_i have the form:

$$N_i(t) = \begin{cases} 0, & 0 \le t < X_i, \\ 1, & t \ge X_i, \end{cases}$$

when $\delta_i = 1$, and $N_i(t) = 0$ for all $t \ge 0$ when $\delta_i = 0$. So (see Appendix, Section A.3):

$$\int_0^\infty \log\{\alpha_i(u, \theta)\}dN_i(u) = \begin{cases} \log\{\alpha_i(X_i, \theta)\}, & \delta_i = 1, \\ 0, & \delta_i = 0. \end{cases} = \delta_i \log\{\alpha_i(X_i, \theta)\}. \qquad (4.13)$$

The trajectories of Y_i have the form:

$$Y_i(t) = \begin{cases} 1, & 0 \le t \le X_i, \\ 0, & t > X_i. \end{cases}$$

So

$$\int_0^\infty Y_i(u)\alpha_i(u)du = \int_0^{X_i} \alpha_i(u)du = -\log\{S_i(X_i, \theta)\}. \qquad (4.14)$$

The equalities (4.9),(4.12)-(4.14) imply that

$$\ln L(\theta) = \sum_{i=1}^{n} \int_0^\infty \log\{\alpha_i(u, \theta)\}dN_i(u) - \sum_{i=1}^{n} \int_0^\infty Y_i(u)\alpha_i(u, \theta)du. \qquad (4.15)$$

$$U_j(\theta) = \sum_{i=1}^{n} \int_0^\infty \frac{\partial}{\partial \theta_j} \log\{\alpha_i(u, \theta)\}dN_i(u) - \sum_{i=1}^{n} \int_0^\infty Y_i(u)\frac{\partial}{\partial \theta_j}\alpha_i(u)du. \qquad (4.16)$$

4.4 Asymptotic properties of the maximum likelihood estimators

The asymptotic properties of the maximum likelihood estimator $\hat{\theta}$ are closely related to the asymptotic properties of the score function $U(\theta)$.

Indeed, denote by θ_0 the true value of the parameter θ. Using Taylor expansion of $n^{-1/2}U_j(\theta)$ around θ_0 in $\theta = \hat{\theta}$, we obtain

$$-U(\theta_0) = U(\hat{\theta}) - U(\theta_0) = \left(\frac{\partial}{\partial \theta_{j'}} U_j(\theta^{(j)}) \right)_{s \times s} (\hat{\theta} - \theta_0), \qquad (4.17)$$

where $\theta^{(j)}$ are on the line segment between $\hat{\theta}$ and θ_0. Set

$$\mathbf{I}(\theta) = (I_{jj'})_{s \times s} = \left(-\frac{\partial}{\partial \theta_{j'}} U_j(\theta) \right)_{s \times s} = \left(-\frac{\partial^2 \ln L(\theta)}{\partial \theta_j \partial \theta_{j'}} \right)_{s \times s}. \qquad (4.18)$$

The equality (4.17) implies that

$$n^{1/2}(\hat{\theta} - \theta_0) = \left(-\frac{1}{n}\frac{\partial}{\partial\theta_{j'}}U_i(\theta^{(j)})\right)^{-1} n^{-1/2}U(\theta_0) =$$

$$\left(-\frac{1}{n}\frac{\partial}{\partial\theta_{j'}}U_j(\theta_0)\right)^{-1} n^{-1/2}\,U(\theta_0) + \Delta = \left(\frac{1}{n}\mathbf{I}(\theta_0)\right)^{-1} n^{-1/2}\,U(\theta_0) + \Delta.$$

$$(4.19)$$

Note that

$$I_{jj'}(\theta) = \sum_{i=1}^{n}\int_{0}^{\infty}\frac{\partial^2}{\partial\theta_j\theta_{j'}}\log\{\alpha_i(u,\theta)\}dN_i(u) - \sum_{i=1}^{n}\int_{0}^{\infty}Y_i(u)\frac{\partial^2}{\partial\theta_j\theta_{j'}}\alpha_i(u,\theta)du.$$

$$(4.20)$$

If $\Delta \xrightarrow{P} 0$ then the asymptotic distributions of the random variables

$$n^{1/2}(\hat{\theta} - \theta_0) \quad \text{and} \quad (\frac{1}{n}\mathbf{I}(\theta_0))^{-1}n^{-1/2}U(\theta_0) \qquad (4.21)$$

are the same.

So if $(\frac{1}{n}\mathbf{I}(\theta_0))^{-1}$ converges in probability to a nonrandom matrix then the asymptotic properties of the maximum likelihood estimator $\hat{\theta}$ can be obtained from the asymptotic properties of the score function $U(\theta)$.

Note that the components (4.16) of the score statistic can be written in the form

$$U_j(\theta) = U_j(\infty, \theta), \qquad (4.22)$$

where

$$U_j(t,\theta) = \sum_{i=1}^{n}\int_{0}^{t}\frac{\partial}{\partial\theta_j}\log\{\alpha_i(u,\theta)\}dM_i(u,\theta), \qquad (4.23)$$

and

$$M_i(t,\theta) = N_i(t) - \int_{0}^{t}Y_i(u)\,\alpha_i(u,\theta)du. \qquad (4.24)$$

Set

$$U_j^{(n)}(t,\theta) = n^{-1/2}U_j(t,\theta), \quad H_{ij}^{(n)}(t,\theta) = n^{-1/2}\frac{\partial}{\partial\theta_j}\log\{\alpha_i(u,\theta)\}. \quad (4.25)$$

Then

$$U_j^{(n)}(t,\theta) = \sum_{i=1}^{n}\int_{0}^{t}H_{ij}^{(n)}(u,\theta)dM_i(u,\theta) \quad (j = 1,\cdots,s). \qquad (4.26)$$

If we are interested not only in the asymptotic distribution of the ML estimator $\hat{\theta}$ but also in goodness-of-fit, it is very useful to have asymptotic properties not only of the random variable $U(\theta)$ but also of the stochastic process $U(\cdot,\theta) = (U_1(\cdot,\theta),\cdots,U_s(\cdot,\theta))^T$.

The score statistics of the form

$$U_j(\tau,\theta) = \sum_{i=1}^{n}\int_{0}^{\tau}H_{ij}(u,\theta)\}dM_i(u,\theta), \qquad (4.27)$$

are obtained and in the case of *semiparametric estimation* using such models as PH, AFT, GPH, CHSS, etc. Semiparametric estimation is used when the baseline function of the model is completely unknown and is treated as an unknown infinite-dimensional parameter. The functions $H_{ij}^{(n)}(u, \theta)$ usually are not deterministic, as here, but (left-continuous) stochastic processes.

The most asymptotic results can be obtained using the fact that the stochastic processes $M_i(t, \theta)$ are *martingales* with respect to the filtration generated by the data (see Appendix, Sections A.5, A.6). Under some assumptions on the processes $H_{ij}^{(n)}(u, \theta)$ the stochastic processes $U_j(t, \theta)$ are also martingales or *local martingales* (see Appendix, Sections A.9, A.10). So the limit distribution of the score statistics can be obtained by applying the *central limit theorem for martingales* (see Appendix, Theorem A.7).

Proofs of the asymptotic properties of the ML estimators under independent right censoring were given by Borgan (1984) and can also be found in Andersen, Borgan, Gill and Keiding (1993, Section VI.1.11). We give only some comments on the conditions of Borgan.

These conditions are:

a) *There exists a neighborhood Θ_0 of the true value θ_0 of θ such that for all i, $\theta \in \Theta_0$ the derivatives of $\alpha_i(u, \theta)$ up to the third order with respect to θ exist and are continuous in θ for $\theta \in \Theta_0$, and the integrals*

$$\int_0^t \alpha_i(u, \theta)du$$

for any finite t may be three times differentiated with respect to $\theta \in \Theta_0$ by interchanging the order of integration and differentiation.

Indeed, these conditions are needed when writing the Taylor expansion (4.17) and differentiating the score function by interchanging the order of integration and differentiation.

b) *There exists a positively definite matrix $\Sigma_0 = (\sigma_{jj'})_{m \times m}$ such that*

$$n^{-1}\sum_{i=1}^{n}\int_0^{\infty} \frac{\partial \log \alpha_i(u, \theta_0)}{\partial \theta_j} \frac{\partial \log \alpha_i(u, \theta_0)}{\partial \theta_{j'}} \alpha_i(u, \theta_0)Y_i(u)du \xrightarrow{\text{P}} \sigma_{jj'}, \quad (4.28)$$

$$n^{-1}\sum_{i=1}^{n}\int_0^{\infty} \left(\frac{\partial \log \alpha_i(u, \theta_0)}{\partial \theta_j}\right)^2 \mathbf{1}_{\{|n^{-1/2}\frac{\partial \log \alpha_i(u, \theta_0)}{\partial \theta_j}| \geq \varepsilon\}} Y_i(v) \alpha_i(u, \theta_0)\, du \xrightarrow{\text{P}} 0.$$

$$(4.29)$$

These are the conditions of the Theorem A.7 for the score function $U(\theta_0)$ to be asymptotically normal. The first condition is just condition on convergence in probability of the predictable covariations of the score process and the second is the Lindeberg condition. Theorem A.7 implies that under these conditions

$$n^{-1/2}U(\theta_0) \xrightarrow{\mathcal{D}} N(0, \Sigma_0); \text{ as } n \to \infty. \tag{4.30}$$

The first condition (4.29) can be written in the following form:

$$n^{-1}\sum_{i=1}^{n}\int_0^\infty \left\{ \frac{\partial^2 \alpha_i(u, \theta_0)}{\partial\theta_j\partial\theta_{j'}} - \frac{\partial^2 \log\{\alpha_i(u, \theta_0)\}}{\partial\theta_j\partial\theta_{j'}} \alpha_i(u, \theta_0) \right\} Y_i(u)du \xrightarrow{P} \sigma_{jj'}. \tag{4.31}$$

The equality (4.20) implies that

$$-n^{-1}I_{jj'}(\theta) = n^{-1}\sum_{i=1}^{n}\int_0^\infty \left\{ \frac{\partial^2 \alpha_i(u, \theta)}{\partial\theta_j\partial\theta_{j'}} - \frac{\partial^2 \log\{\alpha_i(u, \theta)\}}{\partial\theta_j\partial\theta_{j'}} \alpha_i(u, \theta) \right\} Y_i(u)du$$

$$-n^{-1}\sum_{i=1}^{n}\int_0^\infty \frac{\partial^2 \log\{\alpha_i(u, \theta)\}}{\partial\theta_j\partial\theta_{j'}}dM_i(u), \tag{4.32}$$

where M_i are given by (4.24). By (4.31) the first term converges in probability to $\sigma_{jj'}$). If

$$n^{-2}\sum_{i=1}^{n}\int_0^\infty \left(\frac{\partial^2}{\partial\theta_j\partial\theta_{j'}} \log\{\alpha_i(u, \theta_0)\} \right)^2 \alpha_i(u, \theta_0)Y_i(u)du \xrightarrow{P} 0. \tag{4.33}$$

then by Corollary A.6 the second term of the right side of (4.32) converges in probability to zero. So

$$n^{-1}\mathbf{I}(\theta_0) \xrightarrow{P} \Sigma_0,$$

and

$$(n^{-1}\mathbf{I}(\theta_0)^{-1}n^{-1/2}U(\theta_0) \xrightarrow{\mathcal{D}} N(0, \Sigma_0^{-1}) \text{ as } n \to \infty. \tag{4.34}$$

If to the conditions a) and b) to add conditions under which $\Delta \xrightarrow{\mathcal{D}} 0$ then (cf. (4.19))

$$n^{1/2}(\hat{\theta} - \theta_0) \xrightarrow{\mathcal{D}} N(0, \Sigma_0^{-1}).$$

These conditions are related with the third derivatives of the hazard functions in the neighborhood of θ_0 because Δ shows the difference between the second derivatives of the hazard functions (cf. 4.19) at the points $\theta^{(i)}$ and θ_0. So the last group of conditions is

c) *For any n and i there exist measurable functions g_{in} and h_{in} not depending on θ such that for all $t \geq 0$*

$$sup_{\theta\in\Theta_0} \left| \frac{\partial^3 \alpha_i(u, \theta)}{\partial\theta_j\partial\theta_{j'}\partial\theta_{j''}} \right| \leq g_{in}(u), \tag{4.35}$$

and

$$sup_{\theta\in\Theta_0} \left| \frac{\partial^3 \log \alpha_i(u, \theta)}{\partial\theta_j\partial\theta_{j'}\partial\theta_{j''}} \right| \leq h_{in}(u), \tag{4.36}$$

for all j, j', j''. Moreover

$$n^{-1}\sum_{i=1}^{n}\int_0^\infty g_{in}(u)Y_i(u)du, \quad n^{-1}\sum_{i=1}^{n}\int_0^\infty h_{in}(u)\alpha_i(u, \theta_0)Y_i(u)du,$$

$$n^{-1} \sum_{i=1}^{n} \int_{0}^{\infty} \left(\frac{\partial^2}{\partial \theta_j \partial \theta_{j'}} \log\{\alpha_i(u, \theta_0)\} \right)^2 \alpha_i(u, \theta_0) Y_i(u) du$$

all converge in probability to finite quantities as $n \to \infty$, and, for all $\varepsilon > 0$

$$n^{-1} \sum_{i=1}^{n} \int_{0}^{\infty} h_{in}(u) \mathbf{1}_{\{n^{-1/2} h_{in}^{-1/2}(u) \geq \varepsilon\}} \alpha_i(u, \theta_0) Y_i(u) du \xrightarrow{\mathbf{P}} 0.$$

Theorem 4.1. (Borgan (1984)) *Under independent right censoring and the conditions a)-c) , with a probability tending to one, the equation $U(\theta) = 0$ has a solution $\hat{\theta}$ such that $\hat{\theta} \xrightarrow{\mathbf{P}} \theta_0$,*

$$n^{-1/2} U(\theta_0) \xrightarrow{D} N(0, \Sigma_0), \quad n^{1/2}(\hat{\theta} - \theta_0) \xrightarrow{D} N(0, \Sigma_0^{-1}), \qquad (4.37)$$

and the matrix Σ_0 may be consistently estimated by $n^{-1} \mathbf{I}(\hat{\theta})$.

Theorem 4.1 implies that

$$U(\theta_0)^T \mathbf{I}^{-1}(\hat{\theta}) U(\theta_0) \xrightarrow{D} \chi^2(s). \qquad (4.38)$$

and

$$(\hat{\theta} - \theta_0)^T \mathbf{I}(\hat{\theta}) (\hat{\theta} - \theta_0) \xrightarrow{D} \chi^2(s), \qquad (4.39)$$

where $\chi^2(s)$ denotes the chi-square law with s degrees of freedom. The statistic at the left side of (4.38) is called the *score statistic*, and the statistic at the left side of (4.39) is called the *Wald statistic*. (See, for example, Greenwood and Nikulin (1996)).

The Wald statistic is asymptotically equivalent to the likelihood ratio statistic $2(\ln L(\hat{\theta}) - \ln L(\theta_0))$. It is seen from the following considerations. Using Taylor expansion of $\ln(\theta_0)$ around $\hat{\theta}$, we have

$$2 \left(\ln L(\hat{\theta}) - \ln L(\theta_0) \right) = -2U(\hat{\theta}) + (\hat{\theta} - \theta_0)^T \mathbf{I}(\hat{\theta})(\hat{\theta} - \theta_0) + \delta =$$
$$(\hat{\theta} - \theta_0)^T \mathbf{I}(\hat{\theta})(\hat{\theta} - \theta_0) + \delta.$$

It can be shown that under the assumptions of Theorem 4.1 $\delta \xrightarrow{\mathbf{P}} 0$, and hence

$$2 \left(\ln L(\hat{\theta}) - \ln L(\theta_0) \right) \xrightarrow{D} \chi^2(m). \qquad (4.40)$$

So if n is large then the distribution of the the the ML estimator $\hat{\theta}$ is approximated by the normal law:

$$\hat{\theta} \approx N(\theta_0, n \Sigma_0^{-1}), \qquad (4.41)$$

and the distributions of the Wald, score and likelihood ratio statistics are approximated by the chi-square distribution with m degrees of freedom.

The covariance matrix $n \Sigma_0^{-1}$ is estimated by $\mathbf{I}(\hat{\theta})$.

4.5 Approximate confidence intervals

The delta method given in Appendix, Section A.15, gives a method of confidence interval construction for reliability characteristics.

Suppose that $\hat{\theta} = (\hat{\theta}_1, \cdots, \hat{\theta}_s)^T$ is an estimator of $\theta = (\theta_1, \cdots, \theta_s)^T$ and it is known that

$$a_n(\hat{\theta} - \theta) \overset{D}{\to} N_p(0, \Sigma_0^{-1}(\theta)) \quad \text{as} \quad a_n \to \infty. \tag{4.42}$$

For example, $\hat{\theta}$ may be the maximum likelihood estimator.

Suppose that $g : R^p \to R$ is a function verifying the conditions of Theorem A.10. This theorem implies that

$$a_n(g(\theta) - g(\hat{\theta})) \overset{D}{\to} N(0, J_g(\theta)\Sigma_0^{-1}(\theta)J_g^T(\theta)), \tag{4.43}$$

with

$$J_g(\theta) = \left(\frac{\partial g(\theta)}{\partial \theta_1}, \cdots, \frac{\partial g(\theta)}{\partial \theta_s} \right)^T.$$

If n is large then (4.42) implies that

$$\hat{\theta} \approx N_s(\theta, \Sigma^{-1}(\theta)) \tag{4.44}$$

where

$$\Sigma(\theta) = a_n^2 \Sigma_0(\theta).$$

The convergence (4.43) implies that

$$g(\hat{\theta}) \approx N_s(g(\theta), J_g^T(\theta)\Sigma^{-1}(\theta)J_g(\theta)), \tag{4.45}$$

Suppose that $a_n^{-2}\mathbf{I}(\hat{\theta})$ is a consistent estimator of $\Sigma_0(\theta)$.
We have

$$\frac{g(\hat{\theta}) - g(\theta)}{\sigma_g(\hat{\theta})} \approx N(0, 1), \tag{4.46}$$

where

$$\sigma_g^2(\hat{\theta}) = J_g^T(\hat{\theta})\mathbf{I}^{-1}(\hat{\theta})J_g(\hat{\theta}). \tag{4.47}$$

This may be used to construct (see, for example, Bagdonavičius, Nikoulina and Nikulin (1997)) an approximated confidence interval for $g(\theta)$. The most used functions in this book functions

$$g(\theta) = S(t, \theta), \quad g(\theta) = t_p(\theta), \quad g(\theta) = m(\theta),$$

i.e. the survival function, the $p-$quantile and the mean. The survival function takes values in the interval $(0, 1]$, so the approximation by the normal law is improved by using the transformations

$$Q(t, \theta) = \ln \frac{S(t, \theta)}{1 - S(t, \theta)}.$$

Taking into account that

$$(\ln \frac{u}{1 - u})' = \frac{1}{u(1 - u)}$$

and using (4.46), it is obtained that

$$\frac{Q(t, \hat{\theta}) - Q(t, \theta)}{\sigma_Q(\hat{\theta})} \approx N(0, 1),$$

where

$$\sigma_Q(\hat{\theta}) = \frac{1}{S(t,\hat{\theta})(1 - S(t,\hat{\theta}))}\sigma_g(\hat{\theta}). \tag{4.48}$$

It implies that

$$\mathbf{P}\{Q(t,\hat{\theta}) - \sigma_Q(\hat{\theta})w_{1-\alpha/2} \le Q(t,\theta) \le Q(t,\hat{\theta}) + \sigma_Q(\hat{\theta})w_{1-\alpha/2}\} \approx 1 - \alpha,$$

where $w_{1-\alpha/2}$ is the $(1 - \alpha/2)$-quantile of the standard normal law. Solving the inequalities with respect to $S(t,\theta)$ we obtain that an *approximated* $(1-\alpha)$-*confidence interval for the survival function* is

$$\left((1 + \frac{1 - S(t,\hat{\theta})}{S(t,\hat{\theta})}e^{\sigma_Q(\hat{\theta})w_{1-\alpha/2}})^{-1}, \ (1 + \frac{1 - S(t,\hat{\theta})}{S(t,\hat{\theta})}e^{-\sigma_Q(\hat{\theta})w_{1-\alpha/2}})^{-1} \right).$$
$$\tag{4.49}$$

Analogously, if

$$g(\theta) = t_p(\theta) \quad \text{or} \quad g(\theta) = m(\theta)$$

then the function g takes positive values and the approximation is improved by using the transformation

$$K(\theta) = \ln g(\theta).$$

Taking into account that $(\ln u)' = 1/u$ and using (4.46), it is obtained that

$$\frac{K(\hat{\theta}) - K(\theta)}{\sigma_K(\hat{\theta})} \approx N(0,1),$$

where

$$\sigma_K(\hat{\theta}) = \frac{1}{g(\hat{\theta})}\sigma_g(\hat{\theta}). \tag{4.50}$$

It implies an approximate $(1 - \alpha)$-confidence interval for $g(\theta)$:

$$\left(g(\hat{\theta}) \exp\{-\sigma_K(\hat{\theta})w_{1-\alpha/2}\}, \ g(\hat{\theta}) \exp\{\sigma_K(\hat{\theta})w_{1-\alpha/2}\} \right). \tag{4.51}$$

4.6 Some remarks on semiparametric estimation

Going through accelerated life models one can see (see Section 2.11) that generally the cumulative hazard $A_{x(\cdot)}$ can be written as a functional of a baseline function $A(t)$, which does not depend on $x(\cdot)$, and a finite-dimensional parameter θ:

$$A_{x(\cdot)}(t) = g(t, \theta, A(s), x(s), 0 \le s \le t),$$

and

$$\frac{\partial}{\partial \theta} \log\{\alpha_{x(\cdot)}(t)\} = w(t, \theta, A(s), A'(s), A''(s), x(s), 0 \le s \le t).$$

If for a concrete model the function A is completely unknown then this model is semiparameric. It contains parametric submodels with A specified. Under any such submodel the parameter θ can be estimated by the parametric maximum likelihood estimator using the score function (cf. 4.23):

$$U(\theta) = \sum_{i=1}^{n} \int_{0}^{\infty} w(t, \theta, A(s), A'(s), A''(s), x^{(i)}(s), 0 \le s \le t) \times$$

$$\{dN_i(t) - Y_i(t)dg(t, \theta, A(s), x^{(i)}(s), 0 \le s \le t)\}, \tag{4.52}$$

It will be seen in the following sections that if A is unknown then for any fixed θ a consistent estimator $\tilde{A}(\cdot, \theta)$ can be easily obtained using the martingale property of the difference $dN_i(t) - Y_i(t)dg(t, \theta, A(s), x^{(i)}(s), 0 \le s \le t)$, or other considerations. This estimator may be explicit (AFT, PH, and many other models) or defined recurrently (GPH, GAH, and other models). As a rule A is a baseline cumulative hazards and $\tilde{A}$ is some modification of the Nelson-Aalen estimator (Appendix, Section A.13). The weights w do not depend on A, A', A'' for the PH model and depend only on A for some models (GPH, for example). In such a case the unknown function A is replaced in (4.52) by the estimator $\tilde{A}$ and modified score function for estimation of θ is obtained. Under regularity conditions such estimator is semiparametrically efficient in the sense that the asymptotic covariance matrix of $n^{1/2}(\hat{\theta} - \theta)$ under the semiparametric model and under the worst parametric model coincide. Strict definitions of semiparametric efficiency see in Andersen et al (1993) or Bickel et al (1993).

For some models the weight depends also on $\alpha = A'$, $\alpha' = A''$, or both. In such a case two ways are possible. Generally the properties of $\hat{\theta}$ do not depend much on the weight and the first possibility is to replace the optimal weight w by some appropriate simpler weight. An other way is to estimate the baseline hazard α, sometimes α' by some kernel estimators.

The final estimator of A is $\hat{A}(t) = \tilde{A}(t, \hat{\theta})$.

Estimators of reliability characteristics are obtained by replacing θ and A by their estimations in the expressions of these characteristics.

CHAPTER 5

Parametric AFT model

5.1 Parametrization of the AFT model

Let
$$x(\cdot) = (x_0(\cdot), ..., x_m(\cdot))^T,$$
be a possibly time-varying and multidimensional explanatory variable; here $x_0(t) \equiv 1$ and $x_1(\cdot), ..., x_m(\cdot)$ are univariate explanatory variables.

Under the AFT model the survival function under $x(\cdot)$ is

$$S_{x(\cdot)}(t) = S_0 \left(\int_0^t r(\tau) d\tau \right). \qquad (5.1)$$

If the explanatory variables are constant over time then the model (5.1) is written as

$$S_{x(\cdot)}(t) = S_0 \left(r(x)t \right). \qquad (5.2)$$

The function r is parametrized in the following form:

$$r(x) = e^{-\beta^T z}, \qquad (5.3)$$

where $\beta = (\beta_0, \cdots, \beta_m)^T$ is a vector of unknown parameters and

$$z = (z_0, \cdots, z_m)^T = (\varphi_0(x), \cdots, \varphi_m(x))^T$$

is a vector of specified functions φ_i, with $\varphi_0(t) \equiv 1$.

Under the parametrized AFT model the survival function under $x(\cdot)$ is

$$S_{x(\cdot)}(t) = S_0 \left(\int_0^t e^{-\beta^T x(\tau)} d\tau \right), \qquad (5.4)$$

and $x_j(\cdot)$ $(j = 1, \ldots, m)$ are not necessarily the observed explanatory variables. They may be some specified functions $\varphi_j(x)$. Nevertheless, we use the same notation x_j for $\varphi_j(x)$.

If the explanatory variables are constant over time then the model (5.4) is written as

$$S_x(t) = S_0 \left(e^{-\beta^T x} t \right), \qquad (5.5)$$

and the logarithm of the failure time T_x under x may be written as

$$\ln\{T_x\} = \beta^T x + \varepsilon,$$

where the survival function of the random variable ε does not depend on x and is $S(t) = S_0(\ln t)$. Note that *in the case of lognormal failure-time distribution the distribution of ε is normal and we have the standard multiple linear regression model.*

For time-varying explanatory variables the distribution of the random variable

$$R = \int_0^{T_{(x\cdot)}} e^{-\beta^T x(\tau)} d\tau$$

is parameter-free with the survival function $S_0(t)$

Let us discuss the choice of the functions φ_i.

5.1.1 Interval valued explanatory variables

Suppose at first that the explanatory variables are interval-valued (load, temperature, stress, voltage, pressure).

If the model (5.2) holds on E_0, then for all $x_1, x_2 \in E_0$

$$S_{x_2}(t) = S_{x_1}(\rho(x_1, x_2)t), \tag{5.6}$$

where the function $\rho(x_1, x_2) = r(x_2)/r(x_1)$ shows the degree of scale variation. It is evident that $\rho(x, x) = 1$.

Suppose at first that x is one-dimensional. The rate of scale variation with respect is defined by the *infinitesimal characteristic* (see Viertl (1988)):

$$\delta(x) = \lim_{\Delta x \to 0} \frac{\rho(x, x + \Delta x) - \rho(x, x)}{\Delta x} = [log\, r(x)]'. \tag{5.7}$$

So for all $x \in E_0$ the function $r(x)$ is given by the formula:

$$r(x) = r(x_0) \exp\left\{ \int_{x_0}^{x} \delta(v)\, dv \right\}, \tag{5.8}$$

where $x_0 \in E_0$ is a fixed explanatory variable.

Suppose that $\delta(x)$ is *proportional* to a specified function $u(x)$:

$$\delta(x) = \alpha\, u(x).$$

In this case

$$r(x) = e^{-\beta_0 - \beta_1 \varphi_1(x)}, \tag{5.9}$$

where $\varphi_1(x)$ is the primitive of $u(x)$, β_0, β_1 are unknown parameters.

Example 5.1. $\delta(x) = \alpha$, $\varphi_1(x) = x$, i.e. the rate of scale changing is constant. Then

$$r(x) = e^{-\beta_0 - \beta_1 x}. \tag{5.10}$$

It is the *log-linear model*.

Example 5.2. $\delta(x) = \alpha/x$, $\varphi_1(x) = \ln x$. Then

$$r(x) = e^{-\beta_0 - \beta_1 log\, x} = \alpha_1 x^{\beta_1}. \tag{5.11}$$

It is the *power rule model*.

Example 5.3. $\delta(x) = \alpha/x^2$, $\varphi_1(x) = -1/x$. Then

$$r(x) = e^{-\beta_0 - \beta_1/x} = \alpha_1 e^{-\beta_1/x}. \tag{5.12}$$

It is the *Arrhenius model.*

Example 5.4. $\delta(x) = \alpha/x(1-x)$, $\varphi_1(x) = \ln \frac{x}{1-x}$. Then

$$r(x) = e^{-\beta_0 - \beta_1 \ln \frac{x}{1-x}} = \alpha_1 \left(\frac{x}{1-x} \right)^{-\beta_1}, \quad 0 < x < 1. \tag{5.13}$$

It is the *Meeker-Luvalle* model (1995).

The Arrhenius model is used to model product life when the explanatory variable is the temperature, the power rule model - when the explanatory variable is voltage, mechanical loading, the log-linear model is applied in endurance and fatigue data analysis, testing various electronic components (see Nelson (1990)). The model of Meeker-Luvalle is used when x is the proportion of humidity.

If it is not very clear which of the first three models to choose, one can take a larger class of models. For example, all these models are the particular cases of the class of models determined by

$$\delta(x) = \alpha \, x^\gamma$$

with unknown γ or, in terms of the function $r(x)$, by

$$r(x) = \begin{cases} e^{-\beta_0 - \beta_1 (x^\varepsilon - 1)/\varepsilon}, & \text{if } \varepsilon \neq 0; \\ e^{-\beta_0 - \beta_1 \log x}, & \text{if } \varepsilon = 0. \end{cases} \tag{5.14}$$

In this case the parameter ε must be estimated.

The model (5.6) can be generalized. One can suppose that $\delta(x)$ is a linear combination of some specified functions of the explanatory variable:

$$\delta(x) = \sum_{i=1}^{k} \alpha_i \, u_i(x).$$

In such a case

$$r(x) = exp \left\{ -\beta_0 - \sum_{i=1}^{k} \beta_i z_i(x) \right\}, \tag{5.15}$$

where $z_i(x)$ are specified functions of the explanatory variable, $\beta_0, \ldots, \beta_k$ are unknown (possibly not all of them) parameters.

Example 5.5. $\delta(x) = 1/x + \alpha/x^2$.
Then

$$r(x) = e^{-\beta_0 - \beta_1 \log x - \beta_2/x} = \alpha_1 x e^{-\beta_2/x}, \tag{5.16}$$

where $\beta_1 = -1$. It is the *Eyring model*, applied when the explanatory variable x is the temperature.

Example 5.6. $\delta(x) = \sum_{i=1}^{k} \alpha_i/x^i$.
Then

$$r(x) = exp \left\{ -\beta_0 - \beta_1 \log x - \sum_{i=1}^{k-1} \beta_i/x^i \right\}. \tag{5.17}$$

It is the *generalized Eyring model.*

Suppose now that the explanatory variable $x = (x_1, \cdots, x_m)$ is multidimensional.

If there are no interactions between $x_1, \cdots, x_m$ then the model

$$r(x) = exp\left\{-\beta_0 - \sum_{i=1}^{m}\sum_{j=1}^{k_i}\beta_{ij}z_{ij}(x_i)\right\}, \qquad (5.18)$$

could be used; here $z_{ij}(x_i)$ are specified functions, β_{ij} are unknown parameters.

Example 5.7. If the influence of the first explanatory variable is defined by the power rule model and the influence of the second by the Arrhenius model then we have the model

$$r(x_1, x_2) = exp\left\{-\beta_0 - \beta_1 log\, x_1 - \beta_2/x_2\right\}. \qquad (5.19)$$

So $k_1 = k_2 = 1$ here.

If there is interaction between the explanatory variables then complementary terms should be included.

Example 5.8. Suppose that there is interaction between the explanatory variables x_1 and x_2 defined in Example 5.7. Then the model

$$r(x_1, x_2) = exp\left\{-\beta_0 - \beta_1 log\, x_1 - \beta_2/x_2 - \beta_3(log\, x_1)/x_2\right\}. \qquad (5.20)$$

could be considered.

5.1.2 Discrete and categorical explanatory variables.

If the explanatory variables are discrete (number of simultaneous users of a system, number of hardening treatments) then the form of the functions have the same form as in the case of interval-valued explanatory variables, i.e. φ_j may be $\varphi_j(x) = x$, $\ln x$ or $1/x$.

If the jth explanatory variable is categorical (location, manufacturer, design) and take k_j different values, then $x_j(\cdot)$ is understood as a $(k_j - 1)$-dimensional vector

$$x_j(\cdot) = (x_{j1}(\cdot), ..., x_{j,k_j-1}(\cdot))^T,$$

taking k_j different values

$$(0, 0, \cdots, 0)^T, (1, 0, \cdots, 0)^T, (0, 1, 0, \cdots, 0)^T, \cdots, (0, 0, \cdots, 0, 1)^T,$$

and β_j is $(k_j - 1)$-dimensional:

$$\beta_j = (\beta_{j1}, ..., \beta_{j,k_j-1})^T.$$

So, if the jth explanatory variable is categorical, and others are interval-valued or discrete then

$$\beta^T x = \beta_0 + \beta_1 x_1 + \cdots + \beta_{j-1} x_{j-1} + \sum_{l=1}^{k_j-1} \beta_{jl} x_{jl} + \beta_{j+1} x_{j+1} + \cdots + \beta_m x_m. \quad (5.21)$$

The obtained model is equivalent to the model (5.5) with $m + k_j - 2$ univariate explanatory variables. If $k_j = 2$, the explanatory variable x_j is *dichotomous*, taking two values 0 or 1.

5.2 Interpretation of the regression coefficients

Suppose that the explanatory variables are constant over time. Then under the AFT model (5.5) the p-quantile of the failure time T_x is

$$t_p(x) = e^{\beta^T x} S_0^{-1}(1-p), \quad (5.22)$$

so the logarithm

$$\ln\{t_p(x)\} = \beta^T x + c_p \quad (5.23)$$

is a linear function of the regression parameters; here $c_p = \ln(S_0^{-1}(1-p))$.

Let

$$m(x) = \mathbf{E}\{T_x\}$$

be the *mean life* of units under x. Then

$$m(x) = e^{\beta^T x} \int_0^\infty S_0(u) du \quad (5.24)$$

and the logarithm

$$\ln\{m(x)\} = \beta^T x + c \quad (5.25)$$

is also a linear function of the regression parameters; here

$$c = \ln\left\{\int_0^\infty S_0(u) du\right\}.$$

Denote by

$$MR(x,y) = \frac{m(y)}{m(x)} \quad \text{and} \quad QR(x,y) = \frac{t_p(y)}{t_p(x)} \quad (5.26)$$

the ratio of means and quantiles, respectively.

For the AFT model

$$MR(x,y) = QR(x,y) = e^{\beta^T (y-x)}. \quad (5.27)$$

So $e^{\beta^T (y-x)}$ *is the ratio of means, corresponding to the explanatory variables* x *and* y.

Let us consider the interpretation of the parameters β_j under the model(5.5).

5.2.1 Models without interactions

a) *Interval-valued or discrete explanatory variables*

Suppose that the jth explanatory variable x_j is interval-valued or discrete. Then

$$e^{\beta_j} = \frac{e(x_1, ..., x_j + 1, ..., x_m)}{e(x_1, ..., x_j, ..., x_m)} = MR_j, \qquad (5.28)$$

is the ratio of means corresponding to the change of x_j by the unity.

b) *Categorical explanatory variables*

Suppose that $x_j = (x_{j1}, \cdot, ..., x_{j,k_j-1})^T$ is categorical. Its first value is $(0, \cdots, 0)^T$ and the $(i+1)$th value is $(0, \cdots, 0, 1, 0, \cdots, 0)^T$, where the unity is the ith co-ordinate. Then

$$e^{\beta_{ji}} = \frac{e(x_1, ..., x_{j-1}, (0, 0, \cdots, 0, 1, 0, \cdots, 0), x_{j+1}, ..., x_m)}{e(x_1, ..., x_{j-1}, (0, 0, \cdots, 0), x_{j+1}, ..., x_m)} = MR_{ji} \qquad (5.29)$$

is the ratio of means corresponding to the change of x_j from the first to the $(i+1)$th value.

5.2.2 Models with interactions

If the influence of the jth explanatory variable on the mean life is different under various values of other explanatory variables then there is interaction between the explanatory variables and the model must be modified.

a) *Interaction between interval-valued or discrete explanatory variables*

If there are two interval-valued or discrete explanatory variables and there is interaction between them then

$$\beta^T x = \beta_0 + \beta_1 x_1 + \beta_2 x_2 + \beta_3 x_1 x_2. \qquad (5.30)$$

For three explanatory variables

$$\beta^T x = \beta_0 + \beta_1 x_1 + \beta_2 x_2 + \beta_3 x_3 + \beta_4 x_1 x_2 + \beta_5 x_1 x_3 + \beta_6 x_2 x_3 + \beta_7 x_1 x_2 x_3, \qquad (5.31)$$

and so on.

In the case of two explanatory variables the ratio of means

$$MR_2(x_1) = \frac{m(x_1, x_2 + 1)}{m(x_1, x_2)} = e^{\beta_1 + \beta_3 x_1} \qquad (5.32)$$

depends on the value of x_1.

So

$$e^{\beta_1 + \beta_3 x_1} \qquad (5.33)$$

is the ratio of means corresponding to the change of x_2 by the unity, the other explanatory variable being fixed and equal to x_1

b) *Interaction between interval-valued or discrete and categorical explana-tory variables*

Suppose that there are two explanatory variables: x_1 is interval-valued or discrete and x_2 is categorical, with k_2 possible values. Then

$$\beta^T x = \beta_1 x_1 + \sum_{i=1}^{k_2-1} \beta_{2i} x_{2i} + \sum_{i=1}^{k_2-1} \beta_{12i} x_1 x_{2i}, \qquad (5.34)$$

and the mean ratio

$$MR_{2i}(x_1) = \frac{e(x_1, (0, ..., 0, 1, 0, \cdots, 0))}{e(x_1, (0, ..., 0))} = e^{\beta_{2i} + \beta_{12i} x_1} \qquad (5.35)$$

depends on the value of x_1.

So in this example

$$e^{\beta_{2i} + \beta_{12i} x_1}$$

is the ratio of means corresponding to the change of x_2 from the first to the $(i+1)$th value, other explanatory variable being fixed and equal to x_1.

c) *Interaction between categorical explanatory variables*

Suppose that both x_1 and x_2 are categorical with three values for each. Then

$$x_1 = (x_{11}, x_{12})^T, \quad x_2 = (x_{21}, x_{22})^T,$$

and

$$\beta^T x = \beta_{11} x_{11} + \beta_{12} x_{12} + \beta_{21} x_{21} + \beta_{22} x_{22} + \beta_{1121} x_{11} x_{21} +$$
$$\beta_{1122} x_{11} x_{22} + \beta_{1221} x_{12} x_{21} + \beta_{1222} x_{12} x_{22}.$$

In this case the ratio

$$MR_{22}(x_1) = \frac{e(x_1, (1, 0))}{e(x_1, (0, 0))} = e^{\beta_{21} + \beta_{1121} x_{11} + \beta_{1221} x_{12}}$$

depends on the value of $x_1 = (x_{11}, x_{12})^T$.

So

$$e^{\beta_{21} + \beta_{1121} x_{11} + \beta_{1221} x_{12}}$$

is the ratio of means corresponding to the change of x_2 from the first to the second value, other explanatory variable being fixed and equal to $x_1 = (x_{11}, x_{12})^T$.

Generalization is evident if the explanatory variables take three or more values.

5.2.3 Time dependent regression coefficients

Let us consider the AFT model with time dependent explanatory variables (see section 2.8.1):

$$S_{x(\cdot)} = G\left\{ \int_0^t e^{-\beta^T(u)x(u)} \, du \right\}. \qquad (5.36)$$

We shall consider the coefficients $\beta_i(t)$ in the form

$$\beta_i(t) = \beta_i + \gamma_i g_i(t), \quad (i = 1, 2, ..., m),$$

where $g_i(t)$ are some specified deterministic functions or realizations of pre-dictable processes. In such a case the AFT model with time dependent coef-ficients and constant or time dependent explanatory variables can be written in the form (5.4) with different interpretation of the explanatory variables. Indeed, set

$$\theta = (\theta_0, \theta_1, \cdots, \theta_{2m})^T = (\beta_0, \beta_1, \cdots, \beta_m, \gamma_1, \cdots, \gamma_m)^T,$$

$$z(\cdot) = (z_0(\cdot), z_1(\cdot), \cdots, z_{2m}(\cdot))^T =$$

$$(1, x_1(\cdot), \cdots, x_m(\cdot), x_1(\cdot)g_1(\cdot), \cdots, x_m(\cdot)g_m(\cdot))^T. \qquad (5.37)$$

Then

$$\beta^T(u)x(u) = \beta_0 + \sum_{i=1}^m (\beta_i + \gamma_i g_i(t))x_i(t) = \theta^T z(u).$$

So the AFT model with the time dependent regression coefficients can be written in the form

$$S_{z(\cdot)} = G \left\{ \int_0^t e^{-\theta^T z(u)} du \right\}. \qquad (5.38)$$

We have the AFT model where the unknown parameters and the explanatory variables are defined by (5.37).

5.3 FTR data analysis: scale-shape families of distributions

5.3.1 Model and data

Let us consider the AFT model:

$$S_{x(\cdot)}(t) = S_0 \left(\int_0^t e^{-\beta^T x(\tau)} d\tau \right), \qquad (5.39)$$

or the AFT model (5.36) with time dependent regression coefficients, and suppose that S_0 belongs to a specified scale-shape class of survival functions:

$$S_0(t) = G_0 \left\{ (t/\eta)^\nu \right\} \quad (\eta, \nu > 0).$$

For example, if for $t > 0$

$$G_0(t) = e^{-t}, \quad G_0(t) = (1+t)^{-1}, \quad G_0(t) = 1 - \Phi(\ln t),$$

then we obtain the families of the Weibull, loglogistic, lognormal distributions, respectively. Here Φ is the distribution function of the standard normal law.

The parameter η can be included in the coefficient β_0, so suppose that

$$S_0(t; \sigma) = G_0(t^{1/\sigma}), \quad \sigma = 1/\nu.$$

Take notice that if the AFT model with time dependent regression coeffi-cients $\beta_i(t) = \beta_i + \gamma_i g_i(t)$ is considered, then, even in the case of constant explanatory variables, the model (5.36) is reduced to the AFT model (5.38), which is equivalent to (5.39) with time dependent explanatory variables $z(\cdot)$. So all results obtained for the AFT model with time-dependent explanatory variables can be rewritten for the AFT model with time dependent regression coefficients and constant or time dependent explanatory variables. In all formulas

$$m, \quad \beta = (\beta_1, \cdots, \beta_m)^T, \quad x^{(i)}(\cdot) = (x_1^{(i)}(\cdot), \cdots, x_m^{(i)}(\cdot))^T$$

must be replaced by

$$2m, \quad \theta = (\beta_1, \cdots, \beta_m, \gamma_1, \cdots, \gamma_m)^T,$$

$$z^{(i)}(\cdot) = (x_1^{(i)}(\cdot), \cdots, x_m^{(i)}(\cdot), x_1^{(i)}(\cdot)g_1(\cdot), \cdots, x_m^{(i)}(\cdot)g_m(\cdot))^T$$

respectively.

Plan of experiments:

n units are observed. The ith unit is tested under the value

$$x^{(i)}(\cdot) = (x_0^{(i)}(\cdot), ..., x_m^{(i)}(\cdot))^T$$

of a possibly time-varying and multidimensional explanatory variable

$$x(\cdot) = (x_0(\cdot), ..., x_m(\cdot))^T.$$

The data are supposed to be independently right censored.

Let T_i and C_i be the failure and censoring times of the ith unit,

$$X_i = T_i \wedge C_i, \quad \delta_i = \mathbf{1}_{\{T_i \leq C_i\}}.$$

Denote by S_i the survival function $S_{x^{(i)}(\cdot)}$. The model (5.4) may be written in the form

$$S_i(t; \beta, \sigma) = G_0 \left\{ \left(\int_0^t e^{-\beta^T x^{(i)}(u)} du \right)^{1/\sigma} \right\}. \qquad (5.40)$$

If $x^{(i)}$ is constant then

$$S_i(t) = G \left(\frac{\ln t - \beta^T x^{(i)}}{\sigma} \right), \qquad (5.41)$$

where

$$G(u) = G_0(e^u), \quad u \in \mathbf{R}.$$

Note that the distribution of the random variable

$$R_i = \left\{ \int_0^{T_{x^{(i)}}(\cdot)} e^{-\beta^T x^{(i)}(\tau)} d\tau \right\}^{1/\sigma} \qquad (5.42)$$

is parameter-free with the survival function G_0. For the constant $x^{(i)}$:

$$R_i = \left\{ T_{x^{(i)}} e^{-\beta^T x^{(i)}} \right\}^{1/\sigma}.$$

Set

$$g(u) = -G'(u), \quad h(u) = \frac{g(u)}{G(u)} \qquad (5.43)$$

For the Weibull law:

$$G(u) = e^{-e^u}, \quad g(u) = e^u e^{-e^u}, \quad h(u) = e^u, \quad (\ln h)'(u) = 1; \qquad (5.44)$$

for the loglogistic law:

$$G(u) = (1 + e^u)^{-1}, \quad g(u) = \frac{e^u}{(1 + e^u)^2},$$

$$h(u) = e^u (1 + e^u)^{-1}, \quad (\ln h)'(u) = (1 + e^u)^{-1}; \tag{5.45}$$

for the lognormal law:

$$G(u) = 1 - \Phi(u), \quad g(u) = \varphi(u), \quad h(u) = \frac{\varphi(u)}{1 - \Phi(u)}, \quad (\ln h)'(u) = h(u) - u, \tag{5.46}$$

with

$$\varphi(t) = \frac{1}{\sqrt{2\pi}} e^{-t^2/2}.$$

5.3.2 Maximum likelihood estimation of the regression parameters

The likelihood function is

$$L(\beta, \sigma) = \prod_{i=1}^{n} \left\{ \frac{1}{\sigma} e^{-\beta^T x^{(i)}(X_i)} \left(\int_0^{X_i} e^{-\beta^T x^{(i)}(u)} du \right)^{-1} \times \right.$$

$$\left. h \left(\frac{1}{\sigma} \ln \left(\int_0^{X_i} e^{-\beta^T x^{(i)}(u)} du \right) \right) \right\}^{\delta_i} G \left(\frac{1}{\sigma} \ln \left(\int_0^{X_i} e^{-\beta^T x^{(i)}(u)} du \right) \right). \tag{5.47}$$

If $x^{(i)}$ are constant then the likelihood function is

$$L(\beta, \sigma) = \prod_{i=1}^{n} \left\{ \frac{1}{\sigma X_i} h \left(\frac{\ln X_i - \beta^T x^{(i)}}{\sigma} \right) \right\}^{\delta_i} G \left(\frac{\ln X_i - \beta^T x^{(i)}}{\sigma} \right). \tag{5.48}$$

The score functions are

$$U_l(\beta; \sigma) = \frac{\partial \ln L(\beta, \sigma)}{\partial \beta_l} = \frac{1}{\sigma} \sum_{i=1}^{n} z_l^{(i)}(\beta) a_i(\beta, \sigma) + \sum_{i=1}^{n} \delta_i \left(z_l^{(i)}(\beta) - x_l^{(i)}(X_i) \right)$$

$$(l = 0, 1, ..., m),$$

$$U_{m+1}(\beta; \sigma) = \frac{\partial \ln L(\beta, \sigma)}{\partial \sigma} = \frac{1}{\sigma} \sum_{i=1}^{n} \{ v_i(\beta, \sigma) a_i(\beta, \sigma) - \delta_i \}; \tag{5.49}$$

here

$$v_i(\beta, \sigma) = \frac{1}{\sigma} \ln \left(\int_0^{X_i} e^{-\beta^T x^{(i)}(u)} du \right), \quad a_i(\beta, \sigma) = h(v_i(\beta, \sigma)) - \delta_i (\ln h)'(v_i(\beta, \sigma)),$$

$$z_l^{(i)}(\beta) = \frac{\int_0^{X_i} x_l^{(i)}(u) e^{-\beta^T x^{(i)}(u)} du}{\int_0^{X_i} e^{-\beta^T x^{(i)}(u)} du}, \tag{5.50}$$

and the function h is given by (5.43). In the case of constant explanatory variables the score functions are

$$U_l(\beta;\sigma) = \frac{\partial \ln L(\beta,\sigma)}{\partial \beta_l} = \frac{1}{\sigma} \sum_{i=1}^{n} x_l^{(i)} a_i(\beta,\sigma), \quad (l = 0, 1, ..., m),$$

$$U_{m+1}(\beta;\sigma) = \frac{\partial \ln L(\beta,\sigma)}{\partial \sigma} = \frac{1}{\sigma} \sum_{i=1}^{n} \{v_i(\beta,\sigma)a_i(\beta,\sigma) - \delta_i\}, \quad (5.51)$$

where

$$v_i(\beta,\sigma) = \frac{\ln X_i - \beta^T x^{(i)}}{\sigma}, \quad a_i(\beta,\sigma) = h(v_i(\beta,\sigma)) - \delta_i(\ln h)'(v_i(\beta,\sigma)), \quad (5.52)$$

and $(\ln h)'(u)$ is given in (5.43).

The maximum likelihood estimators $\hat{\beta}_j, \hat{\sigma}$, are obtained by solving the system of equations

$$U_l(\beta,\sigma) = 0 \quad (l = 0, 1, ..., m+1).$$

5.3.3 Estimators of the main reliability characteristics

Suppose that $x(\cdot) = (x_1(\cdot), \cdots, x_m(\cdot))^T$ is an arbitrary explanatory variable which may be different from $x^{(i)}(\cdot), (i = 1, \cdots, n)$.

Estimator of the survival function $S_{x(\cdot)}(t)$:

$$\hat{S}_{x(\cdot)}(t) = G_0 \left\{ \left(\int_0^t e^{-\hat{\beta}^T x(u)} du \right)^{1/\hat{\sigma}} \right\}. \quad (5.53)$$

In the case when x is constant, the estimator of the survival function $S_x(t)$ is

$$\hat{S}_x(t) = G \left(\frac{\ln t - \hat{\beta}^T x}{\hat{\sigma}} \right). \quad (5.54)$$

If the AFT model with time dependent regression coefficients is considered then

$$\hat{S}_{z(\cdot)}(t) = G_0 \left\{ \left(\int_0^t e^{-\hat{\theta}^T z(u)} du \right)^{1/\hat{\sigma}} \right\},$$

where

$$\theta = (\beta_1, \cdots, \beta_m, \gamma_1, \cdots, \gamma_m)^T,$$

$$z^{(i)}(\cdot) = (x_1^{(i)}(\cdot), \cdots, x_m^{(i)}(\cdot), x_1^{(i)}(\cdot)g_1(\cdot), \cdots, x_m^{(i)}(\cdot)g_m(\cdot))^T.$$

Estimator of the p-quantile $t_p(x(\cdot))$

The estimator $\hat{t}_p(x(\cdot))$ verifies the equation:

$$G_0 \left\{ \left(\int_0^{\hat{t}_p(x(\cdot))} e^{-\hat{\beta}^T x(u)} du \right)^{1/\hat{\sigma}} \right\} = 1 - p. \qquad (5.55)$$

If x is constant then

$$\hat{t}_p(x) = e^{\hat{\beta}^T x} \{ G_0^{-1}(1-p) \}^{\hat{\sigma}}. \qquad (5.56)$$

Estimator of the mean failure time $m(x(\cdot))$:

$$\hat{m}(x(\cdot)) = \int_0^\infty \hat{S}_{x(\cdot)}(u) du. \qquad (5.57)$$

If x is constant then

$$\hat{m}(x) = \hat{\sigma} e^{\hat{\beta}^T x} \int_0^\infty u^{\hat{\sigma}-1} G_0(u) du \qquad (5.58)$$

Estimators of the mean ratios

The mean ratio $MR(x, y)$ (see (5.27)) is estimated by

$$\widehat{MR}(x, y) = e^{\hat{\beta}^T (y-x)}. \qquad (5.59)$$

1) Models without interactions.

a) Interval or discrete explanatory variable x_j.

The mean ratio MR_j, (see (5.28)), is estimated by

$$\widehat{MR}_j = e^{\hat{\beta}_j}. \qquad (5.60)$$

b) Categorical explanatory variable x_j.

The mean ratio MR_{ji},(see (5.29)), is estimated by

$$\widehat{MR}_{ji} = e^{\hat{\beta}_{ji}}. \qquad (5.61)$$

2) Models with interactions.

a) Interval-valued (discrete) × Interval (discrete) variables.
If two interval (discrete) explanatory variables, for example, x_1 and x_2 interact then the mean ratio $MR_2(x_1)$, (see (5.32)), is estimated by

$$\widehat{MR}_2(x_1) = e^{\hat{\beta}_1 + \hat{\beta}_3 x_1}. \qquad (5.62)$$

b) Interval-valued (discrete) × categorical

If an interval (discrete) explanatory variable x_1 interacts with a k_2-valued categorical explanatory variable x_2, then the mean ratio $MR_{2i}(x_1)$, (see the model (5.35)), is estimated by

$$\widehat{MR}_{2i}(x_1) = e^{\hat{\beta}_{2i} + \hat{\beta}_{12i} x_1}. \tag{5.63}$$

c) Categorical $\times$ categorical explanatory variables.
If two categorical explanatory variables x_1 and x_2 (with, say, 3 possible values) interact, then the mean ratio MR_{22} is estimated by

$$e^{\hat{\beta}_{21} + \hat{\beta}_{1121} x_{11} + \hat{\beta}_{1221} x_{12}}. \tag{5.64}$$

5.3.4 Asymptotic distribution of the regression parameters estimators

Under regularity conditions (see Chapter 4) the distribution of the maximum likelihood estimators $(\hat{\beta}, \hat{\sigma})^T$ for large n is approximated by the normal law:

$$(\hat{\beta}, \hat{\sigma})^T \approx N_{m+2}((\beta, \sigma)^T, \Sigma^{-1}(\beta, \sigma)).$$

The covariance matrix $\Sigma^{-1}(\beta, \sigma)$ is estimated by

$$\mathbf{I}^{-1}(\hat{\beta}, \hat{\sigma}) = (I^{ls}(\hat{\beta}, \hat{\sigma}))_{(m+2) \times (m+2)}, \tag{5.65}$$

where

$$\mathbf{I}(\beta, \sigma) = (I_{lk}(\beta, \sigma))_{(m+2) \times (m+2)} \tag{5.66}$$

is a matrix with the following elements:

$$I_{ls}(\beta, \sigma) = -\frac{\partial^2 \ln L(\beta, \sigma)}{\partial \beta_l \partial \beta_s}, \quad I_{l,m+1}(\beta, \sigma) = -\frac{\partial^2 \ln L(\beta, \sigma)}{\partial \beta_l \partial \sigma},$$

$$I_{m+1,m+1}(\beta, \sigma) = -\frac{\partial^2 \ln L(\beta, \sigma)}{\partial \sigma^2} \quad (l, s = 0, \cdots, m).$$

In particular, the variance of the estimator $\hat{\beta}_j$ is estimated by

$$\hat{\mathbf{Var}}\{\hat{\beta}_j\} = I^{jj}(\hat{\beta}, \hat{\sigma}) \tag{5.67}$$

and

$$\frac{\hat{\beta}_j - \beta_j}{(\hat{\mathbf{Var}} \hat{\beta}_j)^{1/2}} \approx N(0, 1), \quad (j = 0, \cdots, m). \tag{5.68}$$

The expressions for the elements of the matrix $\mathbf{I}(\beta, \sigma)$ are (see notation in (5.50)):

$$I_{ls}(\beta, \sigma) = \frac{1}{\sigma^2} \sum_{i=1}^{n} y_l^{(i)}(\beta) y_s^{(i)}(\beta)) c_i(\beta, \sigma) - \frac{1}{\sigma} \sum_{i=1}^{n} y_{ls}^{(i)}(\beta)(a_i(\beta, \sigma) + \sigma \delta_i),$$

$$I_{l,m+1}(\beta, \sigma) = \frac{1}{\sigma^2} \sum_{i=1}^{n} y_l^{(i)}(\beta)(v_i(\beta, \sigma) c_i(\beta, \sigma) + a_i(\beta, \sigma)), \quad l, s = 0, ..., m.$$

$$I_{m+1,m+1}(\beta, \sigma) = \frac{1}{\sigma} U_{m+1}(\beta, \sigma) + \frac{1}{\sigma^2} \sum_{i=1}^{n} v_i(\beta, \sigma)(v_i(\beta, \sigma) c_i(\beta, \sigma) + a_i(\beta, \sigma)),$$

$$\tag{5.69}$$

where

$$c_i(\beta,\sigma) = h'(v_i(\beta,\sigma)) - \delta_i(\ln h)''(v_i(\beta,\sigma)),$$

$$y_{ls}^{(i)}(\beta) = \frac{\int_0^{X_i} x_l^{(i)}(u)e^{-\beta^T x^{(i)}(u)}du \int_0^{X_i} x_s^{(i)}(u)e^{-\beta^T x^{(i)}(u)}du}{\left(\int_0^{X_i} e^{-\beta^T x^{(i)}(u)}du\right)^2}$$

$$-\frac{\int_0^{X_i} x_l^{(i)}(u)x_s^{(i)}(u)e^{-\beta^T x^{(i)}(u)}du}{\int_0^{X_i} e^{-\beta^T x^{(i)}(u)}du}. \tag{5.70}$$

If $x^{(i)}$ are constant in time then

$$I_{ls}(\beta,\sigma) = \frac{1}{\sigma^2}\sum_{i=1}^{n} x_l^{(i)} x_s^{(i)} c_i(\beta,\sigma),$$

$$I_{l,m+1}(\beta,\sigma) = \frac{1}{\sigma}U_l(\beta,\sigma) + \frac{1}{\sigma^2}\sum_{i=1}^{n} x_l^{(i)} v_i(\beta,\sigma)c_i(\beta,\sigma),$$

$$I_{m+1,m+1}(\beta,\sigma) = \frac{1}{\sigma}U_{m+1}(\beta,\sigma) + \frac{1}{\sigma^2}\sum_{i=1}^{n} v_i(\beta,\sigma)(v_i(\beta,\sigma)c_i(\beta,\sigma) + a_i(\beta,\sigma)).$$

$$\tag{5.71}$$

Take notice that in the case of time dependent regression coefficients $m, \beta, x^{(i)}$ are replaced by $2m, \theta, z^{(i)}$.

5.3.5 Approximate confidence intervals for the main reliability characteristics

The survival functions, quantiles, mean lifetimes, and mean ratios are functions of the parameters β and σ. So the asymptotic distributions and approximate confidence intervals for them are obtained by using the delta method. Forms of approximate confidence intervals for such functions are given in Chapter 4 (the formulas (4.49), (4.51)).

Approximate confidence intervals for the survival functions

The formula (4.49) implies that for any $x(\cdot) = (x_0(\cdot), x_1(\cdot), \cdots, x_m(\cdot)) \in E_0$, $x_0(\cdot) = 1$, an approximate $(1-\alpha)$-confidence interval for the survival function $S_{x(\cdot)}(t)$ is defined by the formula

$$\left(1 + \frac{1 - \hat{S}_{x(\cdot)}(t)}{\hat{S}_{x(\cdot)}(t)} \exp\{\pm\hat{\sigma}_{Q_{x(\cdot)}}(t)w_{1-\alpha/2}\}\right)^{-1}, \tag{5.72}$$

where w_α is the α-quantile of the normal law $N(0,1)$ and (see (4.47), (4.48) and (5.53))

$$\hat{\sigma}_{Q_{x(\cdot)}}^2(t) = \frac{J_{S_{x(\cdot)}(t)}^T \mathbf{I}^{-1}(\hat{\beta},\hat{\sigma}) J_{S_{x(\cdot)}(t)}}{\hat{S}_{x(\cdot)}^2(t)(1 - \hat{S}_{x(\cdot)}(t))^2}, \tag{5.73}$$

$$J^T_{S_{x(\cdot)}(t)} = -\frac{G'(G^{-1}(\hat{S}_{x(\cdot)}(t)))}{\hat{\sigma}^2_{Q_{x(\cdot)}}(t)} \times$$

$$\left(\hat{\sigma}_{Q_{x(\cdot)}}(t) \frac{\int_0^t x^T(u)e^{-\hat{\beta}^T x(u)}\,du}{\int_0^t e^{-\hat{\beta}^T x(u)}\,du}, \ln\{ \int_0^t e^{-\hat{\beta}^T x(u)}\,du\} \right).$$

If x is constant in time, then

$$J^T_{S_x(t)} = -\frac{G'(G^{-1}(\hat{S}_x(t)))}{\hat{\sigma}^2_{Q_x}(t)} \times \left(\hat{\sigma}_{Q_x}(t)x^T, \ln t - \hat{\beta}^T x \right).$$

Approximate confidence intervals for the quantiles

The formula (4.51) implies that an approximate $(1-\alpha)$-confidence interval for the p-quantile $t_p(x(\cdot))$ is

$$\hat{t}_p(x(\cdot)) \exp\{ \pm \hat{\sigma}_{K_p(x(\cdot))} w_{1-\alpha/2} \}, \tag{5.74}$$

where (see (4.47), (4.50) and (5.22))

$$\hat{\sigma}^2_{K_p(x(\cdot))} = \frac{1}{\hat{t}^2_p(x(\cdot))} J^T_{t_p(x(\cdot))} \mathbf{I}^{-1}(\hat{\beta}, \hat{\sigma}) J_{t_p(x(\cdot))},$$

where

$$J^T_{t_p(x(\cdot))} = e^{\hat{\beta}^T x(\hat{t}_p(x(\cdot)))} \left(\int_0^{\hat{t}_p(x(\cdot))} x(u)e^{-\hat{\beta}^T x(u)}\,du, (G_0^{-1}(1-p))^{\hat{\sigma}} \ln G_0^{-1}(1-p) \right).$$

For constant in time x we have

$$J^T_{t_p(x)} = \hat{t}_p(x)(x, \ln G_0^{-1}(1-p)).$$

Approximate confidence interval for the mean lifetime

The formula (4.51) implies that an approximate $(1-\alpha)$-confidence interval for the mean lifetime $m(x(\cdot))$ is

$$\hat{m}(x(\cdot)) \exp \left\{ \pm \frac{\hat{\sigma}_{m(x(\cdot))}(\hat{\beta}, \hat{\sigma})}{\hat{m}(x(\cdot))} w_{1-\alpha/2} \right\}, \tag{5.75}$$

where

$$\hat{\sigma}^2_{m(x(\cdot))} = J^T_{m(x(\cdot))} \mathbf{I}^{-1}(\hat{\beta}, \hat{\sigma}) J_{m(x(\cdot))},$$

$$J_{m(x(\cdot))} = \int_0^\infty J_{S_{x(\cdot)}(t)}\,dt.$$

If $x=const$ then

$$\hat{\sigma}^2_{m(x)}(\hat{\beta}, \hat{\sigma}) = \sum_{l=0}^{m+1} \sum_{s=0}^{m+1} b_l(\hat{\beta}, \hat{\sigma}) I^{ls}(\hat{\beta}, \hat{\sigma}) b_s(\hat{\beta}, \hat{\sigma}),$$

and

$$b_i(\hat\beta, \hat\sigma) = \frac{\partial}{\partial\hat\beta_i}\hat m(x) = \hat m(x)\, x_i \quad (i = 0, 1, ..., m),$$

$$b_{m+1}(\hat\beta, \hat\sigma) = \frac{\partial}{\partial\hat\sigma}\hat m(x) = \frac{\hat m(x)}{\hat\sigma} + \hat\sigma e^{\hat\beta^T x}\int_0^\infty u^{\hat\sigma-1}G_0(u)\ln u\, du.$$

Approximated confidence intervals for the mean ratios

The formula (4.51) implies that an approximate $(1 - \alpha)$-confidence interval for the mean ratio $MR(x, y)$ is:

$$\left(\exp\{\hat\beta^T(y - x) - \hat\sigma_{MR}\}w_{1-\alpha/2},\ \exp\{\hat\beta^T(y - x) + \hat\sigma_{MR}\}w_{1-\alpha/2}\right), \quad (5.76)$$

where

$$\hat\sigma_{MR}^2 = \sum_{l=0}^{m}\sum_{s=0}^{m}(y_l - x_l)I^{ls}(\hat\beta, \hat\sigma)(y_s - x_s). \quad (5.77)$$

5.3.6 Tests for nullity of the regression coefficients

Let us consider the hypothesis

$$H_{k_1,k_2,\cdots,k_l} : \beta_{k_1} = \cdots = \beta_{k_l} = 0, \quad (1 \le k_1 \le k_2 \le \cdots \le k_l). \quad (5.78)$$

Under this hypothesis the explanatory variables $x_{k_1}, \cdots, x_{k_l}$ do not improve the prediction. If $H_{k_1,k_2,\cdots,k_l}$ is verified, these variables are excluded from the model. In particular case, the hypothesis

$$H_{1,2,\cdots,m} : \beta_1 = \cdots = \beta_m = 0$$

means that none of the explanatory variables improve the prediction, i.e. there is no regression. From the practical point of view the most interesting hypothesis is

$$H_k : \beta_k = 0, \quad (k = 1, ..., m).$$

It means that the model with and without the explanatory variable x_k gives the same prediction.

Likelihood ratio test.

Let

$$\hat L = L(\hat\beta, \hat\sigma) = \max_{\sigma,\beta} L(\beta, \sigma) \quad (5.79)$$

be the maximum of the likelihood function under the full model with m explanatory variables and

$$\hat L_{k_1...k_l} = \max_{\sigma,\beta:\beta_{k_1}=...=\beta_{k_l}=0} L(\beta, \sigma) \quad (5.80)$$

be the maximum of the likelihood function under the hypothesis $H_{k_1 \ldots k_l}$ which is the maximum of the likelihood function, corresponding to the model with $(m - l)$ explanatory variables $\{x_1, \ldots, x_m\} \setminus \{x_{k_1}, \ldots, x_{k_l}\}$.

If n is large then the distribution of the likelihood ratio statistic (see (4.40)) :

$$LR^{(*)} = -2\ln \frac{L(\beta, \sigma)}{\hat{L}} \tag{5.81}$$

is approximately chi-square with $m + 2$ degrees of freedom. Under $H_{k_1 \ldots k_l}$ the coefficients $\beta_{k_1}, \cdots, \beta_{k_l}$ are equal to zero in (5.81).

Similarly under $H_{k_1 \ldots k_l}$ the distribution of the likelihood ratio statistic

$$LR^{(**)} = -2\ln \frac{L(\beta, \sigma)}{\hat{L}_{k_1 \ldots k_l}} \tag{5.82}$$

is approximately chi-square with $m - l + 2$ degrees of freedom. The coefficients $\beta_{k_1}, \cdots, \beta_{k_l}$ are equal to zero in (5.82).

It can be shown that the statistics

$$LR^{(**)} \quad \text{and} \quad LR^{(*)} - LR^{(**)}$$

are asymptotically independent.

Thus, under $H_{k_1 \ldots k_l}$ the distribution of the likelihood ratio test statistic

$$LR_{k_1, \ldots, k_l} = -2\ln \frac{\hat{L}_{k_1 \ldots k_l}}{\hat{L}} \tag{5.83}$$

is approximated by the chi-square distribution with $l = (m + 2) - (m - l + 2)$ degrees of freedom when n is large.

The hypothesis $H_{k_1 \ldots k_l} : \beta_{k_1} = \ldots = \beta_{k_l} = 0$ is rejected with the significance level α if

$$LR_{k_1, \ldots, k_l} > \chi^2_{1-\alpha}(l); \tag{5.84}$$

the hypothesis $H_{12 \ldots m} : \beta_1 = \ldots = \beta_m = 0$ is rejected if

$$LR_{1, \ldots, m} > \chi^2_{1-\alpha}(m);$$

the hypothesis $H_k : \beta_k = 0$ is rejected if

$$LR_k > \chi^2_{1-\alpha}(1).$$

The statistic LR_k is often used in stepwise regression procedures when the problem of including or rejecting the explanatory variable x_k is considered.

In the particular case the likelihood ratio test statistics may be used when testing hypothesis about absence of interactions. For example, for the model with

$$\beta^T x = \beta_0 + \beta_1 x_1 + \beta_2 x_2 + \beta_3 x_3 + \beta_4 x_1 x_2 + \beta_5 x_1 x_3 + \beta_6 x_2 x_3$$

the hypothesis $H_{456} : \beta_4 = \beta_5 = \beta_6 = 0$ or $H_5 : \beta_5 = 0$ may be tested.

Wald's tests

Let $A_{k_1..k_l}(\hat{\beta}, \hat{\sigma})$ be the submatrix of $I^{-1}(\hat{\beta}, \hat{\sigma})$ which is in the intersection of the $k_1, ..., k_l$ rows and $k_1, ..., k_l$ columns. Under $H_{k_1,...,k_l}$ and large n the distribution of the statistic (see (4.39))

$$W_{k_1...k_l} = (\hat{\beta}_{k_1}, ..., \hat{\beta}_{k_l})^T A_{k_1...k_l}^{-1}(\hat{\beta}, \hat{\sigma})(\hat{\beta}_{k_1}, ..., \hat{\beta}_{k_l})^T$$

is approximated by the chi-square distribution with l degrees of freedom.

The hypothesis $H_{k_1...k_l}$ is rejected with the significance level α if

$$W_{k_1...k_l} > \chi_{1-\alpha}^2(l),$$

the hypothesis $H_{12...m}$ is rejected if

$$W_{1...m} > \chi_{1-\alpha}^2(m),$$

and the hypothesis H_k is rejected if

$$W_k = \frac{\hat{\beta}_k^2}{I^{kk}(\hat{\beta}, \hat{\sigma})} = \frac{\hat{\beta}_k^2}{\hat{\text{Var}}(\hat{\beta}_k)} > \chi_{1-\alpha}^2(1).$$

Score tests

If n is large then the distribution of the statistic (see (4.38)) :

$$(U_{k_1}(\beta, \sigma), \cdots, U_{k_l}(\beta, \sigma))^T A_{k_1...k_l}(\hat{\beta}, \hat{\sigma}) (U_{k_1}(\beta, \sigma), \cdots, U_{k_l}(\beta, \sigma))$$

is approximately chi-square with l degrees of freedom. Under $H_{k_1...k_l}$ the limit distribution of this statistic does not change if (β, σ) (with the components $\beta_i = 0$ for $i = k_1, \cdots, k_l$) is replaced by $(\tilde{\beta}, \tilde{\sigma})$ verifying the condition

$$L(\tilde{\beta}, \tilde{\sigma}) = \max_{\sigma, \beta: \beta_{k_1} = ... = \beta_{k_l} = 0} L(\beta, \sigma).$$

So the score statistic is

$$U_{k_1...k_l} = (U_{k_1}(\tilde{\beta}, \tilde{\sigma}), \cdots, U_{k_l}(\tilde{\beta}, \tilde{\sigma}))^T A_{k_1...k_l}(\hat{\beta}, \hat{\sigma}) (U_{k_1}(\tilde{\beta}, \tilde{\sigma}), \cdots, U_{k_l}(\tilde{\beta}, \tilde{\sigma})).$$

The hypothesis $H_{k_1...k_l}$ is rejected with the significance level α if

$$U_{k_1...k_l} > \chi_{1-\alpha}^2(l).$$

5.3.7 Residual plots

Let us consider the random variables

$$\hat{R}_i = \left(\int_0^{X_i} e^{-\hat{\beta}^T x^{(i)}(u)} du \right)^{1/\hat{\sigma}},$$

called the *standardized residuals*. They are a particular case of more general *Cox-Snell residuals* (see Cox and Snell(1968)).

If $x^{(i)}$ are constant then

$$\hat{R}_i = \left(e^{-\hat{\beta}^T x^{(i)}} X_i \right)^{1/\hat{\sigma}}.$$

It was noted (see (5.42)) that the random variables

$$R_i = \left(\int_0^{T_i} e^{-\beta^T x^{(i)}(u)} du \right)^{1/\sigma}$$

are i.i.d. with the parameter-free survival function G_0. So if n is large, the random variables

$$\tilde{R}_i = \left(\int_0^{T_i} e^{-\hat{\beta}^T x^{(i)}(u)} du \right)^{1/\hat{\sigma}}$$

are *approximately* i.i.d. with the parameter-free survival function. Not all $\tilde{R}_i$ are observed when there is right censoring. So

$$\hat{R}_i = \tilde{R}_i \quad \text{if} \quad \delta_i = 1,$$

i.e. if the ith unit is not censored. Therefore the plot only of $\hat{R}_i$ for which $\delta_i = 1$ are used. *The points $(X_i, \hat{R}_i)$ with $\delta_i = 1$ are dispersed around a horizontal line in a horizontal band.* Strong departures from linearity or increasing (decreasing) width of the band indicate that the AFT model or parametrization of this model may be false.

Another way of graphical model checking could be comparing the parameter-free survival function G_0 and its Kaplan-Meier (1958) estimator (see Appendix, Section A13) obtained from the residuals $\hat{\varepsilon}_i$.

In practice the residuals are at first transformed. Indeed, take notice that the random variables with a specified parameter-free distribution G_0 may be transformed to the random variables with any other parameter-free distribution. So the residuals $\hat{\varepsilon}_i$ are transformed in such a way that the Kaplan-Meier estimator obtained from these transformed residuals estimates not G_0 but the survival function of some standard distribution such as uniform on $(0,1)$ or the standard exponential.

The random variables

$$\hat{\varepsilon}_i^u = 1 - G_0(\hat{R}_i)$$

are called the *uniformly standardized residuals.*

Denote by

$$S_U(t) = t \mathbf{1}_{(0,1)}(t)$$

the survival function of the uniform distribution on $(0,1)$. The random variables

$$\varepsilon_i^u = 1 - G_0(R_i)$$

are i.i.d. with the survival function S_U. So when n is large then

$$(\hat{\varepsilon}_1^u, \delta_1), \cdots, (\hat{\varepsilon}_n^u, \delta_n)$$

can be interpreted as right censored data corresponding to the survival distribution S_U.

Denote by $\hat{S}_U$ the Kaplan-Meier (1958) estimator (see Appendix) of S_U. Under the AFT model and for i suh that $\delta_i = 1$ the points

$$(u_i, v_i) = (\hat{\varepsilon}_i^u, 1 - \hat{S}(\hat{\varepsilon}_i^u)), \quad (i = 1, \cdots, n)$$

should be dispersed around the line

$$v = u, \quad u \in [0, 1].$$

Strong departure from linearity indicates that the AFT model or its parametrization may be false.

Alternatively, the distribution G_0 can be transformed to the standard exponential distribution $\mathcal{E}(1)$ with the survival function

$$S_{\mathcal{E}}(t) = e^{-t}\mathbf{1}_{(0,\infty)}(t).$$

The random variables

$$\hat{\varepsilon}_i^{\mathcal{E}} = -\ln[1 - G_0(\hat{R}_i)], \quad (i = 1, \cdots, n),$$

are called the *exponentially standardized residuals*. When n is large then

$$(\hat{\varepsilon}_i^{\mathcal{E}}, \delta_i), \quad , (i = 1, \cdots, n),$$

can be interpreted as right censored data corresponding to the survival distribution $S_{\mathcal{E}}$. Denote by $\hat{S}_{\mathcal{E}}(t)$ the Kaplan-Meier estimator of $S_{\mathcal{E}}(t)$. Under the AFT model and for i such that $\delta_i = 1$ the points

$$(u_i, v_i) = (\hat{\varepsilon}_i^{\mathcal{E}}, -\ln[\hat{S}_{\mathcal{E}}(\hat{\varepsilon}_i^{\mathcal{E}})]), \quad (i = 1, \cdots, n)$$

should be dispersed around the line

$$v = u, \quad u \geq 0.$$

Several examples of residual plots from actual data can be found in Meeker and Escobar (1998).

5.4 FTR data analysis: generalized Weibull distribution

5.4.1 Model

Families of scale-shape distributions such as Weibull, loglogistic, lognormal do not give $\cup$-shaped hazard rates.

A family which allows all possible forms of the hazard rate (constant, increasing, decreasing, $\cap$-shaped, $\cup$-shaped) is the *generalized Weibull distribution* with the survival function

$$S_0(t) = \exp\left\{1 - (1 + (t/\theta)^\nu)^\gamma\right\} \quad (\gamma, \theta, \nu > 0).$$

If the AFT model with baseline generalized Weibull distribution is used, the parameter θ can be included in the coefficient β_0, so we suppose that

$$S_0(t) = \exp\left\{1 - (1 + t^\nu)^\gamma\right\} \quad (\gamma, \nu > 0). \tag{5.85}$$

Under the AFT model the survival function $S_{x^{(i)}}(\cdot)$ has the form

$$S_i(t; \beta, \nu, \gamma) = \exp\left\{1 - \left(1 + \left(\int_0^t e^{-\beta^T x^{(i)}(u)} du\right)^\nu\right)^\gamma\right\}, \qquad (5.86)$$

and for the constant $x^{(i)}$

$$S_i(t; \beta, \nu, \gamma) = \exp\left\{1 - \left(1 + (e^{-\beta^T x^{(i)}} t)^\nu\right)^\gamma\right\}. \qquad (5.87)$$

Set

$$f_i(t, \beta, \gamma) = \int_0^t e^{-\beta^T x^{(i)}(u)} du.$$

5.4.2 Maximum likelihood estimation for the regression parameters

Similarly as in 5.1.5 we obtain the likelihood function

$$L(\beta, \nu, \gamma) = \prod_{i=1}^n \left\{\nu\gamma \left(1 + (f_i(X_i, \beta, \gamma))^\nu\right)^{\gamma-1} e^{-\beta^T x^{(i)}(X_i)} (f_i(X_i, \beta, \gamma))^{\nu-1}\right\}^{\delta_i}$$

$$\times \exp\left\{1 - (1 + (f_i(X_i, \beta, \gamma))^\nu)^\gamma\right\}. \qquad (5.88)$$

If $x^{(i)}$ are constant then

$$L(\beta, \nu, \gamma) = \prod_{i=1}^n \left\{\nu\gamma e^{-\nu\beta^T x^{(i)}} X_i^{\nu-1} \left(1 + (e^{-\beta^T x^{(i)}} X_i)^\nu\right)^{\gamma-1}\right\}^{\delta_i} \times$$

$$\exp\left\{1 - \left(1 + (e^{-\beta^T x^{(i)}} X_i)^\nu\right)^\gamma\right\}. \qquad (5.89)$$

In the case of constant explanatory variables the score functions are

$$U_l(\beta, \nu, \gamma) = \frac{\partial \ln L(\beta, \nu, \gamma)}{\partial \beta_l} = \nu \sum_{i=1}^n x_l^{(i)}(\gamma \omega_i(\beta, \nu, \gamma) - \delta_i u_i(\beta, \nu, \gamma))$$

$$(l = 0, \cdots, m),$$

$$U_{m+1}(\beta, \nu, \gamma) = \frac{\partial \ln L(\beta, \nu, \gamma)}{\partial \nu} =$$

$$\frac{D}{\nu} - \frac{1}{\nu} \sum_{i=1}^n (\gamma \omega_i(\beta, \nu, \gamma) - \delta_i u_i(\beta, \nu, \gamma)) \ln z_i(\beta, \nu),$$

$$U_{m+2}(\beta, \nu, \gamma) = \frac{\partial \ln L(\beta, \nu, \gamma)}{\partial \gamma} = \frac{D}{\gamma} - \sum_{i=1}^n \{(1 + z_i(\beta, \nu))^\gamma - \delta_i\} \ln(1 + z_i(\beta, \nu))\},$$

$$(5.90)$$

where

$$D = \sum_{i=1}^n \delta_i, \quad z_i(\beta, \nu) = \left(e^{-\beta^T x^{(i)}} X_i\right)^\nu,$$

$$u_i(\beta,\nu,\gamma) = 1 + (\gamma - 1)\frac{z_i(\beta,\nu)}{1 + z_i(\beta,\nu)},$$

$$\omega_i(\beta,\nu,\gamma) = (1 + z_i(\beta,\nu))^{\gamma-1}z_i(\beta,\nu). \qquad (5.91)$$

When the explanatory variables are time-varying the formulas are slightly more complicated.

The maximum likelihood estimators $\hat{\beta}$, $\hat{\nu}$, $\hat{\gamma}$ are obtained by solving the system of equations

$$U_l(\beta,\nu,\gamma) = 0 \quad (l = 0, 1, ..., m+2).$$

5.4.3 Estimators of the main reliability characteristics

Suppose that $x(\cdot)$ is an arbitrary explanatory variable.

Estimator of the survival function $S_{x(\cdot)}(t)$:

$$\hat{S}_{x(\cdot)}(t) = \exp\left\{1 - \left(1 + (\int_0^t e^{-\hat{\beta}^T x(u)}du)^{\hat{\nu}}\right)^{\hat{\gamma}}\right\}. \qquad (5.92)$$

If x is constant then

$$\hat{S}_x(t) = \exp\left\{1 - \left(1 + (e^{-\hat{\beta}^T x}t)^{\hat{\nu}}\right)^{\hat{\gamma}}\right\}. \qquad (5.93)$$

Estimator of the p-quantile

The estimator $\hat{t}_p(x(\cdot))$ verifies the equation:

$$\hat{S}_{x(\cdot)}(\hat{t}_p(x(\cdot))) = 1 - p. \qquad (5.94)$$

If x is constant then

$$\hat{t}_p(x) = \{(1 - \ln(1 - p))^{1/\hat{\gamma}} - 1\}^{1/\hat{\nu}}e^{\hat{\beta}^T x}. \qquad (5.95)$$

Estimator of the mean failure time $m(x(\cdot))$:

$$\hat{m}(x(\cdot)) = \int_0^\infty \hat{S}_{x(\cdot)}(u)du. \qquad (5.96)$$

If x is constant then

$$\hat{m}(x) = e^{\hat{\beta}^T x + 1}\int_0^\infty \exp\{-(1 + u^{\hat{\nu}})^{\hat{\gamma}}\}du. \qquad (5.97)$$

Estimators of the mean ratios

Estimators of the mean ratios have the same forms (5.59)-(5.64) as in the case of the scale-shape family of distributions.

5.4.4 *Asymptotic distribution of the regression parameters*

The distribution of the maximum likelihood estimators $(\hat{\beta}, \hat{\nu}, \hat{\gamma})^T$ for large n is approximated by the normal law:

$$(\hat{\beta}, \hat{\nu}, \hat{\gamma})^T \approx N_{m+3}((\beta, \nu, \gamma)^T, \Sigma^{-1}(\beta, \nu, \gamma)). \tag{5.98}$$

The covariance matrix $\Sigma^{-1}(\beta, \nu, \gamma)$ is estimated by

$$\mathbf{I}^{-1}(\hat{\beta}, \hat{\nu}, \hat{\gamma}) = (I^{ls}(\hat{\beta}, \hat{\nu}, \hat{\gamma}))_{(m+3)\times(m+3)}, \tag{5.99}$$

where

$$\mathbf{I}(\beta, \nu, \gamma) = (I_{lk}(\beta, \nu, \gamma))_{(m+3)\times(m+3)} \tag{5.100}$$

is a matrix of minus second partial derivatives of $\ln L(\beta, \nu, \gamma)$ with respect to its arguments. Under constant explanatory variables we have (the indices l ($l = 0, \cdots, m$), $m+1$ and $m+2$ mean the derivatives with respect β_l, ν and γ, respectively):

$$I_{ls}(\beta, \nu, \gamma) =$$

$$\nu^2 \sum_{i=1}^{n} \frac{x_l^{(i)} x_s^{(i)}}{1 + z_i(\beta, \nu)} \left\{ \gamma \omega_i(\beta, \nu, \gamma) \left(1 + \gamma z_i(\beta, \nu)\right) - \delta_i \left(u_i(\beta, \nu, \gamma) - 1\right) \right\},$$

$$I_{l,m+1}(\beta, \nu, \gamma) = -\frac{1}{\nu} U_l(\beta, \nu, \gamma) -$$

$$\sum_{i=1}^{n} x_l^{(i)} \frac{\ln z_i(\beta, \nu)}{1 + z_i(\beta, \nu)} \left\{ \gamma \omega_i(\beta, \nu, \gamma) \left(1 + \gamma z_i(\beta, \nu)\right) - \delta_i \left(u_i(\beta, \nu, \gamma) - 1\right) \right\},$$

$$I_{l,m+2}(\beta, \nu, \gamma) =$$

$$-\nu \sum_{i=1}^{n} x_l^{(i)} \left\{ \omega_i(\beta, \nu, \gamma) \left(1 + \gamma \ln(1 + z_i(\beta, \nu))\right) - \delta_i \frac{z_i(\beta, \nu)}{1 + z_i(\beta, \nu)} \right\},$$

$$(l, s = 0, ..., m).$$

$$I_{m+1,m+1}(\beta, \nu, \gamma) =$$

$$\frac{1}{\nu} U_{m+1}(\beta, \nu, \gamma) + \frac{1}{\nu^2} \sum_{i=1}^{n} \frac{\ln^2 z_i(\beta, \nu)}{1 + z_i(\beta, \nu)} \left\{ \gamma \omega_i(\beta, \nu, \gamma)(1 + \gamma z_i(\beta, \nu)) \right.$$

$$\left. -\delta_i \left(u_i(\beta, \nu, \gamma) - 1\right) \right\} + \frac{1}{\nu^2} \sum_{i=1}^{n} \ln z_i(\beta, \nu) \left\{ \gamma \omega_i(\beta, \nu, \gamma) - \delta_i u_i(\beta, \nu, \gamma) \right\},$$

$$I_{m+1,m+2}(\beta, \nu, \gamma) =$$

$$\frac{1}{\nu} \sum_{i=1}^{n} \frac{z_i(\beta, \nu) \ln z_i(\beta, \nu)}{1 + z_i(\beta, \nu)} \left\{ (1 + z_i(\beta, \nu))^\gamma + \gamma(1 + z_i(\beta, \nu))^\gamma \ln(1 + z_i(\beta, \nu)) - \delta_i \right\},$$

$$I_{m+2,m+2}(\beta,\nu,\gamma) =$$

$$\frac{D}{\gamma^2} + \sum_{i=1}^{n}(1+z_i(\beta,\nu))^\gamma \ln^2(1+z_i(\beta,\nu)), \qquad (5.101)$$

In particular, the variances of the estimators $\hat{\beta}_j, \hat{\nu}, \hat{\gamma}$ are estimated by

$$\hat{\mathbf{Var}}\{\hat{\beta}_j\} = I^{jj}(\hat{\beta},\hat{\nu},\hat{\gamma}), \quad \hat{\mathbf{Var}}\{\hat{\nu}\} = I^{m+1,m+1}(\hat{\beta},\hat{\nu},\hat{\gamma}),$$

$$\hat{\mathbf{Var}}\{\hat{\gamma}\} = I^{m+2,m+2}(\hat{\beta},\hat{\nu},\hat{\gamma}).$$

5.4.5 Approximate confidence intervals for the main reliability characteristics

Approximate $(1-\alpha)$ confidence intervals for the survival function, the p-quantile and the mean life under any x are obtained as in the case of the scale-shape families of distributions with the only difference that the functions Q and K contain $m+3$ parameters instead of $m+2$.

Approximate confidence intervals for the survival functions

For any $x = (x_0,x_1,\cdots,x_m) \in E_0$, $x_0 = 1$, an approximate $(1-\alpha)$-confidence interval for the survival function $S_x(t)$ is defined by the formula

$$\left(1 + \frac{1-\hat{S}_x(t)}{\hat{S}_x)(t)}\exp\{\pm\hat{\sigma}_{Q_x}w_{1-\alpha/2}\}\right)^{-1},$$

where w_α is the α-quantile of the normal law $N(0,1)$ and

$$\hat{\sigma}^2_{Q_x} = \frac{1}{(1-\hat{S}_x(t))^2}\sum_{l=0}^{m+2}\sum_{s=0}^{m+2}a_l(\hat{\beta},\hat{\nu},\hat{\gamma})I^{ls}(\hat{\beta},\hat{\nu},\hat{\gamma})a_s(\hat{\beta},\hat{\nu},\hat{\gamma}),$$

where

$$a_l(\hat{\beta},\hat{\nu},\hat{\gamma}) = -a_{m+1}(\hat{\beta},\hat{\nu},\hat{\gamma})\,\nu\,x_l/(\ln t - \hat{\beta}^T x) \quad (l=0,\cdots,m),$$

$$a_{m+1}(\hat{\beta},\hat{\nu},\hat{\gamma}) = -\gamma\,(e^{-\hat{\beta}^T x}t)^{\hat{\nu}}\left(1+(e^{-\hat{\beta}^T x}t)^{\hat{\nu}}\right)^{\hat{\gamma}-1}(\ln t - \hat{\beta}^T x),$$

$$a_{m+2}(\hat{\beta},\hat{\nu},\hat{\gamma}) = -(1-\ln S_x(t))\ln\{1+(e^{-\hat{\beta}^T x}t)^{\hat{\nu}}\}.$$

Approximate confidence intervals for the quantiles

An approximate $(1-\alpha)$-confidence interval for the p-quantile $t_p(x)$ is

$$\hat{t}_p(x)\exp\{\pm\hat{\sigma}_{K_p(x)}w_{1-\alpha/2}\},$$

where

$$\hat{\sigma}^2_{K_p(x)} = \sum_{l=0}^{m+2}\sum_{s=0}^{m+2}b_l(\hat{\beta},\hat{\nu},\hat{\gamma})\,I^{ls}(\hat{\beta},\hat{\nu},\hat{\gamma})\,b_s(\hat{\beta},\hat{\nu},\hat{\gamma}),$$

where
$$b_l(\hat\beta,\hat\nu,\hat\gamma) = x_l \quad (l = 0,1,...,m), \quad b_{m+1}(\hat\beta,\hat\nu,\hat\gamma) = 1,$$
$$b_{m+2}(\hat\beta,\hat\nu,\hat\gamma) =$$
$$-\frac{1}{\nu\gamma^2}\ln(1-\ln(1-p))(1-\ln(1-p))^{1/\hat\gamma}\left\{(1-\ln(1-p))^{1/\hat\gamma}\right\}^{-1}.$$

Approximate confidence interval for the mean lifetime

An approximate $(1-\alpha)$-confidence interval for the mean lifetime $m(x)$ is
$$\hat m(x)\exp\left\{\pm\frac{\hat\sigma_{m(x)}(\hat\beta,\hat\sigma)}{\hat m(x)}w_{1-\alpha/2}\right\},$$

where
$$\hat\sigma^2_{m(x)} = \sum_{l=0}^{m+2}\sum_{s=0}^{m+2} c_l(\hat\beta,\hat\nu,\hat\gamma)\,I^{ls}(\hat\beta,\hat\nu,\hat\gamma)\,c_s(\hat\beta,\hat\nu,\hat\gamma),$$

and
$$c_l(\hat\beta,\hat\nu,\hat\gamma) = x_l m(x) \quad (l=0,1,...,m),$$
$$c_{m+1}(\hat\beta,\hat\nu,\hat\gamma) = -\gamma e^{\hat\beta^T x+1}\int_0^\infty u^{\hat\nu}(1+u^{\hat\nu})^{\hat\gamma-1}\exp\{-(1+u^{\hat\nu})^{\hat\gamma}\}\ln u\,du,$$
$$c_{m+2}(\hat\beta,\hat\nu,\hat\gamma) = -e^{\hat\beta^T x+1}\int_0^\infty (1+u^{\hat\nu})^{\hat\gamma}\exp\{-(1+u^{\hat\nu})^{\hat\gamma}\}\ln(1+u^\nu)\,du.$$

Approximated confidence intervals for the mean ratios

An approximate $(1-\alpha)$-confidence interval for the mean ratio $MR(x,y)$ is:
$$\left(\exp\{\hat\beta^T(y-x)-\hat\sigma_{MR}\}w_{1-\alpha/2}, \exp\{\hat\beta^T(y-x)+\hat\sigma_{MR}\}w_{1-\alpha/2}\right),$$

where
$$\hat\sigma^2_{MR} = \sum_{l=0}^{m}\sum_{s=0}^{m}(y_l-x_l)I^{ls}(\hat\beta,\hat\sigma)(y_s-x_s).$$

5.4.6 Tests for nullity of the regression coefficients

All results of Section 5.3.6 are evidently reformulated for consideration in this section.

Likelihood ratio tests

The hypothesis $H_{k_1...k_l} : \beta_{k_1} = ... = \beta_{k_l} = 0$ is rejected with the significance level α if
$$LR_{k_1,...,k_l} > \chi^2_{1-\alpha}(l), \tag{5.102}$$

where

$$LR_{k_1,\ldots,k_l} = -2\ln\left\{\frac{\max_{\nu,\gamma,\beta:\beta_{k_1}=\ldots=\beta_{k_l}=0} L(\beta,\nu,\gamma)}{\max_{\beta,\nu,\gamma} L(\beta,\nu,\gamma)}\right\} \qquad (5.103)$$

Wald's tests

The hypothesis $H_{k_1\ldots k_l}$ is rejected with the significance level α if

$$W_{k_1\ldots k_l} > \chi^2_{1-\alpha}(l), \qquad (5.104)$$

where

$$W_{k_1\ldots k_l} = (\hat{\beta}_{k_1},\ldots,\hat{\beta}_{k_l})^T A^{-1}_{k_1\ldots k_l}(\hat{\beta},\hat{\nu},\hat{\gamma})(\hat{\beta}_{k_1},\ldots,\hat{\beta}_{k_l})^T, \qquad (5.105)$$

and $A_{k_1\ldots k_l}(\hat{\beta},\hat{\nu},\hat{\gamma})$ is the submatrix of $I^{-1}(\hat{\beta},\hat{\nu},\hat{\gamma})$ which is in the intersection of the $k_1,\ldots,k_l$ rows and $k_1,\ldots,k_l$ columns.

Score tests

The hypothesis $H_{k_1\ldots k_l}$ is rejected with the significance level α if

$$U_{k_1\ldots k_l} > \chi^2_{1-\alpha}(l),$$

where

$$U_{k_1\ldots k_l} =$$

$$(U_{k_1}(\tilde{\beta},\tilde{\nu}\tilde{\gamma}),\cdots,U_{k_l}(\tilde{\beta},\tilde{\nu},\tilde{\gamma}))^T A_{k_1\ldots k_l}(\hat{\beta},\hat{\nu},\hat{\gamma})\,(U_{k_1}(\tilde{\beta},\tilde{\nu},\tilde{\gamma}),\cdots,U_{k_l}(\tilde{\beta},\tilde{\nu},\tilde{\gamma})),$$

and $(\tilde{\beta},\tilde{\nu},\tilde{\gamma}))$ verifies the equality

$$L(\tilde{\beta},\tilde{\nu},\tilde{\gamma})) = \max_{\nu,\gamma,\beta:\beta_{k_1}=\ldots=\beta_{k_l}=0} L(\beta,\nu,\gamma).$$

5.4.7 Residual plots

All discussion of Section 5.3.7 holds and for the case of the generalized baseline distribution with the only difference that the standartized residuals are defined as

$$\hat{R}_i = \left\{1 + \left(\int_0^{X_i} e^{-\hat{\beta}^T x^{(i)}(u)}du\right)^{\nu}\right\}^{\gamma} - 1, \qquad (5.106)$$

and, when data is sufficiently large, are interpreted as failure or censoring times from right censored data from the standard exponential distribution. So here $\hat{R}_i^E = \hat{R}_i$ and $\hat{R}_i^U = 1 - e^{-\hat{R}_i}$.

If $x^{(i)}$ are constant then

$$\hat{R}_i = \left\{1 + \left(e^{-\hat{\beta}^T x^{(i)}} X_i\right)^{\nu}\right\}^{\gamma} - 1. \qquad (5.107)$$

5.5 FTR data analysis: exponential distribution

5.5.1 Model

Suppose that the baseline survival distribution is exponential:

$$S_0(t) = e^{-t/\theta}.$$

The parameter θ can be included in the coefficient β_0, so we suppose that

$$S_0(t) = e^{-t}.$$

Under the AFT model the survival function $S_i = S_{x^{(i)}}(\cdot)$ and the hazard rate $\alpha_i = \alpha_{x^{(i)}}(\cdot)$ are

$$S_i(t) = \exp\left\{-\int_0^t \exp\{-\beta^T x^{(i)}(u)\}du\right\}, \quad \alpha_i(t) = e^{-\beta^T x^{(i)}(t)}. \qquad (5.108)$$

Under constant $x^{(i)}$:

$$S_i(t) = \exp\{-\exp(-\beta^T x^{(i)})t\}, \quad \alpha_i(t) = e^{-\beta^T x^{(i)}}. \qquad (5.109)$$

5.5.2 Maximum likelihood estimation of the regression parameters

The likelihood function is

$$L(\beta) = \prod_{i=1}^n \{\alpha_i(X_i)\}^{\delta_i} S_i(X_i) =$$

$$\exp\left\{-\sum_{i=1}^n \left(\delta_i \beta^T x^{(i)}(X_i) + \int_0^{X_i} \exp\{-\beta^T x^{(i)}(u)\}du\right)\right\}. \qquad (5.110)$$

For constant $x^{(i)}$:

$$L(\beta) = \exp\left\{-\sum_{i=1}^n (\delta_i \beta^T x^{(i)} + e^{-\beta^T x^{(i)}} X_i)\right\}. \qquad (5.111)$$

The score functions are

$$U_l(\beta) = \frac{\partial \ln L(\beta)}{\partial \beta_l} = -\sum_{i=1}^n \left(\delta_i x_l^{(i)}(X_i) - \int_0^{X_i} x_l^{(i)}(u) \exp\{-\beta^T x^{(i)}(u)\}du\right). \qquad (5.112)$$

For constant $x^{(i)}$:

$$U_l(\beta) = \frac{\partial \ln L(\beta)}{\partial \beta_l} = -\sum_{i=1}^n x_l^{(i)}(\delta_i - e^{-\beta^T x^{(i)}} X_i). \qquad (5.113)$$

Denote by $\hat{\beta}$ the maximum likelihood estimator of β.

5.5.3 Estimators of the main reliability characteristics

The estimators of the survival function $S_x(t)$, the p-quantile $t_p(x)$ and the mean $m(x)$ are

$$\hat{S}_x(t) = \exp\{-e^{-\hat{\beta}^T x} t\}, \quad \hat{t}_p(x) = -e^{-\hat{\beta}^T x} \ln(1-p), \quad \hat{m}(x) = e^{\hat{\beta}^T x}.$$

(5.114)

5.5.4 Asymptotic properties of the regression parameters

The law of $\hat{\beta}$ for large n is approximated by the normal law:

$$\hat{\beta} \approx N(\beta, \Sigma^{-1}(\beta)).$$

The covariance matrix $\Sigma^{-1}(\beta)$ is estimated by

$$\mathbf{I}^{-1}(\hat{\beta}) = (I^{ls}(\hat{\beta}))_{(m+1)\times(m+1)}$$

where

$$\mathbf{I}(\hat{\beta}) = (I_{ls}(\hat{\beta})), \quad (l, s = 0, ..., m),$$

and

$$I_{ls}(\beta) = -\frac{\partial^2 \ln L(\beta)}{\partial \beta_l \partial \beta_s} = \sum_{i=1}^{n} \int_0^{X_i} x_l^{(i)}(u) x_s^{(i)}(u) \exp\{-\beta^T x^{(i)}(u)\} du. \quad (5.115)$$

For constant $x^{(i)}$:

$$I_{ls}(\beta) = \sum_{i=1}^{n} x_l^{(i)} x_s^{(i)} \exp\{-\beta^T x^{(i)}\} X_i. \quad (5.116)$$

5.5.5 Approximate confidence intervals for the main reliability characteristics

Set

$$\hat{u} = (\sum_{l=0}^{m} \sum_{s=0}^{m} x_l x_s I^{ls}(\hat{\beta}))^{1/2}. \quad (5.117)$$

Approximate $(1 - \alpha)$-confidence interval for $S_x(t)$:

$$\left(1 + \frac{1 - \hat{S}_x(t)}{\hat{S}_x(t)} \exp\{\pm\hat{\sigma}_{Q_x} w_{1-\alpha/2}\}\right)^{-1}. \quad (5.118)$$

where

$$\hat{\sigma}_{Q_x} = \frac{\ln\{\hat{S}_x(t)\}}{1 - \hat{S}_x(t)} \hat{u}.$$

Approximate $(1 - \alpha)$ confidence interval for $t_p(x)$:

$$\hat{t}_p(x) \exp\{\pm\hat{u}\, w_{1-\alpha/2}\}. \tag{5.119}$$

Approximate $(1 - \alpha)$-confidence interval for the mean $m(x)$:

$$\hat{m}(x) \exp\{\pm\hat{u}\, w_{1-\alpha/2}\}. \tag{5.120}$$

5.5.6 Tests for nullity of the regression coefficients

All results of Section 5.3.6 are evidently reformulated for consideration in this section.

Likelihood ratio tests

The hypothesis $H_{k_1\dots k_l} : \beta_{k_1} = \dots = \beta_{k_l} = 0$ is rejected with the significance level α if

$$LR_{k_1,\dots,k_l} > \chi^2_{1-\alpha}(l), \tag{5.121}$$

where

$$LR_{k_1,\dots,k_l} = -2\ln\left\{\frac{\max_{\beta:\beta_{k_1}=\dots=\beta_{k_l}=0} L(\beta)}{\max_\beta L(\beta)}\right\} \tag{5.122}$$

Wald's tests

The hypothesis $H_{k_1\dots k_l}$ is rejected with the significance level α if

$$W_{k_1\dots k_l} > \chi^2_{1-\alpha}(l), \tag{5.123}$$

where

$$W_{k_1\dots k_l} = (\hat{\beta}_{k_1}, \dots, \hat{\beta}_{k_l})^T A^{-1}_{k_1\dots k_l}(\hat{\beta})(\hat{\beta}_{k_1}, \dots, \hat{\beta}_{k_l})^T, \tag{5.124}$$

and $A_{k_1..k_l}(\hat{\beta})$ is the submatrix of $I^{-1}(\hat{\beta})$ which is in the intersection of the $k_1, \dots, k_l$ rows and $k_1, \dots, k_l$ columns.

Score tests

The hypothesis $H_{k_1\dots k_l}$ is rejected with the significance level α if

$$U_{k_1\dots k_l} > \chi^2_{1-\alpha}(l),$$

where

$$U_{k_1\dots k_l} = (U_{k_1}(\tilde{\beta}), \cdots, U_{k_l}(\tilde{\beta}))^T A_{k_1\dots k_l}(\hat{\beta})\, (U_{k_1}(\tilde{\beta}), \cdots, U_{k_l}(\tilde{\beta})),$$

and $\tilde{\beta}$ verifies the equality

$$L(\tilde{\beta}) = \max_{\beta:\beta_{k_1}=\dots=\beta_{k_l}=0} L(\beta).$$

5.5.7 Residual plots

All discussion of Section 5.3.7 holds and for the case of the generalized baseline distribution with the only difference that the standartized residuals are defined as

$$\hat{R}_i = \int_0^{X_i} e^{-\hat{\beta}^T x^{(i)}(u)} du, \tag{5.125}$$

and, when data is sufficiently large, are interpreted as failure or censoring times from right censored data from the standard exponential distribution. So here $\hat{R}_i^E = \hat{R}_i$ and $\hat{R}_i^U = 1 - e^{-\hat{R}_i}$.

If $x^{(i)}$ are constant then

$$\hat{R}_i = e^{-\hat{\beta}^T x^{(i)}} X_i. \tag{5.126}$$

5.6 Plans of experiments in accelerated life testing

The purpose of ALT is to give estimators of the main reliability characteristics under usual (design) stress using data of accelerated experiments when units are tested at higher than usual stress conditions.

A stress $x_2(\cdot)$ is higher than a stress $x_1(\cdot)$, $x_2(\cdot) > x_1(\cdot)$, if for any $t \geq 0$ the inequality $S_{x_1(\cdot)}(t) \geq S_{x_2(\cdot)}(t)$ holds and exists $t_0 > 0$ such that $S_{x_1(\cdot)}(t_0) > S_{x_2(\cdot)}(t_0)$.

Denote by $x_0 = (x_{00}, x_{01}, \cdots, x_{0m})$, $x_{00} = 1$, the usual stress.

If the AFT model holds on a set of stresses E then for any $x(\cdot) \in E$:

$$S_{x(\cdot)}(t) = S_0 \left(\int_0^t r\{x(\tau)\} d\tau \right). \tag{5.127}$$

If $x(\tau) \equiv x = const$ than

$$S_x(t) = S_0(r(x)t). \tag{5.128}$$

Generally accelerated life testing experiments are done under an one-dimensional stress (m=1), sometimes under two-dimensional $(m = 2)$.

Several plans of experiments may be used in ALT.

5.6.1 First plan of experiments

Let $x_1, ..., x_k$ be constant over time accelerated stresses:

$$x_0 < x_1 < ... < x_k;$$

here $x_i = (x_{i0}, x_{i1}, \cdots, x_{im})$, $x_{i0} = 1$. The usual stress x_0 is not used during experiments.

k groups of units are tested. The ith group of n_i units, $\sum_{i=1}^k n_i = n$, is tested under the stress x_i. The data can be complete or independently right censored.

If the form of the function r is completely unknown and this plan of experiments is used, the function S_{x_0} can not be estimated even if it is supposed to know a parametric family to which belongs the distribution $S_{x_0}(t)$.

For example, if $S_0(t) = e^{-(t/\theta)^\alpha}$ then for constant stresses

$$S_x(t) = \exp\left\{-\left(\frac{r(x)}{\theta}t\right)^\alpha\right\}.$$

Under the given plan of experiments the parameters

$$\alpha, \frac{r(x_1)}{\theta}, \cdots, \frac{r(x_k)}{\theta}$$

and the functions $S_{x_1}(t), ..., S_{x_k}(t)$ may be estimated. Nevertheless, the function $r(x)$ being completely unknown, the parameter $r(x_0)$ can not be written as a known function of these estimated parameters. So $r(x_0)$ and, consequently, $S_{x_0}(t)$ can not be estimated.

Thus, the function r must be chosen from some class of functions. Usually the model

$$S_x(t) = S_0(e^{-\beta^T x}t), \tag{5.129}$$

with $x = (x_0, \cdots, x_m)^T$, $\beta = (\beta_0, \cdots, \beta_m)^T$ is used, where x_j may be not the stress components but some functions $\varphi_j(x)$ of them. The form of the functions $\varphi_j(x)$ is discussed in Section 5.1.

The choice of the functions $\varphi_j(x)$ is very important in accelerated life testing because the *usual stress is not in the region of the stresses used in the experiment*, and bad choice of the model may give bad estimators of the reliability characteristics under the usual stress.

5.6.2 Second plan of experiments

In step-stress accelerated life testing the plan of experiments is as follows.

n units are placed on test at an initial low stress and if it does not fail in a predetermined time t_1, the stress is increased and so on. Thus, all units are tested under the step-stress

$$x(\tau) = \begin{cases} x_1, & 0 \leq \tau < t_1, \\ x_2, & t_1 \leq \tau < t_2, \\ \cdots & \cdots \\ x_k, & t_{k-1} \leq \tau < t_k; \end{cases} \tag{5.130}$$

here $x_j = (x_{j0}, \cdots, x_{jm})^T$, $x_{j0} = 1$, $t_0 = 0, t_k = \infty$.

In this case the function $r(x)$ should be also parametrized because, even when the usual stress is used until the moment t_1, the data of failures occurring after this moment do not give any information about the reliability under the usual stress when the function $r(x)$ is unknown.

Thus, the model

$$S_{x(\cdot)}(t) = S_0\left(\int_0^t e^{-\beta^T x(\tau)} d\tau\right) \tag{5.131}$$

should be used. For step-stresses it is written in the form (cf.(2.24)):if $t \in$

$[t_{i-1}, t_i), i = 1, \cdots, k$

$$S_{x(\cdot)}(t) = S_0 \left\{ \mathbf{1}_{\{i>1\}} \sum_{j=1}^{i-1} e^{-\beta^T x_j}(t_j - t_{j-1}) + e^{-\beta^T x_i}(t - t_{i-1}) \right\}. \quad (5.132)$$

As in the case of the first plan, x_{jl} may be not the stress components but some specified functions of them and good choice of these functions is also very important.

5.6.3 Third plan of experiments

Application of the first two plans may not give satisfactory results because assumptions on the form of the function $r(x)$ are done. These assumptions can not be statistically verified because of lack of experiments under the usual stress.

If the function $r(x)$ is completely unknown, and the coefficient of variation (defined as the ratio of the standard deviation and the mean) of failure times is not too large, the following plan of experiments may be used.

Suppose that the failure time under the usual stress x_0 takes large values and most of the failures occur after the moment t_2 given for the experiment.

Two groups of units are tested:

a) The first group of n_1 units under a constant accelerated stress x_1;

b) The second group of n_2 units under a step-stress: time t_1 under x_1, and after this moment under the usual stress x_0 until the moment t_2, i.e. under the stress:

$$x_2(\tau) = \begin{cases} x_1, & 0 \leq \tau \leq t_1, \\ x_0, & t_1 < \tau \leq t_2. \end{cases}$$

Units use much of their resources until the moment t_1 under the accelerated stress x_1, so after the switch-up failures occur in the interval $[t_1, t_2]$ even under usual stress. The figure 5.1 illustrates this phenomenon. We denote by f_{x_0}, f_{x_1}, and f_{x_2} the densities of failure times under stresses x_0, x_1, and x_2, respectively. The area under f_{x_0} in the interval $[0, t_2]$ is very small but the area under f_{x_2} in the interval $[t_1, t_2]$ (when the units of the second group are under the usual stress x_0) is large.

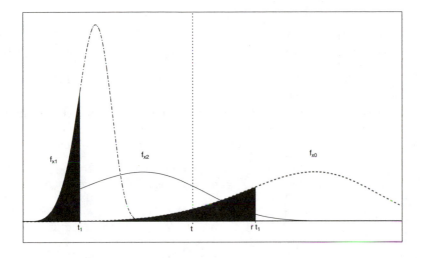

Figure 5.1

The AFT model implies that

$$S_{x_1}(u) = S_{x_0}(ru),$$

where $r = r(x_1)/r(x_0)$, and

$$S_{x_2(\cdot)}(u) = \begin{cases} S_{x_0}(ru), & 0 \leq u \leq t_1, \\ S_{x_0}(rt_1 + u - t_1), & t_1 < u \leq t_2, \end{cases}$$

or, shortly,

$$S_{x_2(\cdot)}(t) = S_{x_0}(r(u \wedge t_1) + (u - t_1) \vee 0), \qquad (5.133)$$

with $a \wedge b = min(a, b)$ and $a \vee b = max(a, b)$.

It will be shown in the next chapter that if the third plan is used, and both functions S_{x_0} and $r(x)$ are completely unknown, semiparametric estimation of S_{x_0} is possible.

The third plan may be modified. The moment t_1 may be chosen as random. The most natural is to choose t_1 as the moment when the failures begin to occur.

5.6.4 Fourth plan of experiments

If the failure-time distribution under the usual stress is exponential, exact confidence intervals can be obtained using the following plan of experiments.

k groups of units are observed. The ith group of n_i units is tested under one-dimensional constant stress $x^{(i)}$ until the r_ith failure ($r_i \leq n_i$) (type II *censoring). The failure moments of the ith group are*

$$T_{i1} \leq \cdots \leq T_{ir_i} \quad (i = 1, 2, \cdots, k).$$

5.7 Parametric estimation in ALT under the AFT model

Suppose that the baseline survival function S_0 in the AFT model (5.1) belongs to a specified class of distributions.

Each of the above mentioned plans of experiments is a particular case of the plan of experiments considered in Section 5.3.1.

The first plan is the particular case of that general plan taking

$$x^{(i)} = \begin{cases} x_1, & i = 1, \cdots, n_1, \\ x_2, & i = n_1 + 1, \cdots, n_1 + n_2, \\ \cdots & \cdots \\ x_k, & i = \sum_{j=1}^{k-1} n_j + 1, \cdots, n, \end{cases} \quad (5.134)$$

where $x_j = (x_{j0}, \cdots, x_{jm}) \ (i = 1, \cdots, k)$.

Second plan: $x^{(i)}(\cdot) = x(\cdot)$ for any $i = 1, \cdots, n$, where

$$x(\tau) = \begin{cases} x_1, & 0 \leq \tau < t_1, \\ x_2, & t_1 \leq \tau < t_2, \\ \cdots & \cdots \\ x_k, & t_{m-1} \leq \tau < t_k, \end{cases} \quad (5.135)$$

and $x_j = (x_{j0}, \cdots, x_{jm}) \ (i = 1, \cdots, k)$.

Note the difference. In the case of the first plan the constant stress x_j is used *all time of the experiment for the jth group of units*. In the case of the second plan the constant stress x_j is used *in the interval $[t_{j-1}, t_j)$ for all units*.

Third plan:

$$x^{(i)}(\tau) = x_1, \ \tau \geq 0, \quad (i = 1, \cdots, n_1);$$

$$x^{(i)}(\tau) = \begin{cases} x_1, & \tau < t_1, \\ x_0, & \tau \in [t_1, t_2], \end{cases} \quad (i = n_1 + 1, \cdots, n), \quad (5.136)$$

where $x_j = (x_{j0}, \cdots, x_{jm}) \ (i = 1, 2)$.

5.7.1 Maximum likelihood estimation

First plan:

The likelihood function may have the forms (5.48) (shape-scale families of distributions), (5.89) (generalized Weibull distribution), (5.111) (exponential distribution). The score functions have, respectively, the forms (5.51), (5.90), and (5.113).

Second plan:

The likelihood function may have the forms (5.47) (shape-scale families of distribution), (5.89) (generalized Weibull distribution), (5.111) (exponential distribution). The score functions for the first and the third case have the forms (5.49) and (5.113).Take notice that all integrals in the expressions of the likelihood and score functions are sums, because the stress is piecewise constant. For example, if $X_i \in [t_{j-1}, t_j)$, $(j = 1, \cdots, k)$, then

$$\int_0^{X_i} e^{-\beta^T x^{(i)}(u)} du = \mathbf{1}_{\{j>1\}} \sum_{s=1}^{j-1} e^{-\beta^T x_s}(t_s - t_{s-1}) + e^{-\beta^T x_j}(X_i - t_{j-1}),$$

and for $l = 1, \cdots, m$

$$\int_0^{X_i} x_l^{(i)}(u) e^{-\beta^T x^{(i)}(u)} du = \mathbf{1}_{\{j>1\}} \sum_{s=1}^{j-1} x_{sl} \, e^{-\beta^T x_s}(t_s - t_{s-1}) + e^{-\beta^T x_j}(X_i - t_{j-1}).$$

Third plan:

The function $r(x)$ is not parametrized and its form is supposed to be completely unknown. So we can not use expressions obtained for the parametrized model in the previous sections.

Let us consider this plan when the failure time distribution under the usual stress belongs to a family of shape-scale distributions:

$$S_{x_0}(t) = S_0\left((t/\theta)^\alpha\right).$$

Under the AFT model

$$S_{x(\cdot)}(t) = S_0\left\{\left(\int_0^t r\{x(\tau)\}d\tau/\theta\right)^\alpha\right\},$$

where $r(x_0) = 1$. It implies :

$$S_{x_1}(t) = S_0\left(\left(\frac{rt}{\theta}\right)^\alpha\right), \tag{5.137}$$

$$S_{x_2(\cdot)}(t) = S_0\left\{((r(t_1 \wedge t) + (t - t_1) \vee 0)/\theta)^\alpha\right\}, \tag{5.138}$$

where $r = r(x_1)$.

Set

$$\rho = \ln r, \quad \psi = \ln\theta, \quad S(t) = S_0(e^t), \quad f(t) = -S'(t), \quad \lambda(t) = f(t)/S(t). \tag{5.139}$$

Then

$$S_{x_1}(t) = S(\alpha(\ln t + \rho - \psi)); \tag{5.140}$$

$$S_{x_2(\cdot)}(t) = \begin{cases} S(\alpha(\ln t + \rho - \psi)), & t \le t_1, \\ S(\alpha(\ln(e^\rho t_1 + t - t_1) - \psi)), & t > t_1. \end{cases} \tag{5.141}$$

Denote by r_2 the number of failures of the second group until the moment t_1. The likelihood function

$$L = \prod_{j=1}^{n_1} f(\alpha(\ln T_{1j} + \rho - \psi))\frac{\alpha}{T_{1j}} \prod_{j=1}^{r_2} f(\alpha(\ln T_{2j} + \rho - \psi))\frac{\alpha}{T_{2j}} \times$$

$$\prod_{j=r_2+1}^{m_2} f(\alpha(\ln{(e^\rho t_1 + T_{2j} - t_1)} - \psi)) \frac{\alpha}{e^\rho t_1 + T_{2j} - t_1} \times$$

$$S^{n_2-m_2}(\alpha(\ln{(e^\rho t_1 + t_2 - t_1)} - \psi)). \qquad (5.142)$$

The maximum likelihood estimators verify the system of equations

$$U_i(\alpha, \rho, \psi) = 0 \quad (i = 1, 2, 3),$$

where

$$U_1(\alpha, \rho, \psi) = \frac{\partial \ln L}{\partial \alpha} = \sum_{j=1}^{n_1} (\ln f)'(c(T_{1j})) \frac{c(T_{1j})}{\alpha} + \frac{n_1 + m_2}{\alpha}$$

$$+ \sum_{j=1}^{r_2} (\ln f)'(c(T_{2j})) \frac{c(T_{2j})}{\alpha} + \sum_{j=r_2+1}^{m_2} (\ln f)'(d(T_{2j})) \frac{d(T_{2j})}{\alpha} - (n_2 - m_2)\lambda(d(t_2)) \frac{d(t_2)}{\alpha},$$

$$U_2(\alpha, \rho, \psi) = \frac{\partial \ln L}{\partial \rho} = \sum_{j=1}^{n_1} (\ln f)'(c(T_{1j}))\alpha + \sum_{j=1}^{r_2} (\ln f)'(c(T_{2j}))\alpha$$

$$+ \sum_{j=r_2+1}^{m_2} (\ln f)'(d(T_{2j})) \frac{\alpha e^\rho t_1}{e^\rho t_1 + T_{2j} - t_1} -$$

$$\sum_{j=r_2+1}^{m_2} \frac{e^\rho t_1}{e^\rho t_1 + T_{2j} - t_1} - (n_2 - m_2)\lambda(d(t_2)) \frac{\alpha e^\rho t_1}{e^\rho t_1 + t_2 - t_1},$$

$$U_3(\alpha, \rho, \psi) = \frac{\partial \ln L}{\partial \psi} = -\alpha \left[\sum_{j=1}^{n_1} (\ln f)'(c(T_{1j})) + \sum_{j=1}^{r_2} (\ln f)'(c(T_{2j})) \right.$$

$$\left. + \sum_{j=r_2+1}^{m_2} (\ln f)'(d(T_{2j})) - (n_2 - m_2)\lambda(d(t_2)) \right], \qquad (5.143)$$

$$c(u) = \alpha(\ln u + \rho - \psi), \quad d(u) = \alpha(\ln{(e^\rho t_1 + u - t_1)} - \psi).$$

In the case of Weibull, loglogistic, and lognormal distributions:

$$(\ln f)'(t) = e^t; \quad (\ln f)'(t) = \frac{1 - e^t}{1 + e^t}; \quad (\ln f)'(t) = -t, \qquad (5.144)$$

respectively, and

$$\lambda(t) = e^t; \quad \lambda(t) = (1 + e^{-t})^{-1}; \quad \lambda(t) = \frac{\varphi(t)}{1 - \Phi(t)}, \qquad (5.145)$$

respectively.

Fourth plan:

Suppose that the failure-time distribution under the usual stress is exponential, i.e. the model (5.109) is considered.

k groups of units are observed. The ith group of n_i units is tested under one-dimensional constant stress $x^{(i)}$ until the r_ith failure ($r_i \leq n_i$) (type II censoring). The failure moments of the ith group are

$$T_{i1} \leq \cdots \leq T_{ir_i} \quad (i = 1, 2, \cdots, k).$$

The survival and the probability density functions under stress $x^{(i)}$ are

$$S_{x^{(i)}}(t, \beta) = \exp\{-e^{\beta^T x^{(i)}} t\}, p_{x^{(i)}}(t, \beta) = e^{\beta^T x^{(i)}} \exp\{-e^{\beta^T x^{(i)}} t\},$$

$$x^{(i)} = (1, x_{i1}), \quad \beta = (\beta_0, \beta_1).$$

The likelihood function

$$L(\beta) = \prod_{i=1}^{k} \prod_{j=1}^{r_i} p_{x^{(i)}}(T_{ij}, \beta) \, S_{x^{(i)}}^{n_i - r_i}(T_{ir_i}, \beta) =$$

$$\prod_{i=1}^{k} r_i e^{\beta^T x^{(i)}} \exp\{-e^{\beta^T x^{(i)}} U_i\},$$

where

$$U_i = \sum_{j=1}^{r_i} T_{ij} + (n_i - r_i) T_{ir_i}.$$

The score function is

$$U(\beta) = \sum_{i=1}^{k} x^{(i)} (r_i - e^{\beta^T x^{(i)}} U_i).$$

Set $r = \sum_{i=1}^{k} r_i$. The ML estimator $\hat{\beta}_1$ of the parameter β_1 verifies the equation

$$r - \frac{\sum_{i=1}^{k} r_i x_{1i} \sum_{i=1}^{k} e^{\hat{\beta}_1 x_{1i}} U_i}{\sum_{i=1}^{k} x_{1i} e^{\hat{\beta}_1 x_{1i}} U_i} = 0, \tag{5.146}$$

and

$$e^{\hat{\beta}_0} = \frac{r}{\sum_{i=1}^{k} e^{\hat{\beta}_1 x_{1i}} u_i}. \tag{5.147}$$

5.7.2 *Estimators of the main reliability characteristics under the usual stress*

First and second plans

Estimators of the survival function $S_{x_0}(t)$, p-quantile $t_p(x_0)$, and the mean failure time $m(x_0)$ under the usual stress x_0 are calculated using the formulas (5.54), (5.56), and (5.58) (shape-scale families of distributions), (5.93), (5.95), and (5.97) (generalized Weibull distribution), (5.114) (exponential distribution), and taking $x = x_0$ in all formulas.

Choosing specified family of shape-scale distributions and parametrization of $r(x)$, concrete formulas of estimators are obtained.

Example 5.9. If T_{x_0} has the Weibull distribution, i.e.

$$S_{x_0}(t) = e^{-(t/\theta)^\nu}, \quad t \geq 0,$$

and the Arrhenius parametrization is choosed (for example, stress is temperature), i.e.

$$r(x) = e^{-\beta_0 - \beta_1/x},$$

then $G(t) = \exp\{\exp(-t)\}$,

$$\hat{S}_{x_0}(t) = \exp\left\{-\exp\{\frac{\ln t - \hat{\beta}_0 - \hat{\beta}_1/x_0}{\hat{\sigma}}\}\right\}, \quad \hat{t}_p(x_0) = e^{\hat{\beta}_0 + \hat{\beta}_1/x_0}(-\ln(1-p))^{\hat{\sigma}}.$$

Example 5.10. If T_{x_0} has the loglogistic distribution, i.e.

$$S_{x_0}(t) = (1 + (t/\theta)^\nu)^{-1}, \quad t \geq 0,$$

and the power rule parametrization is chosen (for example, stress is voltage), i.e.

$$r(x) = e^{-\beta_0 - \beta_1 \ln x},$$

then

$$G(t) = (1 + e^t)^{-1},$$

So

$$\hat{S}_{x_0}(t) = \left[1 + \exp\left(\frac{\ln t - \hat{\beta}_0 - \hat{\beta}_1 \ln x_0}{\hat{\sigma}}\right)\right]^{-1}, \quad \hat{t}_p(x_0) = e^{\hat{\beta}_0 + \hat{\beta}_1 \ln x_0}\left(\frac{p}{1-p}\right)^{\hat{\sigma}}.$$

Example 5.11. If T_{x_0} has the lognormal law and the Eyring parametrization is chosen, i.e.

$$r(x) = e^{-\beta_0 - \beta_1 \ln x - \beta_2/x},$$

then $G(t) = 1 - \Phi(t)$, and

$$\hat{S}_{x_0}(t) = 1 - \Phi\left(\frac{\ln t - \hat{\beta}_0 - \hat{\beta}_1 \ln x_0 - \hat{\beta}_2/x_0}{\hat{\sigma}}\right),$$

$$\hat{t}_p(x_0) = e^{\hat{\beta}_0 + \hat{\beta}_1 \ln x_0 + \hat{\beta}_2/x_0 + \hat{\sigma}\Phi^{-1}(p)}.$$

Example 5.12. Suppose that the time-to-failure T_{x_0} has the Weibull law and the stress $x = (x_1, x_2)^T$ is bidimensional (for example, voltage and temperature) and the parametrization (5.19) is chosen. Then

$$\hat{S}_{x_0}(t) = \exp\left\{-\exp\left\{\frac{\ln t - \hat{\beta}_0 - \hat{\beta}_1 \ln x_{10} - \hat{\beta}_2/x_{20}}{\hat{\sigma}}\right\}\right\},$$

$$\hat{t}_p(x_0) = e^{\hat{\beta}_0 + \hat{\beta}_1 \ln x_{10} + \hat{\beta}_2/x_{20}}\{-\ln(1-p)\}^{\hat{\sigma}}.$$

Third plan

If $\hat{\alpha}$, $\hat{\rho}$, $\hat{\psi}$ are the maximum likelihood estimators of α, ρ, ψ then the estimators of the survival function $S_{x_0}(t)$, the p-quantile $t_p(x_0)$, and the mean life $m(x_0)$ are

$$\hat{S}_{x_0}(t) = S(\hat{\alpha}(\ln t - \hat{\psi})), \quad \hat{t}_p(x_0) = \exp\{\hat{\psi} + \frac{1}{\hat{\alpha}}S^{-1}(1-p)\},$$

$$\hat{m}(x_0) = \int_0^\infty \hat{S}_{x_0}(t)dt.$$

In the case of Weibull, loglogistic, and lognormal distributions

$$S^{-1}(p) = \ln(-\ln(1-p)); \; S^{-1}(p) = -\ln(\frac{1}{p} - 1); \; S^{-1}(p) = \Phi^{-1}(1-p),$$

respectively.

Fourth plan

Estimators of the survival function $S_{x_0}(t)$, p-quantile $t_p(x_0)$, and the mean failure time $m(x_0)$ under the usual stress x_0 are calculated using the formulas (5.114) (and taking $x = x_0$) in all formulas.

5.7.3 Asymptotic distribution of the regression parameters

First plan

a) Shape-scale families of distributions

Under regularity conditions (see Chapter 4) the distribution of the maximum likelihood estimators $(\hat{\beta}, \hat{\sigma})^T$ for large n is approximated by the normal law:

$$(\hat{\beta}, \hat{\sigma})^T \approx N_{m+2}((\beta, \sigma)^T, \Sigma^{-1}(\beta, \sigma)), \qquad (5.148)$$

and the covariance matrix $\Sigma^{-1}(\beta, \sigma)$ is estimated by the random matrix

$$\mathbf{I}^{-1}(\hat{\beta}, \hat{\sigma}) = (I^{ls}(\hat{\beta}, \hat{\sigma}))_{(m+2) \times (m+2)}, \qquad (5.149)$$

where

$$\mathbf{I}(\beta, \sigma) = (I_{lk}(\beta, \sigma))_{(m+2) \times (m+2)} \qquad (5.150)$$

is a matrix with the elements given by (5.71).

b) Generalized Weibull family of distributions

The distribution of the maximum likelihood estimators $(\hat{\beta}, \hat{\nu}, \hat{\gamma})^T$ for large n is approximated by the normal law:

$$(\hat{\beta}, \hat{\nu}, \hat{\gamma})^T \approx N_{m+3}((\beta, \nu, \gamma)^T, \Sigma^{-1}(\beta, \nu, \gamma)). \qquad (5.151)$$

The covariance matrix $\Sigma^{-1}(\beta, \nu, \gamma)$ is estimated by

$$\mathbf{I}^{-1}(\hat{\beta}, \hat{\nu}, \hat{\gamma}) = (I^{ls}(\hat{\beta}, \hat{\nu}, \hat{\gamma}))_{(m+3) \times (m+3)}, \qquad (5.152)$$

where

$$\mathbf{I}(\beta, \nu, \gamma) = (I_{lk}(\beta, \nu, \gamma))_{(m+3) \times (m+3)} \qquad (5.153)$$

is a matrix with the elements given by (5.101).

c) Exponential distribution

The law of $\hat{\beta}$ for large n is approximated by the normal law:

$$\hat{\beta} \approx N(\beta, \Sigma^{-1}(\beta)). \tag{5.154}$$

The covariance matrix $\Sigma^{-1}(\beta)$ is estimated by

$$\mathbf{I}^{-1}(\hat{\beta}) = (I^{ls}(\hat{\beta}))_{(m+1)\times(m+1)}$$

where

$$\mathbf{I}(\hat{\beta}) = (I_{ls}(\hat{\beta})), \quad (l, s = 0, ..., m),$$

is a matrix with the elements given by (5.116).

Second plan

a) Shape-scale families of distributions

The distribution of the maximum likelihood estimators $(\hat{\beta}, \hat{\sigma})^T$ for large n is approximated by the normal law (5.148), the covariance matrix $\Sigma^{-1}(\beta, \sigma)$ is estimated by $\mathbf{I}^{-1}(\hat{\beta}, \hat{\sigma})$, and the elements of $\mathbf{I}(\beta, \sigma)$ are given by (5.69).

b) Exponential distribution

The distribution of the maximum likelihood estimator $\hat{\beta}$ for large n is approximated by the normal law (5.154), the covariance matrix $\Sigma^{-1}(\beta)$ is estimated by $\mathbf{I}^{-1}(\hat{\beta})$, and the elements of $\mathbf{I}(\beta)$ are given by (5.115).

Third plan

Shape-scale families of distributions

The distribution of the maximum likelihood estimators $(\hat{\alpha}, \hat{\rho}, \hat{\psi})$ for large n is approximated by the normal law:

$$(\hat{\alpha}, \hat{\rho}, \hat{\psi}) \approx N_3((\hat{\alpha}, \hat{\rho}, \hat{\psi}), \Sigma^{-1}(\hat{\alpha}, \hat{\rho}, \hat{\psi})),$$

and the covariance matrix $\Sigma^{-1}(\hat{\alpha}, \hat{\rho}, \hat{\psi})$ is estimated by the random matrix

$$\mathbf{I}^{-1}(\hat{\alpha}, \hat{\rho}, \hat{\psi}) = \left\| I^{ls}(\hat{\alpha}, \hat{\rho}, \hat{\psi}) \right\|_{3\times3}, \tag{5.155}$$

and

$$\mathbf{I}(\alpha, \rho, \psi) = \| I_{ij}(\alpha, \rho, \psi) \|_{3\times3} \tag{5.156}$$

is a symmetric matrix with the following elements :

$$I_{11} = -\frac{\partial^2 \ln L}{\partial \alpha^2} = -\frac{1}{\alpha^2} \left\{ \sum_{j=1}^{n_1} (\ln f)''(c(T_{1j}))[c(T_{1j})]^2 - n_1 - m_2 \right.$$

$$\left. + \sum_{j=1}^{r_2} (\ln f)''(c(T_{2j}))[c(T_{2j})]^2 + \sum_{j=r_2+1}^{m_2} (\ln f)''(d(T_{2j}))[d(T_{2j})]^2 - \right.$$

$$(n_2 - m_2)\lambda'(d(t_2))[d(t_2)]^2\},$$

$$I_{12} = I_{21} = -\frac{\partial^2 \ln L}{\partial \alpha \partial \rho} = -\sum_{j=1}^{n_1} (\ln f)''(c(T_{1j}))c(T_{1j}) - \sum_{j=1}^{r_2} (\ln f)''(c(T_{2j}))$$

$$-\sum_{j=r_2+1}^{m_2} (\ln f)''(d(T_{2j}))d(T_{2j}) + (n_2 - m_2)\lambda'(d(t_2))\frac{e^{\rho}t_1}{e^{\rho}t_1 + t_2 - t_1}$$

$$-\frac{1}{\alpha}U_2(\alpha, \rho, \psi) - \frac{1}{\alpha}\sum_{j=r_2+1}^{m_2} \frac{e^{\rho}t_1}{e^{\rho}t_1 + T_{2j} - t_1},$$

$$I_{13} = I_{31} = -\frac{\partial^2 \ln L}{\partial \alpha \partial \psi} = \sum_{j=1}^{n_1} (\ln f)''(c(T_{1j}))c(T_{1j}) + \sum_{j=1}^{r_2} (\ln f)''(c(T_{2j}))c(T_{2j})$$

$$+\sum_{j=r_2+1}^{m_2} (\ln f)''(d(T_{2j}))d(T_{2j}) - (n_2 - m_2)\lambda'(d(t_2))d(t_2) - \frac{1}{\alpha}U_3(\alpha, \rho, \psi),$$

$$I_{22} = -\frac{\partial^2 \ln L}{\partial \rho^2} = -\alpha^2 \sum_{j=1}^{n_1} (\ln f)''(c(T_{1j})) - \alpha^2 \sum_{j=1}^{r_2} (\ln f)''(c(T_{2j}))$$

$$-\alpha^2 \sum_{j=r_2+1}^{m_2} (\ln f)''(d(T_{2j})) \left(\frac{e^{\rho}t_1}{e^{\rho}t_1 + T_{2j} - t_1}\right)^2 -$$

$$\sum_{j=r_2+1}^{m_2} [\alpha(\ln f)'(d(T_{2j})) - 1]\frac{e^{\rho}t_1(T_{2j} - t_1)}{(e^{\rho}t_1 + T_{2j} - t_1)^2}$$

$$+(n_2 - m_2)\lambda'(d(t_2)) \left(\frac{\alpha e^{\rho}t_1}{e^{\rho}t_1 + t_2 - t_1}\right)^2 + (n_2 - m_2)\lambda(d(t_2))\frac{\alpha e^{\rho}t_1(t_2 - t_1)}{(e^{\rho}t_1 + t_2 - t_1)^2},$$

$$I_{23} = I_{32} = -\frac{\partial^2 \ln L}{\partial \rho \partial \psi} = \alpha^2 \left\{\sum_{j=1}^{n_1} (\ln f)''(c(T_{1j})) + \sum_{j=1}^{r_2} (\ln f)''(c(T_{2j}))\right.$$

$$\left.+\sum_{j=r_2+1}^{m_2} (\ln f)''(d(T_{2j}))\frac{e^{\rho}t_1}{e^{\rho}t_1 + T_{2j} - t_1} - (n_2 - m_2)\lambda'(d(t_2))\frac{e^{\rho}t_1}{(e^{\rho}t_1 + t_2 - t_1)}\right\},$$

$$I_{33} = -\frac{\partial^2 \ln L}{\partial \psi^2} = -\alpha^2 \left[\sum_{j=1}^{n_1} (\ln f)''(c(T_{1j})) + \sum_{j=1}^{r_2} (\ln f)''(c(T_{2j}))\right.$$

$$\left.+\sum_{j=r_2+1}^{m_2} (\ln f)''(d(T_{2j})) - (n_2 - m_2)\lambda'(d(t_2))\right]. \qquad (5.157)$$

In the case of Weibull, loglogistic, and lognormal distributions

$$(\ln f)''(t) = e^t; \quad (\ln f)''(t) = \frac{-2e^t}{(1+e^t)^2}; \quad (\ln f)''(t) = -1, \quad (5.158)$$

respectively, and

$$\lambda'(t) = e^t; \quad \lambda'(t) = \frac{e^t}{(1+e^t)^2}; \quad \lambda'(t) = -t\frac{\varphi(t)}{1-\Phi(t)} + \left(\frac{\varphi(t)}{1-\Phi(t)}\right)^2, \quad (5.159)$$

respectively. See also, for example, Meeker and Hahn (1978).

5.7.4 Confidence intervals for the main reliability characteristics

First and second plans

Approximate confidence intervals for the survival function $S_{x_0}(t)$, p-quantile $t_p(x_0)$, and the mean failure time $m(x_0)$ under the usual stress x_0 are given in Section 5.3.5 (shape-scale families of distributions), Section 5.4.5 (generalized Weibull distribution), Section 5.5.5 (exponential distribution). Take $x = x_0$ in all formulas.

Third plan

Approximate $(1-\alpha)$ confidence interval for $S_{x_0}(t)$

$$\left(1 + \frac{1-\hat{S}_{x_0}(t)}{\hat{S}_{x_0}(t)} \exp\{\mp\hat{\sigma}_{Q_0} w_{1-\alpha/2}\}\right)^{-1},$$

where

$$\hat{\sigma}_{Q_0} = -\frac{S'(S^{-1}(\hat{S}_{x_0}(t)))}{\hat{S}_{x_0}(t)(1-\hat{S}_{x_0}(t))} \times$$

$$\left((\ln t - \hat{\psi})^2 I^{11}(\hat{\alpha},\hat{\rho},\hat{\psi}) - 2\hat{\alpha}(\ln t - \hat{\psi})I^{13}(\hat{\alpha},\hat{\rho},\hat{\psi}) + \hat{\alpha}^2 I^{33}(\hat{\alpha},\hat{\rho},\hat{\psi})\right)^{1/2}.$$

$$\hat{\sigma}_{K_p}^2 = \left(\frac{S^{-1}(1-p)}{\alpha^2}\right)^2 I^{11} - \frac{S^{-1}(1-p)}{\alpha^2}I^{13} + I^{33}.$$

Approximate $(1-\alpha)$ confidence interval for $t_p(x_0)$

$$\hat{t}_p(x_0) \exp\{\pm\hat{\sigma}_{K_p} w_{1-\alpha/2}\},$$

$$\hat{\sigma}_{K_p}^2 = \left(\frac{S^{-1}(1-p)}{\alpha^2}\right)^2 I^{11} - \frac{S^{-1}(1-p)}{\alpha^2}I^{13} + I^{33}.$$

Fourth plan

It is easy to show (cf. Lawless (1982)) that the random variables

$$X_1^2 = 2e^{\beta^T z^{(1)}} U_1, \cdots, X_k^2 = 2e^{\beta^T z^{(k)}} U_k$$

are independent and the random variable X_i^2 is chi-square distributed with $2r_i$ degrees of freedom.

Set

$$A_i = e^{\hat{\beta}_i - \beta_i} \ (i = 0, 1).$$

The equalities (5.146) and (5.147) imply that A_1 verifies the equation

$$r - \frac{\sum_{i=1}^{k} r_i z_{1i} \sum_{i=1}^{k} A_1^{z_{1i}} X_i^2}{\sum_{i=1}^{k} z_{1i} A_1^{z_{1i}} X_i^2} = 0,$$

and

$$A_0 = \frac{2r}{\sum_{i=1}^{k} A_1^{z_{1i}} X_i^2}.$$

The last equalities imply that the vector (A_0, A_1) is a function of $X_1^2, \cdots, X_1^2$. So the distribution of this vector is parameter-free.

Generating values of k independent chi-squared random variables $X_1^2, \cdots, X_k^2$, the α-quantiles C_α of the random variable

$$C = \frac{e^{\hat{\beta} z^{(0)}}}{e^{\beta z^{(0)}}} = A_0 A_1^{z_{10}}$$

can be obtained. So the $(1 - \alpha)$ confidence interval for $e^{\beta x^{(0)}}$ is

$$\left(e^{\hat{\beta} x^{(0)}} / C_{1-\alpha/2} \, , \ e^{\hat{\beta} x^{(0)}} / C_{\alpha/2} \right).$$

Hence,

$(1 - \alpha)$- *confidence interval for the survival function $S_{x_0}(t)$:*

$$\left(\exp\{-\frac{e^{\hat{\beta}^T x^{(0)}}}{C_{\alpha/2}} t\} \, , \ \exp\{-\frac{e^{\hat{\beta}^T x^{(0)}}}{C_{1-\alpha/2}} t\} \right).$$

$(1 - \alpha)$- *confidence interval for the p-quantile $t_p(x^{(0)})$:*

$$\left(-C_{\alpha/2} e^{-\hat{\beta}^T x^{(0)}} \ln(1 - p) \, , \ -C_{1-\alpha/2} e^{-\hat{\beta}^T x^{(0)}} \ln(1 - p) \right).$$

$(1 - \alpha)$- *confidence interval for the mean $m(x^{(0)})$:*

$$\left(C_{\alpha/2} e^{-\hat{\beta}^T x^{(0)}} \, , \ C_{1-\alpha/2} e^{-\hat{\beta}^T x^{(0)}} \ln(1 - p) \right).$$

The results of this chapter were explained also in Bagdonavičius and Nikulin (1994, 1995), Gerville-Réache and Nikoulina (1999,2000), Bagdonavičius, Gerville-Réache, Nikoulina and Nikulin (2000), Bagdonavičius, Gerville-Réache and Nikulin (2001), Glaser (1984), Hirose (1987,1993,1997a,b), Khamis and Higgins (1998), Klein and Basu (1981), Nelson and Macarthur (1992), Rodrigues, Bolfarine and Lourada-Neto (1993).

CHAPTER 6

Semiparametric AFT model

6.1 FTR data analysis

6.1.1 Introduction

Let us consider the AFT model:

$$S_{x(\cdot)}(t) = S_0 \left(\int_0^t e^{-\beta^T x(u)} du \right), \tag{6.1}$$

and suppose that the baseline survival function S_0 is completely unknown. Note that the parameter $\beta = (\beta_1, \cdots, \beta_m)^T$ and the vector of the explanatory variables $x(\cdot) = (x_1(\cdot), \cdots, x_m(\cdot))^T$ have not $m+1$ (as in the case of parametric models) but m coordinates, because the term $\beta_0^T x_0(t) \equiv \beta_0$ may be included in the unknown function S_0. So $x_1(\cdot), \cdots, x_m(\cdot)$ are one-dimensional explanatory variables.

Suppose that n units are observed. The ith unit is tested under the explanatory variable $x^{(i)}(\cdot) = (x_1^{(i)}(\cdot), ..., x_m^{(i)}(\cdot))^T$.

The data are supposed to be independently right censored and may be presented in the form

$$(X_i, \delta_i, x^{(i)}(t), 0 \le t \le X_i), \quad (i = 1, \cdots, n)$$

or

$$(N_i(t), Y_i(t), x^{(i)}(t), 0 \le t \le \sup\{s : Y_i(s) > 0\}), \quad (i = 1, \cdots, n).$$

Denote by S_i and α_i the survival and the hazard rate functions under $x^{(i)}(\cdot)$. Under the AFT model

$$\alpha_i(t; \beta) = \alpha_0 \left\{ \int_0^t e^{-\beta^T x^{(i)}(u)} du \right\} e^{-\beta^T x^{(i)}(t)},$$

where $\alpha_0 = -S_0'/S_0$ is the baseline hazard function. If $x^{(i)}$ is constant then

$$\alpha_i(t, \beta) = \alpha_0 \left(e^{-\beta^T x^{(i)}} t \right) e^{-\beta^T x^{(i)}}.$$

Take notice that if the AFT model

$$S_{x(\cdot)}(t) = S_0 \left(\int_0^t e^{-\beta^T (u) x(u)} du \right),$$

with time dependent regression coefficients $\beta_i(t) = \beta_i + \gamma_i g_i(t)$ is considered, then even in the case of constant explanatory variables it is reduced to the

AFT model (6.1) with explanatory variables $z(\cdot)$ instead of $x(\cdot)$, where

$$z(\cdot) = (x_1(\cdot), \cdots, x_m(\cdot), x_1(\cdot)g_1(\cdot), \cdots, x_m(\cdot)g_m(\cdot))^T$$

So all results obtained for the AFT model with time-dependent explanatory variables can be rewritten for the AFT model with time dependent regression coefficients and constant or time dependent explanatory variables. In all formulas

$$m, \quad \beta = (\beta_1, \cdots, \beta_m)^T, \quad x^{(i)}(\cdot) = (x_1^{(i)}(\cdot), \cdots, x_m^{(i)}(\cdot))^T$$

must be replaced by

$$2m, \quad \theta = (\beta_1, \cdots, \beta_m, \gamma_1, \cdots, \gamma_m)^T,$$

$$z^{(i)}(\cdot) = (x_1^{(i)}(\cdot), \cdots, x_m^{(i)}(\cdot), x_1^{(i)}(\cdot)g_1(\cdot), \cdots, x_m^{(i)}(\cdot)g_m(\cdot))^T$$

respectively.

6.1.2 Semiparametric estimation of the regression parameters

If S_0 is specified then the parametric maximum likelihood estimators of the parameters β_j are obtained by solving the system of equations

$$U_j(\beta) = 0, \quad (j = 1, \cdots, m),$$

with (cf. (4.22))

$$U_j(\beta) = \sum_{i=1}^n \int_0^\infty W_j^{(i)}(u, \beta) \{dN_i(u) - Y_i(u)\, dA_0\{f_i(u, \beta)\}\}, \qquad (6.2)$$

where

$$f_i(t, \beta) = \int_0^t e^{-\beta^T x^{(i)}(v)} dv, \quad A_0(t) = -\ln\{S_0(t)\}, \quad W_j^{(i)}(t, \beta) =$$

$$\frac{\partial}{\partial \beta_j} \log\{\alpha_i(t, \beta)\} = -x_j^{(i)}(t) - \frac{\alpha_0'(f_i(t, \beta))}{\alpha_0(f_i(t, \beta))} \int_0^t x_j^{(i)}(u) e^{-\beta^T x^{(i)}(u)} du.$$

If S_0 is unknown then the score functions U_j depend not only on β but also on unknown functions A_0, α_0, and α_0'.

The idea of semiparametric estimation of β is to replace the unknown baseline cumulative hazard function A_0 in (6.2) by its efficient estimator (still depending on β), and the weight function $W_j^{(i)}(t, \beta)$ by a suitable function which does not depend on the unknown baseline functions α_0 and α_0'. In such a way the obtained modified score function does not contain unknown infinite-dimensional parameters A_0, α_0 and α_0' and contains only the finite-dimensional parameter β.

The optimal weights depend on the derivative of the baseline hazard rate for which estimation is complicated when the law is unknown. So simplest weights

$$W_j^{(i)}(t, \beta) = x_j^{(i)}(t) \qquad (6.3)$$

may be chosen. They are optimal when the baseline distribution is exponential and do not depend on unknown parameters. The greater the stress at the moment t the greater the weight at this moment.

Different weights do not influence much the efficiency of the estimators but good choice of an estimator $\tilde{A}_0(t, \beta)$ (which should replace the baseline cumulative hazard function A_0 in (6.2)) is crucial.

The idea of the estimator $\tilde{A}_0(t, \beta)$ construction is as follows.

The random variables

$$T_i^*(\beta) = f_i(T_i, \beta) = \int_0^{T_i} e^{-\beta^T x^{(i)}(v)} dv$$

have the same survival function S_0. Set

$$C_i^*(\beta) = f_i(C_i, \beta), \quad X_i^*(\beta) = f_i(X_i, \beta) = T_i^*(\beta) \wedge C_i^*(\beta),$$

$$\delta_i^* = \mathbf{1}_{\{T_i^*(\beta) \leq C_i^*(\beta)\}} = \delta_i.$$

Then the pairs

$$(X_1^*(\beta), \delta_1^*), \cdots, (X_n^*(\beta), \delta_n^*) \tag{6.4}$$

may be considered as censored data from an experiment, in which all n units have the same survival function S_0 and cumulative hazard function A_0. The function A_0 may be estimated by the Nelson-Aalen estimator $\tilde{A}_0(t, \beta)$(see Appendix, Section A14). This estimator depends on β because the data (6.4) depend on β, and has the form

$$\tilde{A}_0(t, \beta) = \int_0^t \frac{dN^*(u, \beta)}{Y^*(u, \beta)} = \sum_{i=1}^n \int_0^t \frac{dN_i(g_i(u, \beta))}{\sum_{l=1}^n Y_l(g_l(u, \beta))}, \tag{6.5}$$

where $g_i(u, \beta)$ is the function inverse to $f_i(u, \beta)$ with respect to the first argument, and

$$N^*(t, \beta) = \sum_{i=1}^n N_i^*(t, \beta), \quad N_i^*(t, \beta) = \mathbf{1}_{\{X_i^*(\beta) \leq t, \delta_i^* = 1\}} = N_i(g_i(t, \beta)),$$

$$Y^*(t, \beta) = \sum_{i=1}^n Y_i^*(t, \beta), \quad Y_i^*(t, \beta) = \mathbf{1}_{\{X_i^*(\beta) \geq t\}} = Y_i(g_i(t, \beta)).$$

Replacing the function A_0 by $\tilde{A}_0(t, \beta)$ and using the weight (6.3) in (6.2), modified score functions $\tilde{U}_j(\beta)$, depending only on β, are obtained. The random vector

$$\tilde{U}(\beta) = (\tilde{U}_1(\beta), \cdots, \tilde{U}_m(\beta))^T$$

has the form:

$$\tilde{U}(\beta) = \sum_{i=1}^n \int_0^\infty x^{(i)}(u) \left(dN_i(u) - Y_i(u) \, d\tilde{A}_0\{f_i(u, \beta)\} \right) =$$

$$\sum_{i=1}^n \int_0^\infty \{x^{(i)}(u) - \bar{x}(f_i(u, \beta), \beta)\} dN_i(u), \tag{6.6}$$

where

$$\bar{x}(v, \beta) = \frac{\sum_{j=1}^{n} x^{(j)}(g_j(v, \beta)) Y_j(g_j(v, \beta))}{\sum_{j=1}^{n} Y_j(g_j(v, \beta))}. \tag{6.7}$$

Note that it is calculated very simply:

$$\tilde{U}(\beta) = \sum_{i=1}^{n} \delta_i \{x^{(i)}(X_i) - \bar{x}(f_i(X_i, \beta), \beta)\}.$$

If $x^{(i)}$ are constant then

$$\tilde{U}(\beta) = \sum_{i=1}^{n} \delta_i \{x^{(i)} - \bar{x}(e^{-\beta^T x^{(i)}} X_i, \beta)\}, \tag{6.8}$$

where

$$\bar{x}(u, \beta) = \frac{\sum_{j=1}^{n} x^{(j)} \mathbf{1}_{\{\ln X_j - \beta^T x^{(j)} \geq \ln u\}}}{\sum_{j=1}^{n} \mathbf{1}_{\{\ln X_j - \beta^T x^{(j)} \geq \ln u\}}}.$$

We obtained the score function identical to the score function for the rank estimator given by Lin and Geyer (1992).

If the parameter β is one-dimensional then the function (6.8) may have jumps only at the points

$$\frac{\ln X_l - \ln X_{l'}}{x^{(l)} - x^{(l')}} \quad (x^{(l)} \neq x^{(l')}),$$

and is constant between them. If β is multi-dimensional then the function $U(\beta)$ may have jumps only on the hyper-planes

$$\beta^T (x^{(l)} - x^{(l')}) = \ln X_l - \ln X_{l'},$$

and is constant in the regions bounded by them.

The values of the random vector $\tilde{U}(\beta)$ are dispersed around zero but may be not equal to zero for all values of the parameter β. So an estimator of the parameter β may be obtained by minimizing the distance of $\tilde{U}(\beta)$ from zero.

Denote by $B_0 \subset \mathbf{R}^m$ a set such that for any $\tilde{\beta} \in B_0$ the value $||\tilde{U}(\tilde{\beta})||$ is minimizing the distance $||\tilde{U}(\beta)||$ of $\tilde{U}(\beta)$ from zero:

$$\tilde{\beta} = \text{Argmin}_{\beta \in \mathbf{R}^m} ||\tilde{U}(\beta)||.$$

When n is finite, the set B_0 contains an infinite number of elements (see comments after the formula (6.8)).

To have an unique estimator of the parameter β one may do as follows.

The value $\exp\{\beta^T x\}$ (and $\beta^T x$) shows the influence of the constant explanatory variable x on the lifetimes of units. Suppose that $x^{(0)}$ is a standard value of the explanatory variable vector (for example the mean of all $x^{(i)}$ in FTR analysis or the usual stress in ALT). An unique estimator of the parameter β may be defined as

$$\hat{\beta} = \text{Argmax}_{\tilde{\beta} \in B_0}(\tilde{\beta}^T x^{(0)}) \quad \text{or} \quad \hat{\beta} = \text{Argmin}_{\tilde{\beta} \in B_0}(\tilde{\beta}^T x^{(0)}),$$

or

$$\hat{\beta} = \left(\text{Argmin}_{\tilde{\beta} \in B_0} (\tilde{\beta}^T x^{(0)}) + \text{Argmax}_{\tilde{\beta} \in B_0} (\tilde{\beta}^T x^{(0)}) \right) / 2.$$

6.1.3 Estimators of the main reliability characteristics

Suppose that $x(\cdot)$ is an arbitrary explanatory variable which may be different from $x^{(i)}(\cdot), (i = 1, \cdots, n)$.

Estimator of the survival function $S_{x(\cdot)}(t)$:

It was noted that the pairs (6.4) may be considered as censored data from an experiment, in which all n units have the same survival function S_0. So at first the baseline survival function S_0 is estimated by the Kaplan-Meier estimator $\tilde{S}_0(t, \beta)$(see Appendix, Chapter A14). This estimator depends on β because the data (6.4) depend on β, and is the product integral (see Appendix, Chapter A15)

$$\tilde{S}_0(t, \beta) = \prod_{0 \leq s \leq t} \left(1 - d\tilde{A}_0(s, \beta) \right) =$$

$$\prod_{j:\delta_j=1, f_j(X_j,\beta) \leq t} \left(1 - \frac{1}{\sum_{l=1}^{n} \mathbf{1}_{\{f_l(X_l,\beta) \geq f_j(X_j,\beta)\}}} \right). \qquad (6.9)$$

The formula (6.1) implies an estimator of the survival function $S_{x(\cdot)}(t)$:

$$\hat{S}_{x(\cdot)}(t) = \tilde{S}_0 \left(\int_0^t e^{-\hat{\beta}^T x(u)} du, \hat{\beta} \right) =$$

$$\prod_{j:\delta_j=1, f_j(X_j,\hat{\beta}) \leq f_{x(\cdot)}(t,\hat{\beta})} \left(1 - \frac{1}{\sum_{l=1}^{n} \mathbf{1}_{\{f_l(X_l,\hat{\beta}) \geq f_j(X_j,\hat{\beta})\}}} \right), \qquad (6.10)$$

where

$$f_{x(\cdot)}(t, \hat{\beta}) = \int_0^t e^{-\hat{\beta}^T x(u)} du. \qquad (6.11)$$

If x and $x^{(i)}$ are constant then

$$\hat{S}_x(t) = \prod_{j:\delta_j=1, X_j \leq t\, e^{\hat{\beta}^T (x^{(j)} - x)}} \left(1 - \frac{1}{\sum_{l=1}^{n} \mathbf{1}_{\{X_l \geq X_j \, \exp\{\hat{\beta}^T (x^{(l)} - x^{(j)})\}\}}} \right). \qquad (6.12)$$

Estimator of the p-quantile $t_p(x(\cdot))$

:

$$\hat{t}_p(x(\cdot)) = \sup\{t : \hat{S}_{x(\cdot)}(t) \geq 1 - p\}. \qquad (6.13)$$

Estimator of the mean failure time $m(x(\cdot))$:

$$\hat{m}(x(\cdot)) = -\int_0^\infty u\, d\hat{S}_{x(\cdot)}(u) = \sum_{j=1}^n \delta_j\, g_{x(\cdot)}(f_j(X_j, \hat{\beta}), \hat{\beta}) \times$$

$$\{\tilde{S}_0(f_j(X_j, \hat{\beta})-, \hat{\beta}) - \tilde{S}_0(f_j(X_j, \hat{\beta}), \hat{\beta})\}, \qquad (6.14)$$

where $g_{x(\cdot)}(t, \beta)$ is the function inverse to $f_{x(\cdot)}(t, \beta)$ with respect to the first argument, $X_0 = 0$.

If x is constant then

$$\hat{m}(x) = \sum_{j=1}^n \delta_j\, e^{\hat{\beta}^T(x - x^{(j)})} X_j\{\tilde{S}_0(e^{-\hat{\beta}^T(x^{(j-1)})} X_j-, \hat{\beta}) - \tilde{S}_0(e^{-\hat{\beta}^T(x^{(j)})} X_j, \hat{\beta})\}$$

$$(6.15)$$

The estimator of the mean may be underestimated if the last $X_i^*(\hat{\beta})$ is the censoring time.

Estimators of the mean ratios

The mean ratio $MR(x, y)$ (see (5.26)) is estimated by

$$\hat{M}R(x, y) = e^{\hat{\beta}^T(y-x)}. \qquad (6.16)$$

6.1.4 Asymptotic distribution of the regression parameters estimators

Tsiatis (1990), Wei et al (1990), Ying (1993) (constant explanatory variables), Robins and Tsiatis (1992), Lin and Ying (1995) (time-dependent explanatory variables) give asymptotic properties of the regression parameters for random right censored data. In all above mentioned papers, boundedness of the density of the censoring variable is required.

In the case of accelerated life testing when type one censoring is generally used, this condition does not hold. Bagdonavičius and Nikulin (2000b) give asymptotic properties of the estimators under the third plan of experiments (see Section 5.6.3).

The limit distribution of the score function $\tilde{U}$ is obtained using the Theorem A7 because the score function (6.6) can be written as an integral of a predictable process with respect to a martingale:

$$\tilde{U}(\beta_0) = \sum_{i=1}^n \int_0^\infty \{x^{(i)}(u) - \bar{x}(f_i(u, \beta_0), \beta_0)\} dM_i(u, \beta_0).$$

Under regularity conditions and constant explanatory variables

$$n^{-1/2}\tilde{U}(\beta_0) \xrightarrow{D} N(0, B),$$

where

$$B = \int_0^\infty \varphi(u, \beta_0)\, dA_0(u),$$

$$\varphi(u, \beta) = P \lim n^{-1} \sum_{i=1}^{n} \{x^{(i)}(g_i(u, \beta)) - \bar{x}(u, \beta)\}^{\otimes 2} Y_i(g_i(u, \beta)).$$

$P \lim$ denotes the limit in probability. If $\beta \neq \beta_0$ then

$$n^{-1}\tilde{U}(\beta) \xrightarrow{P} U(\beta),$$

where $U(\beta)$ is non-random function, $U(\beta_0) = 0$.

If the score function $\tilde{U}(\beta)$ would be differentiable then using the Taylor series expansion (cf. (4.17)), the result

$$n^{1/2}(\hat{\beta} - \beta_0) = (n^{-1}\frac{\partial}{\partial\beta}\tilde{U}(\beta_0))^{-1}n^{-1/2}\tilde{U}(\beta_0) + o_P(1)$$

could be obtained under regularity conditions. Unfortunately, the score function $\tilde{U}(\beta)$ is a discrete function of β and the limit distribution of the estimator $\hat{\beta}$ can not be obtained using the Taylor series expansion. Instead, under regularity conditions (see the above mentioned papers) the matrix of derivatives of the differentiable limit function $U(\beta)$ can be used and the following result holds:

$$n^{1/2}(\hat{\beta} - \beta_0) = (\frac{\partial}{\partial\beta}U(\beta_0))^{-1}n^{-1/2}\tilde{U}(\beta_0) + o_P(1).$$

Set $C = (\frac{\partial}{\partial\beta}U(\beta_0))$. So

$$n^{1/2}(\hat{\beta} - \beta_0) = C^{-1}n^{-1} \sum_{i=1}^{n} \int_{0}^{\infty} \{x^{(i)}(u) - \bar{x}(f_i(u, \beta_0), \beta_0)\} dM_i(u, \beta_0) + o_p(1).$$

$$(6.17)$$

Hence the asymptotic law of β:

$$n^{1/2}(\hat{\beta} - \beta_0) \xrightarrow{D} N(0, C^{-1}BC^{-1}).   \qquad (6.18)$$

Expression of the matrix C under constant explanatory variables is

$$C = \int_{0}^{\infty} u \, \varphi(u, \beta_0) \, d\alpha_0(u) + B.$$

The asymptotic covariance matrix of $\hat{\beta}$ involves the derivative of the baseline hazard rate.

Estimator of the matrix B is evident

$$\hat{B} = n^{-1} \sum_{i=1}^{n} \int_{0}^{\infty} \{x^{(i)} - \bar{x}(u, \hat{\beta})\}^{\otimes 2} \mathbf{1}_{\{\ln X_i - \hat{\beta}^T x^{(i)} \geq \ln u\}} d\tilde{A}_0(u, \hat{\beta}).$$

The matrix C could be estimated by

$$\hat{C} = n^{-1} \sum_{i=1}^{n} \int_{0}^{\infty} u \, \{x^{(i)} - \bar{x}(u, \hat{\beta})\}^{\otimes 2} \mathbf{1}_{\{\ln X_i - \hat{\beta}^T x^{(i)} \geq \ln u\}} \tilde{\alpha}'_0(u, \hat{\beta}) du + \hat{B},$$

where

$$\tilde{\alpha}'_0(t, \hat{\beta}) = b^{-2} \int_{0}^{\infty} K'\{(t - u)/b\} d\tilde{A}_0(u)$$

is a kernel function estimator of the derivative $\alpha'_0(t)$. Conditions on K, α_0, and the censoring mechanism should be established for consistency of the estimator $\hat{C}$.

Robins and Tsiatis (1992) propose other way: to estimate the matrix C by numerical derivatives of $n^{-1}\tilde{U}(\beta)$ at $\hat{\beta}$:

$$\hat{C} = (\hat{C}_{ij})_{m \times m}, \quad \hat{C}_{ij} = \frac{1}{2} n^{-1/2} \frac{\tilde{U}_i(\hat{\beta} + n^{-1/2} c_j e_j) - \tilde{U}_i(\hat{\beta} - n^{-1/2} c_j e_j)}{c_j},$$

where $e_1 = (1, 0, \cdots, 0), \cdots, e_m = (0, 0, \cdots, 1)$, and $c_j > 0$ are fixed constants. Unfortunately, as noted by Lin and Ying (1995), this approach can yield rather different estimators for varying step sizes and will be unreliable in finite sample since $\tilde{U}_i(\beta)$ are neither continuous nor monotone.

6.1.5 Tests for nullity of the regression coefficients.

Let us consider the hypothesis

$$H_{k_1, k_2, \cdots, k_l} : \beta_{k_1} = \cdots = \beta_{k_l} = 0, \quad (1 \le k_1 < k_2 < \cdots < k_l). \tag{6.19}$$

Set (see (6.10),(6.14) for the notations f_i, g_i)

$$g_{ji}(\beta) = g_j\{f_i(X_i, \beta), \beta\},$$

and

$$\mathbf{I}(\beta) = (I_{sr}(\beta))_{m \times m}$$

the matrix with the elements

$$I_{sr}(\beta) = \sum_{i=1}^{n} \delta_i \left\{ \sum_{j=1}^{n} x_s^{(j)}\{g_{ji}(\beta)\} x_r^{(j)}\{g_{ji}(\beta)\} Y_j\{g_{ji}(\beta)\} \sum_{j=1}^{n} Y_j\{g_{ji}(\beta)\} - \right.$$

$$\left. \sum_{j=1}^{n} x_s^{(j)}\{g_{ji}(\beta)\} Y_j\{g_{ji}(\beta)\} \sum_{j=1}^{n} x_r^{(j)}\{g_{ji}(\beta)\} Y_j\{g_{ji}(\beta)\} \right\} / \left(\sum_{j=1}^{n} Y_j\{g_{ji}(\beta)\} \right)^2.$$

Lin and Ying (1995) proved that under the hypothesis $H^0_{k_1, k_2, \cdots, k_l} : \beta_{k_1} = \beta^0_{k_1}, \cdots, \beta_{k_l} = \beta^0_{k_l}$, random right censoring and some smoothness conditions on the explanatory variables, the baseline density, its derivative, and the density of the censoring variables, the score statistic

$$\inf_{\beta_{k_1} = \beta^0_{k_1}, \cdots, \beta_{k_l} = \beta^0_{k_l}, |\hat{\beta}_s - \beta_s| \le \varepsilon, s \ne k_1, \cdots, k_l} \tilde{U}^T(\beta) I^{-1}(\hat{\beta}) \tilde{U}(\beta)$$

is asymptotically chi-square distributed with l degrees of freedom, for every fixed $\varepsilon > 0$. So the hypothesis $H_{k_1, k_2, \cdots, k_l}$ is rejected with the approximate significance level α if

$$W_{k_1 \ldots k_l} > \chi^2_{1-\alpha}(l);$$

here

$$W_{k_1 \ldots k_l} = \inf_{\beta_{k_1} = \cdots = \beta_{k_l} = 0, |\hat{\beta}_s - \beta_s| \le \varepsilon, s \ne k_1, \cdots, k_l} \tilde{U}^T(\beta) I^{-1}(\hat{\beta}) \tilde{U}(\beta).$$

If $x^{(i)}$ are constant then

$$I_{sr}(\beta) =$$

$$\sum_{i=1}^{n} \delta_i \frac{\sum_{j=1}^{n} x_s^{(j)} x_r^{(j)} a_{ji}(\beta) \sum_{j=1}^{n} a_{ji}(\beta) - \sum_{j=1}^{n} x_s^{(j)} a_{ji}(\beta) \sum_{j=1}^{n} x_r^{(j)} a_{ji}(\beta)}{\left(\sum_{j=1}^{n} a_{ji}(\beta)\right)^2},$$

where

$$a_{ji}(\beta) = \mathbf{1}_{\{X_j \geq e^{\beta^T(x^{(j)} - x^{(i)})} X_i\}}.$$

6.2 Semiparametric estimation in ALT

Suppose that the baseline survival function S_0 in the AFT model (5.1) is completely unknown. In Section 5 several plans of ALT were discussed.

The first plan of experiments (Section 5.6.1) may be used when the function r is parametrized and the model has the form (6.1).

The second plan (Section 5.6.2) can not be used because when S_0 is unknown and all units are tested under the same step-stress (5.131), then the score function (6.6) does not depend on β.

If the third plan is used (Section 5.6.3), semiparametric estimation of the survival function under the usual stress is possible even when both functions S_0 and r are unknown.

The fourth plan (Section 5.6.4) is used when the baseline distribution is exponential. So it is not used when S_0 is completely unknown.

6.2.1 First plan of experiments: parametrized r and unknown S_0

The first plan is the particular case of the plan of experiments considered in 6.1.1, when $x^{(i)}$ have the form (5.134). So the score function has the form (6.6).

Estimators of the main reliability characteristics under the usual stress $x^{(0)}$ are defined by formulas (6.11)-(6.15), taking $x^{(0)}$ instead of x in all formulas.

6.2.2 Third plan of experiments: unspecified S_0 and r

Suppose that the AFT model (5.1) holds. Redefining $r(x) := r(x)/r(x_0)$, this model may be written:

$$S_{x(\cdot)}(t) = S_{x^{(0)}} \left(\int_0^t r\{x(\tau)\} d\tau \right). \tag{6.20}$$

Take notice that then $r(x^{(0)}) = 1$.

Suppose that the functions r and $S_0(t)$ (or, equivalently, r and $S_{x^{(0)}}(t)$) are completely unknown, and the third plan of experiments is considered.

Let $x^{(1)} > x^{(0)}$ be an accelerated constant stress.

Two groups of units are tested. The first group of n_1 units is tested under the accelerated stress $x^{(1)}$ and a complete sample

$$T_{11} \leq ... \leq T_{1n_1}$$

is obtained. The second group of n_2 units is tested under the step-stress

$$x^{(2)}(u) = \begin{cases} x^{(1)}, & 0 \leq u \leq t_1, \\ x^{(0)}, & t_1 < u \leq t_2, \end{cases}$$

and a type I censored sample

$$T_{21} \leq ... \leq T_{2m_2}$$

is obtained $(m_2 \leq n_2)$.

Here m_2 is the random number of failures of units of the second group.

The moment t_1 is supposed to be such that after the switch-up from stress $x^{(1)}$ to stress $x^{(0)}$, a sufficiently large number of failures under the usual stress $x^{(0)}$ can be obtained in $[t_1, t_2]$.

Set

$$S_i = S_{x^{(i)}}, \quad \alpha_i = \alpha_{x^{(i)}}, \quad A_i = -\ln S_i, \quad (i = 0, 1, 2),$$

$$f_i(u, r) = \begin{cases} ru, & i = 1, \\ r(u \wedge t_1) + (u - t_1) \vee 0, & i = 2, \end{cases}$$

$$g_i(u, r) = \begin{cases} u/r, & i = 1, \\ r(u \wedge t_1) + (u - rt_1) \vee 0, & i = 2. \end{cases}$$

Under the AFT model $S_i(u) = S_0(f_i(u, r))$. Analogously as in the case of the first plan (cf.(6.5)) define the Nelson-Aalen type estimator

$$\tilde{A}_0(s, r) = \sum_{i=1}^{2} \int_0^t \frac{dN_i(g_i(u, r))}{\sum_{l=1}^n Y_l(g_l(u, r))} =$$

$$\int_0^s \frac{dN_1(u/r) + dN_2((u/r) \wedge t_1 + (u - rt_1) \vee 0)}{Y_1(u/r) + Y_2((u/r) \wedge t_1 + (u - rt_1) \vee 0)}, \tag{6.21}$$

and the Kaplan-Meier type estimator

$$\tilde{S}_0(s, r) = \prod_{(i,j) \in B(s)} \left(1 - \frac{1}{Y_1(T_{1i}) + Y_2(t_1 \wedge T_{1i} + r((T_{1i} - t_1) \vee 0))} \right) \times$$

$$\left(1 - \frac{1}{Y_2(T_{2j}) + Y_1(t_1 \wedge T_{2j} + (\frac{T_{2j} - t_1}{r}) \vee 0)} \right), \tag{6.22}$$

where

$$B(s) = \{(i, j) | rT_{1i} \leq s, r(T_{2j} \wedge t_1) + (T_{2j} - t_1) \vee 0 \leq s\}.$$

The modified score function (cf.(6.6)) is

$$\tilde{U}(\beta) = \sum_{i=1}^{2} \int_0^\infty x^{(i)}(u) \left(dN_i(u) - Y_i(u) \, d\tilde{A}_0\{f_i(u, \beta)\} \right) =$$

$$(x^{(0)} - x^{(1)}) \int_{t_1}^{t_2} \frac{Y_1(t_1 + \frac{u-t_1}{r})dN_2(u) - Y_2(u)dN_1(t_1 + \frac{u-t_1}{r})}{Y_1(t_1 + \frac{u-t_1}{r}) + Y_2(u)}$$

The score function is defined as follows:

$$\hat{U}(r) = \int_{t_1}^{t_2} \frac{Y_1(t_1 + \frac{u-t_1}{r})dN_2(u) - Y_2(u)dN_1(t_1 + \frac{u-t_1}{r})}{Y_1(t_1 + \frac{u-t_1}{r}) + Y_2(u)}, \tag{6.23}$$

or

$$\hat{U}(r) = \sum_{j:T_{2j}>t_1} \frac{Y_1(t_1 + \frac{T_{2j}-t_1}{r})}{Y_1(t_1 + \frac{T_{2j}-t_1}{r}) + Y_2(T_{2j})} -$$

$$\sum_{j:T_{1j}>t_1} \frac{Y_2(t_1 + r(T_{1j} - t_1))}{Y_1(T_{1j}) + Y_2(t_1 + r(T_{1j} - t_1))}.$$

It is an increasing step function, $\hat{U}(0) < 0$, $\hat{U}(\infty) > 0$ with the probability 1.
  The estimator of the parameter r :

$$\hat{r} = \hat{U}^{-1}(0) = \sup\{r : \hat{U}(r) \le 0\}. \tag{6.24}$$

The following initial estimator may be taken:

$$\tilde{r} = (T_{2m_2} - t_1)/(T_{1k} - t_1),$$

where k satisfies the inequalities

$$\frac{k-1}{n_1} < \frac{m_2}{n_2} \le \frac{k}{n_1}.$$

It is implied by the equality (cf. (2.33))

$$r = (t_{x_2(\cdot)}(p) - t_1)/(t_{x_1}(p) - t_1).$$

The survival function $S_{x^0}(t)$ is estimated by:

$$\hat{S}_{x^0}(t) = \hat{S}_0(t) = \tilde{S}_0(t, \hat{r}). \tag{6.25}$$

The quantile $t_p(x^{(0)})$ is estimated by

$$\hat{t}_p(x^{(0)}) = \sup\{t : \hat{S}_{x^{(0)}}(t) \ge 1 - p\} \tag{6.26}$$

and the mean time-to-failure $m(x^{(0)}) = \mathbf{E}(T_{x^{(0)}})$ by

$$\hat{m}(x^{(0)}) = -\int_0^\infty u d\hat{S}_{x^{(0)}}(u). \tag{6.27}$$

6.2.3 Asymptotic properties of the estimators

Let us consider asymptotic properties of the estimator $\hat{r}$. First we obtain the asymptotic distribution of the function $\hat{U}(r)$ which may be written in the following way:

$$\hat{U}(r) = \int_{t_1}^{t_2} \frac{\hat{S}_2(u-)d\hat{S}_1(t_1 + \frac{u-t_1}{r}) - \hat{S}_1(t_1 + \frac{u-t_1}{r}-)d\hat{S}_2(u)}{(n_1/n)\hat{S}_1(t_1 + \frac{u-t_1}{r}-) + (n_2/n)\hat{S}(u-)}, \tag{6.28}$$

where $\hat{S}_i$ is the empirical survival function for the ith group of units:

$$\hat{S}_i(u) = 1 - N_i(u)/n_i, \quad u \geq 0, \quad i = 1, 2.$$

Suppose that $n_1/n \to l_1 \in (0, 1)$, $n_2/n \to l_2 = 1 - l_1$. Then (Appendix, Section A.13)

$$a_n(\hat{S}_i - S_i) \xrightarrow{D} V_i \quad \text{on} \quad D[0, s], \tag{6.29}$$

where

$$s \in [0, \infty) \ (i = 1) \quad \text{or} \quad s \in [0, t_2] \ (i = 2),$$

V_i is a mean zero Gaussian process such that for all $0 \leq s_1 \leq s_2 \leq s$:

$$\textbf{Cov}\,(V_i(s_1), V_i(s_2)) = l_{3-i}(1 - S_i(s_1))S_i(s_2), \ (i = 1, 2). \tag{6.30}$$

Set

$$U(r) = \int_{t_1}^{t_2} \frac{S_2(u)dS_1(t_1 + \frac{u-t_1}{r}) - S_1(t_1 + \frac{u-t_1}{r})dS_2(u)}{l_1 S_1(t_1 + \frac{u-t_1}{r}) + l_2 S_2(u)}. \tag{6.31}$$

Proposition 6.1. *Suppose that S_i are absolutely continuous and $S_i(u) > 0$ for all $u \geq 0$. Then*

$$a_n(\hat{U}(r) - U(r)) \xrightarrow{D} W(r), \tag{6.32}$$

where

$$W(r) = \int_{t_1}^{t_2} Z_2(u)dV_1(t_1 + \frac{u - t_1}{r}) - Z_1(u)dV_2(u) -$$

$$W_1(u)d(l_1 S_1(t_1 + \frac{u - t_1}{r}) + l_2 S_2(u)),$$

$$Z_1(u) = S_1(t_1 + \frac{u - t_1}{r})/(l_1 S_1(t_1 + \frac{u - t_1}{r}) + l_2 S_2(u)),$$

$$Z_2(u) = (1 - l_1 Z_1(u))/l_2,$$

$$W_1(u) = \frac{S_2(u)V_1(t_1 + \frac{u-t_1}{r}) - S_1(t_1 + \frac{u-t_1}{r})V_2(u)}{(l_1 S_1(t_1 + \frac{u-t_1}{r}) + l_2 S_2(u))^2}.$$

Proof. Set

$$\hat{Z}_1(u) = \frac{\hat{S}_1(t_1 + \frac{u-t_1}{r}-)}{l_1 \hat{S}_1(t_1 + \frac{u-t_1}{r}-) + l_2 \hat{S}_2(u-)}, \quad \hat{Z}_2(u) = (1 - l_1 \hat{Z}_1(u))/l_2.$$

Using (6.29) and the functional delta method (see Appendix, Theorem A.10) we obtain

$$a_n(\hat{Z}_1(\cdot) - Z_1(\cdot)) \xrightarrow{D} l_2 W_1(\cdot), \quad a_n(\hat{Z}_2(\cdot) - Z_2(\cdot)) \xrightarrow{D} -l_1 W_1(\cdot). \tag{6.33}$$

Noting that

$$a_n(\hat{U}(r) - U(r)) =$$

$$a_n \left\{ \int_{t_1}^{t_2} \hat{Z}_2(u)d\hat{S}_1(t_1 + \frac{u - t_1}{r}) - Z_2(u)d\hat{S}_1(t_1 + \frac{u - t_1}{r}) - \right.$$

$$\int_{t_1}^{t} \hat{Z}_1(u)d\hat{S}_2(u) - Z_1(u)d\hat{S}_2(u) \Big\},$$

using (6.29), (6.33) and the functional delta method for stochastic integrals (Theorem A.11), we obtain (6.32).

Denote by r_0 the true value of r under the AFT model. Under the AFT model $U(r_0) = 0$.

Proposition 6.2. *Under assumptions of the proposition 6.1 and the AFT model*

$$a_n \hat{U}(r_0) \xrightarrow{D} W(r_0), \tag{6.34}$$

and

$$\mathbf{E}\{W(r_0)\} = 0, \quad \mathbf{Var}\{W(r_0)\} = S_2(t_1) - S_2(t_2). \tag{6.35}$$

Proof. Taking into account the equality $S_1(t_1 + (u - t_1)/r) = S_2(u)$ for $u \geq t_1$ we obtain

$$W(r_0) = V_1(t_1 + \frac{t_2 - t_1}{r_0}) - V_1(t_1) - V_2(t_2) + V_2(t_1) -$$

$$\int_{t_1}^{t} [V_1(t_1 + \frac{u - t_1}{r_0}) - V_2(u)]d\ln S_2(u).$$

Calculation of $\mathbf{Var}\{W(r_0)\}$ is tedious but elementary.

Remark 6.1. The proposition 4.2 implies that if n_1 and n_2 are large, then

$$\mathbf{P}\{-u_{1-\frac{\alpha}{2}} < \hat{U}(r_0) < u_{1-\frac{\alpha}{2}}\} \approx 1 - \alpha,$$

where

$$u_{1-\frac{\alpha}{2}} = w_{1-\frac{\alpha}{2}} \left\{ \frac{n}{n_1 n_2} (\hat{S}_2(t_1) - \hat{S}_2(t_2)) \right\}^{1/2},$$

$w_{1-\frac{\alpha}{2}}$ is the $(1 - \alpha/2)$ quantile of the standard normal distribution, and

$$\hat{S}_2(s) = \hat{S}_0(\hat{r}(s \wedge t_1) + (s - t_1) \vee 0)$$

is the estimator of the survival function $S_2 = S_{x_2}$.

So the approximating $(1 - \alpha)$ confidence interval for r_0 is $(\underline{r}, \bar{r})$, where

$$\underline{r} = \sup\{r : \hat{U}(r) \geq u_{1-\frac{\alpha}{2}}\}, \quad \bar{r} = \inf\{r : \hat{U}(r) \leq -u_{1-\frac{\alpha}{2}}\}. \tag{6.36}$$

Proposition 6.3. *Suppose that the densities $f_i = -S_i'$ are continuous and positive on $[0, \infty[$ $(i = 1, 2)$. Then*

$$a_n(\hat{r} - r_0) \xrightarrow{D} \frac{W(r_0)}{U'(r_0)}. \tag{6.37}$$

Proof. Under the assumption of the proposition the function $U(r)$ is differentiable and decreasing on $[0, \infty[$ and $U(r_0) = 0$.

Taking into account that $\hat{r} = \hat{U}^{-1}(0)$, $r_0 = U^{-1}(0)$,

$$a_n(\hat{U}(r_0) - U(r_0)) = a_n \hat{U}(r_0) \xrightarrow{D} W(r_0)$$

and using the functional delta method (Theorem A12) we obtain (6.37).

Remark 6.2. It can be shown that

$$U'(r_0) = -(t_2 - t_1)\alpha_0(r_0t_1 + t_2 - t_1) S_0(r_0t_1 + t_2 - t_1) +$$

$$\int_0^{t_2-t_1} v\,\alpha_0(r_0t_1 + v)dS_0(r_0t_1 + v), \qquad (6.38)$$

where $\alpha_0(u) = -S_0'(u)/S_0(u)$ is the failure rate under the usual stress.

Let us consider the asymptotic properties of the estimator

$$\hat{A}_0 = \tilde{A}_0(s, \hat{r}).$$

It can be written in the form : for $s \in [0, \hat{r}t_1 + t - t_1]$

$$\hat{A}_0(s) = -\int_0^s \frac{k_1 d\hat{S}_1(u/\hat{r}) + k_2 d\hat{S}_2((u/\hat{r}) \wedge t_1 + (u - \hat{r}t_1) \vee 0)}{k_1 \hat{S}_1(u/\hat{r}-) + k_2 \hat{S}_2((u/\hat{r}) \wedge t_1 + (u - \hat{r}t_1) \vee 0-)} \qquad (6.39)$$

and for $s > \hat{r}t_1 + t_2 - t_1$

$$\hat{A}_0(s) = \hat{A}_0(\hat{r}t_1 + t_2 - t_1) - \int_{\hat{r}t_1+t_2-t_1}^s \frac{d\hat{S}_1(u/\hat{r})}{\hat{S}_1(u/\hat{r}-)}, \qquad (6.40)$$

where $k_1 = n_1/n$, $k_2 = n_2/n$.

Proposition 6.4. *Under the assumptions of the proposition 6.3*

$$a_n(\hat{A}_0(s) - A_0(s)) \xrightarrow{D} -\frac{W(r_0)}{r_0 U'(r_0)} \{l_1 s\alpha_0(s) + l_2(s \wedge (r_0t_1))\alpha_0(s \wedge (r_0t_1 + t_2 - t_1))\} -$$

$$\frac{l_1 V_1(s/r_0)}{S_0(s)} - \frac{l_2 V_2((\frac{s}{r_0} \wedge t_1 + (s - r_0t_1) \vee 0) \wedge t_2)}{S_0(s \wedge (r_0t_1 + t_2 - t_1))}. \qquad (6.41)$$

Proof.

Using the functional delta method (Theorem A13) we obtain for all $s \geq 0$ the convergence

$$a_n\{\hat{S}_1(\cdot/\hat{r}) - S_1(\cdot/r_0)\} \xrightarrow{D} Q_1(\cdot) \quad \text{on} \quad D(0, s) \quad \text{as} \quad n_1 \to \infty, \qquad (6.42)$$

where

$$Q_1(u) = -\frac{W(r_0)}{U'(r_0)} \frac{u}{r_0} S_0'(u) + V_1(u/r_0).$$

Analogously,

$$a_n\{\hat{S}_2(\frac{\cdot}{\hat{r}} \wedge t_1 + (\cdot - \hat{r}t_1) \vee 0) - S_2(\frac{\cdot}{r_0} \wedge t_1 + (\cdot - r_0t_1) \vee 0)\} \xrightarrow{D} Q_2(\cdot), \qquad (6.43)$$

$$\text{on} \quad D(0, r_0t_1 + t_2 - t_1), \quad \text{as} \quad n_2 \to \infty,$$

where

$$Q_2(u) = -\frac{W(r_0)}{U'(r_0)}(\frac{u}{r_0} \wedge t_1)S_0'(u) + V_2(\frac{u}{r_0} \wedge t_1 + (u - r_0t_1) \vee 0).$$

The results (6.42) and (6.43) imply

$$a_n\{[k_1\hat{S}_1(\cdot/\hat{r}-) + k_2\hat{S}_2(\frac{\cdot}{\hat{r}} \wedge t_1 + (\cdot - \hat{r}t_1) \vee 0-)]^{-1} -$$

$$[l_1 S_1(\cdot/r_0) + l_2 S_2(\frac{\cdot}{r_0} \wedge t_1 + (\cdot - r_0 t_1) \vee 0)]^{-1}\} \xrightarrow{D} -\frac{l_1 Q_1(\cdot) + l_2 Q_2(\cdot)}{S_0^2(\cdot)}$$

on $D[0, r_0 t_1 + t_2 - t_1]$ as $n_i \to \infty$, $k_i \to l_i$ $(i = 1, 2)$.

Using the functional delta method for stochastic integrals we obtain

$$a_n(\hat{A}_0(s) - A_0(s)) \xrightarrow{D} -\frac{l_1 Q_1(s) + l_2 Q_2(s)}{S_0(s)} \quad \text{as} \quad n_i \to \infty, \quad k_i \to l_i \ (i = 1, 2).$$

$$(6.44)$$

The right side of (6.44) may be written in the form (6.41).
If $s > r_0 t_1 + t_2 - t_1$, we obtain

$$a_n(\hat{A}_0(s) - A_0(s)) \xrightarrow{D} -\frac{l_1 Q_1(r_0 t_1 + t_2 - t_1) + l_2 Q_2(r_0 t_1 + t_2 - t_1)}{S_0(r_0 t_1 + t_2 - t_1)} -$$

$$\frac{Q_1(s)}{S_0(s)} + \frac{Q_1(r_0 t_1 + t_2 - t_1)}{S_0(r_0 t_1 + t_2 - t_1)}$$

which may be written in the form (6.41).

Corollary 6.1. Under the assumptions of the proposition for all $s \geq 0$

$$a_n(\hat{A}_0(s) - A_0(s)) \xrightarrow{D} N(0, \sigma_s^2), \quad a_n(\hat{S}_0(s) - S_0(s)) \xrightarrow{D} N(0, \sigma_s^2 S_0^2(s)), \quad (6.45)$$

$$a_n(\hat{t}_p(x_0) - t_p(x_0)) \xrightarrow{D} N(0, \sigma_s^2/\alpha_0^2(t_p(x_0))), \quad (6.46)$$

where

$$\sigma_s^2 =$$

$$\left(\frac{l_1 s\alpha_0(s) + l_2(s \wedge (r_0 t_1))\alpha_0(s \wedge (r_0 t_1 + t_2 - t_1))}{r_0 U'(r_0)}\right)^2 (S_0(t_1) - S_0(r_0 t_1 + t_2 - t_1))$$

$$+ l_1 l_2 \left(\frac{l_1(1 - S_0(s))}{S_0(s)} + \frac{l_2(1 - S_0(s \wedge (r_0 t_1 + t_2 - t_1)))}{S_0(s \wedge (r_0 t_1 + t_2 - t_1))}\right). \quad (6.47)$$

Proof. Take $s \in [0, r_0 t_1 + t_2 - t_1]$. Then

$$\mathbf{E}\{W(r_0)(l_1 V_1(s/r_0) + l_2 V_2((s/r_0) \wedge t_1 + (s - r_0 t_1) \vee 0))\} = 0.$$

So the expression of σ_s^2 is obtained by using expressions of the variances of $W(r_0)$, $V_1(s/r_0)$ and $V_2((s/r_0) \wedge t_1 + (s - r_0 t_1) \vee 0)$ done by the formulas (6.35) and (6.30) and from the result (6.41).
Take $s > r_0 t_1 + t_2 - t_1$. Then

$$\mathbf{E}\left\{W(r_0)\left(\frac{l_2 V_2(t_2)}{S_0(r_0 t_1 + t_2 - t_1)} + \frac{V_1(s/r_0)}{S_0(s)}\right)\right\} = 0.$$

The limit distribution of the estimator $\hat{S}_0(t)$ is obtained by the delta method using the fact that

$$\hat{S}_0(s) = \mathop{\pi}_{0 \leq u \leq s}(1 - d\hat{A}_0(u)).$$

The limit distribution of $\hat{t}_p(x_0)$ is obtained by using Theorem A.12.

Replacing the unknown functions S_0 and α_0 by their estimators $\hat{S}_0$ and $\hat{\alpha}_0$ in (6.47), we obtain an estimator $\hat{\sigma}_s^2$ of σ_s^2. The function α_0 is estimated by

$$\hat{\alpha}_0(s) = b^{-1}\int_0^\infty K(\frac{s-u}{b})d\hat{A}_0(u) =$$

$$\hat{\alpha}_0(s) = b^{-1}\int_0^\infty K\left(\frac{s-u}{b}\right)\frac{dN_1(u/\hat{r}) + dN_2((u/\hat{r})\wedge t_1 + (u - \hat{r}t_1)\vee 0)}{Y_1(u/\hat{r}) + Y_2[(u/\hat{r})\wedge t_1 + (u - \hat{r}t_1)\vee 0]} =$$

$$b^{-1}\left\{\sum_{j=1}^{n_1}\frac{K\left(\frac{s-\hat{r}T_{1j}}{b}\right)}{Y_1(T_{1j}) + Y_2[T_{1j}\wedge t_1 + \hat{r}(T_{1j} - t_1)\vee 0]} + \right.$$

$$\left.\sum_{j=1}^{m_2}\frac{K(\frac{s-\hat{r}(T_{2j}\wedge t_1)-(T_{2j}-t_1)\vee 0}{b})}{Y_1(T_{2j}\wedge t_1 + \frac{T_{2j}-t_1}{\hat{r}}\vee 0) + Y_2(T_{2j})}\right\},$$

where K is the kernel function with the window size b. The derivative $U'(r_0)$ (see (6.47)) can also be estimated by the numerical derivative

$$\frac{1}{2}n^{-1/2}\frac{\tilde{U}(\hat{r} + n^{-1/2}c) - \tilde{U}(\hat{r} - n^{-1/2}c)}{c},$$

where c is a positive constant.

6.2.4 Approximate confidence intervals

Approximate $(1-\alpha)$-confidence interval for $S_{x_0}(s)$:

$$\left\{1 + \frac{1 - \hat{S}_0(s)}{\hat{S}_0(s)}\exp\{\pm\frac{\hat{\sigma}_s w_{1-\frac{\alpha}{2}}}{a_n(1 - \hat{S}_0(s))}\}\right\}^{-1}, \qquad (6.48)$$

where $\hat{\sigma}_s$ is obtained by replacing the unknown functions S_0 and α_0 by their estimators $\hat{S}_0$ and $\hat{\alpha}_0$ in the expression (6.47).

Approximate confidence interval for the quantile $t_p(x_0)$:

$$\hat{t}_p(x_0)\exp\left\{\mp\frac{\hat{\sigma}_s w_{1-\frac{\alpha}{2}}}{a_n\,\hat{t}_p(x_0)\,\hat{\alpha}_0\{\hat{t}_p(x_0)\}}\right\}. \qquad (6.49)$$

6.2.5 Case of random moments of switch-up of stresses

The considered plan of experiments can be modified. In the preliminary experiments the time interval in which failures begin to appear is unknown, and the moment t_1, when switch-up from $x^{(1)}$ to $x^{(0)}$ is done, is difficult to choose.

A better plan of experiments is to switch up at the random moment, for example, at the moment of, say, kth failure of the tested group, i.e. when failures really begin to appear.

Suppose that two groups of units are tested: the first group of n_1 units are tested under the constant in time accelerated stress x_1 and the complete sample $T_{11}, \cdots, T_{1n_1}$ or in terms of order statistics

$$T_{1,(1)} \leq \cdots \leq T_{1,(n_1)}$$

is obtained. The second group of n_2 units is tested under stress x_1 until the kth failure and after this moment under usual stress x_0 until the moment t, i.e. under stress

$$x_2(u) = \begin{cases} x_1 & \text{if} \quad 0 \leq u < T_{2,(k)} \\ x_0 & \text{if} \quad T_{2,(k)} \leq u \leq t \end{cases}$$

and the type I censored sample

$$T_{2,(1)} \leq \cdots T_{2,(k)} \leq \cdots \leq T_{2,(m_2)}$$

is obtained ($m_2 \leq n_2$).

The purpose, as before, is to estimate the reliability function $S_0 = S_{x^{(0)}}$, the p-quantile $t_p(x^{(0)})$, and the mean value $m(x_0)$ under usual stress $x^{(0)}$.

Denote by $S_1 = S_{x^{(1)}}$ and $S_2^{(n_2)} = S_{x^{(2)}(\cdot)}^{(n_2)}$ the reliability functions of the times-to-failure for the units of the first and the second groups, respectively, by $A_2^{(n_2)} = -ln S_2^{(n_2)}$-the cumulative failure rate function for the units of the second group. Then

$$S_2^{(n_2)}(\tau) = - \int_0^\infty \mathbf{P}\left(T_{2i} > \tau | T_{2,(k)} = u\right) \, dS_{T_{2,(k)}}(u),$$

where T_{2j} are the times to failure under stress $x^{(2)}(\cdot)$ (possibly not all observed), $j = 1, \cdots, n_2$. We suppose that T_{2j} are absolutely continuous with positive densities on $[0, \infty)$. Then there exist a survival function S_2 such that

$$n_2^{1/2}(S_2^{(n)} - S_2) \xrightarrow{\mathcal{D}} 0 \quad \text{as} \quad n_2 \to \infty.$$

Denote by $N_1(u)$, $N_2(u)$ the numbers of observed failures in the interval $[0, u]$ and by $Y_1(u)$, $Y_2(u)$-the numbers of units at risk just before the moment u for the units of the first and the second group, respectively, $r = r(x)$. Note that $r(x^{(0)}) = 1$. If the AFT model holds, the moment u under stress x_0 is equivalent to the moments u/r and $(u/r) \wedge T_{2,(k)} + (u - rT_{2,(k)}) \vee 0$ under stresses $x^{(1)}$ and $x^{(2)}(\cdot)$, respectively. So the estimator (depending on r) of the cumulative hazard rate function $A_0 = -ln\{S_{x^{(0)}}\}$ is

$$\tilde{A}_0(s, r) = \int_0^s \frac{dN_1\left(\frac{u}{r}\right) + dN_2\left(\frac{u}{r} \wedge T_{2,(k)} + (u - rT_{2,(k)}) \vee 0\right)}{Y_1\left(\frac{u}{r}\right) + Y_2\left(\frac{u}{r} \wedge T_{2,(k)} + (u - rT_{2,(k)}) \vee 0\right)}.$$

The estimator of the reliability function S_0 is the product-integral

$$\tilde{S}_0(s, r) = \prod_{0 \leq u \leq s}(1 - d\tilde{A}_0(u, r)).$$

Then the estimators for S_1 and S_2 are

$$\tilde{S}_1(s, r) = \tilde{S}_0(rs, r), \quad \tilde{S}_2(s, r) = \tilde{S}_0(r(s \wedge T_{2,(k)}) + (s - T_{2,(k)}) \vee 0, r).$$

Under the AFT model the moment u under stress $x_2(\cdot)$ is equivalent to the moment

$$u \wedge T_{2,(k)} + \frac{u - T_{2,(k)}}{r} \vee 0$$

under stress x_1. The score function is defined as

$$\hat{U}(r) =$$

$$\int_0^\infty \frac{Y_1(u \wedge T_{2,(k)} + \frac{u - T_{2,(k)}}{r} \vee 0)dN_2(u) - Y_2(u)dN_1(u \wedge T_{2,(k)} + \frac{u - T_{2,(k)}}{r} \vee 0)}{Y_1(u \wedge T_{2,(k)} + \frac{u - T_{2,(k)}}{r} \vee 0) + Y_2(u)}$$

$$= \sum_{j=1}^{m_2} \frac{Y_1(T_{2,(j)} \wedge T_{2,(k)} + \frac{T_{2,(j)} - T_{2,(k)}}{r} \vee 0)}{Y_1(T_{2,(j)} \wedge T_{2,(k)} + \frac{T_{2,(j)} - T_{2,(k)}}{r} \vee 0) + Y_2(T_{2,(j)})} -$$

$$\sum_{j=1}^{n_1} \frac{Y_2(T_{1,(j)} \wedge T_{2,(k)} + r(T_{1,(j)} - T_{2,(k)}) \vee 0)}{Y_1(T_{1,(j)}) + Y_2(T_{1,(j)} \wedge T_{2,(k)} + r(T_{1,(j)} - T_{2,(k)}) \vee 0)}.$$

The function $\hat{U}$ is an increasing step function of r. If we assume that times to failure are absolutely continuous random variables with supports on $[0, \infty[$, then we have that $\hat{U}(0) < 0$, $\hat{U}(\infty) > 0$ with the probability 1.

The estimator of the parameter r is defined in the following manner:

$$\hat{r} = \hat{U}^{-1}(0) = \sup \{r : \hat{U}(r) \le 0\}.$$

The cumulative failure-rate function A_0 and the survival function S_0 under the usual stress can be estimated by

$$\hat{A}_0(s) = \tilde{A}_0(s, \hat{r}), \quad \hat{S}_0(s) = \tilde{S}_0(s, \hat{r}).$$

The estimators $\hat{A}_0(s)$ and $\hat{S}_0(s)$ are

$$\hat{A}_0(s) = \int_0^s \frac{dN_1(\frac{u}{\hat{r}}) + dN_2\left[\frac{u}{\hat{r}} \wedge T_{2,(k)} + (u - \hat{r}T_{2,(k)}) \vee 0\right]}{Y_1(\frac{u}{\hat{r}}) + Y_2\left[\frac{u}{\hat{r}} \wedge T_{2,(k)} + (u - \hat{r}T_{2,(k)}) \vee 0\right]} =$$

$$\sum_{j:\, T_{1,(j)} \le \frac{s}{\hat{r}}} \frac{1}{Y_1(T_{1,(j)}) + Y_2[T_{1,(j)} \wedge T_{2,(k)} + \hat{r}(T_{1,(j)} - T_{2,(k)}) \vee 0]} +$$

$$\sum_{j:\, T_{2,(j)} \le \frac{s}{\hat{r}} \wedge T_{2,(k)} + (s - \hat{r}T_{2,(k)}) \vee 0} \frac{1}{Y_1[T_{2,(j)} \wedge T_{2,(k)} + \frac{T_{2,(j)} - T_{2,(k)}}{\hat{r}} \vee 0] + Y_2(T_{2,(j)})}.$$

and

$$\hat{S}_0(s) = \prod_{(i,j) \in B(s)} \left(1 - \frac{1}{Y_1(T_{1,(i)}) + Y_2(T_{1,(i)} \wedge T_{2,(k)} + \hat{r}(T_{1,(i)} - T_{2,(k)}) \vee 0)}\right) \times$$

$$\left(1 - \frac{1}{Y_2(T_{2,(j)}) + Y_1(T_{2,(j)} \wedge T_{2,(k)} + \frac{T_{2,(j)} - T_{2,(k)}}{\hat{r}} \vee 0)}\right),$$

where

$$B(s) = \{(i,j) | \hat{r}T_{1,(i)} \le s;\ \hat{r}(T_{2,(j)} \wedge T_{2,(k)}) + (T_{2,(j)} - T_{2,(k)}) \vee 0 \le s\}.$$

The estimator of the mean time to failure under the normal stress $x^{(0)}$:

$$\hat{m}(x^{(0)}) = -\int_0^\infty s d\hat{S}_0(s) = -\sum_{i=1}^{n_1} \hat{r} T_{1,(i)} \Delta \hat{S}_0(\hat{r} T_{1,(i)}) - \sum_{i=1}^{r_2} \hat{r} T_{2,(i)} \Delta \hat{S}_0(\hat{r} T_{2,(i)})$$

$$- \sum_{i=r_2+1}^{m_2} (T_{2,(i)} - T_{2,(k)} + \hat{r} T_{2,(k)}) \Delta \hat{S}_0(T_{2,(i)} - T_{2,(k)} + \hat{r} T_{2,(k)}).$$

The asymptotic properties of estimators are given in Bagdonavičius and Nikulin (1999b). See also Duchesne (2000), Duchesne and Lawless (2000), etc.

CHAPTER 7

The Cox or PH model

7.1 Introduction

Let
$$x(\cdot) = (x_1(\cdot), ..., x_m(\cdot))^T,$$

be a possibly time-varying and multidimensional explanatory variable; here $x_1(\cdot), ..., x_m(\cdot)$ are one-variate explanatory variables.

Under the PH or Cox model the hazard rate under $x(\cdot)$ is

$$\alpha_{x(\cdot)}(t) = r\{x(t)\}\,\alpha(t). \qquad (7.1)$$

We shall consider only semiparametric estimation, because under the assumption that the survival distribution under a specified value of the explanatory variable $x \in E_1$ belongs to any family considered in Chapter 1 (with exception of the exponential and Weibull families) the survival distribution under any other value $y \in E_1$ does not belong to the same family. In the case of the exponential or Weibull distributions under constant explanatory variables the PH model coincides with the AFT model. Parametric estimation for the AFT model is considered in Chapter 5.

7.2 Parametrization of the PH model

The function r is parametrized in the following form:

$$r(x) = e^{\beta^T z}, \qquad (7.2)$$

where $\beta = (\beta_1, \cdots, \beta_m)^T$ is a vector of unknown parameters and

$$z = (z_1, \cdots, z_m)^T = (\varphi_1(x), \cdots, \varphi_m(x)^T$$

is a vector of specified functions φ_i. In what follows we use the same notation x_j for $z_j = \varphi_j(x)$.

The parametrized PH model has the form

$$\alpha_{x(\cdot)}(t) = e^{\beta^T x(t)}\,\alpha(t). \qquad (7.3)$$

The baseline hazard rate function $\alpha(t)$ is supposed to be unknown.

Let us discuss the choice of the functions φ_i.

7.3 Interpretation of the regression coefficients

Suppose that the explanatory variables are constant over time.

7.3.1 Interval valued explanatory variables

Suppose at first that the explanatory variables are interval-valued (load, temperature, stress, voltage, pressure).

If the model (7.3) holds on E_0, then for all $x_1, x_2 \in E_0$ the hazard ratio is

$$HR(x_1, x_2) = \alpha_{x_2}(t)/\alpha_{x_1}(t) = \rho(x_1, x_2), \qquad (7.4)$$

where $\rho(x_1, x_2) = r(x_2)/r(x_1)$. It is evident that $\rho(x, x) = 1$.

Suppose at first that x is one-dimensional. The speed of hazard rate variation with respect to x is defined by the *infinitesimal characteristic*:

$$\delta(x) = \lim_{\Delta x \to 0} \frac{\rho(x, x + \Delta x) - \rho(x, x)}{\Delta x} = [log\, r(x)]'. \qquad (7.5)$$

So for all $x \in E_0$ the function $r(x)$ is given by the formula:

$$r(x) = r(x_0) \exp \left\{ \int_{x_0}^{x} \delta(v)\, dv \right\}, \qquad (7.6)$$

where $x_0 \in E_0$ is a fixed explanatory variable.

Suppose that $\delta(x)$ is *proportional* to a specified function $u(x)$:

$$\delta(x) = \alpha\, u(x).$$

In this case

$$r(x) = e^{\beta_0 + \beta_1 \varphi_1(x)}, \qquad (7.7)$$

where $\varphi_1(x)$ is the primitive of $u(x)$, β_0, β_1 are unknown parameters.

So we have the model

$$\alpha_x(t) = e^{\beta_0 + \beta_1 \varphi_1(x)}\, \alpha(t).$$

Taking into consideration that the function $\alpha(t)$ is unknown, the parameter β_0 can be included in this function and the model

$$\alpha_x(t) = e^{\beta_1 \varphi_1(x)}\, \alpha(t) \qquad (7.8)$$

is obtained.

Example 7.1. $\delta(x) = \alpha$, $\varphi_1(x) = x$. Then

$$r(x) = e^{\beta_1 x}. \qquad (7.9)$$

It is the *log-linear model*.

Example 7.2. $\delta(x) = \alpha/x$, $\varphi_1(x) = \ln x$. Then

$$r(x) = e^{\beta_1 log\, x} = x^{\beta_1}. \qquad (7.10)$$

Example 7.3. $\delta(x) = \alpha/x^2$, $\varphi_1(x) = 1/x$. Then

$$r(x) = e^{\beta_1/x}. \tag{7.11}$$

Example 7.4. $\delta(x) = \alpha/x(1-x)$, $\varphi_1(x) = \ln \frac{x}{1-x}$. Then

$$r(x) = e^{\beta_1 \ln \frac{x}{1-x}} = \left(\frac{x}{1-x}\right)^{\beta_1}, \quad 0 < x < 1. \tag{7.12}$$

The infinitesimal characteristic $\delta(x)$ may be taken as a linear combination of some specified functions of the explanatory variable:

$$\delta(x) = \sum_{i=1}^{k} \alpha_i \, u_i(x).$$

In such a case

$$r(x) = exp\left\{\sum_{i=1}^{k} \beta_i z_i(x)\right\}, \tag{7.13}$$

where $z_i(x)$ are specified functions of the explanatory variable, $\beta_1, \ldots, \beta_k$ are unknown (possibly not all of them) parameters.

Example 7.5. $\delta(x) = \sum_{i=1}^{k} \alpha_i/x^i$.

Then

$$r(x) = exp\left\{\beta_1 log\, x + \sum_{i=1}^{k-1} \beta_i/x^i\right\}. \tag{7.14}$$

Suppose now that the explanatory variable $x = (x_1, \cdots, x_m)^T$ is multidimensional.

If there is no interaction between $x_1, \cdots, x_m$ then the model

$$r(x) = exp\left\{\sum_{i=1}^{m}\sum_{j=1}^{k_i} \beta_{ij} z_{ij}(x_i)\right\}, \tag{7.15}$$

could be used; here $z_{ij}(x_i)$ are specified functions, β_{ij} are unknown parameters.

Example 7.6. If the influence of the first explanatory variable is defined as in Example 7.2 and the influence of the second as in Example 7.3 then we have the model

$$r(x_1, x_2) = exp\left\{\beta_1 log\, x_1 + \beta_2/x_2\right\}. \tag{7.16}$$

So $k_1 = k_2 = 1$ here.

If there is interaction between the explanatory variables then complementary terms should be included.

Example 7.7. Suppose that there is interaction between the explanatory variables x_1 and x_2 defined in Example 7.6. Then the model

$$r(x_1, x_2) = exp\left\{\beta_1 log\, x_1 + \beta_2/x_2 + \beta_3 log\, x_1/x_2\right\}. \qquad (7.17)$$

could be considered.

7.3.2 Discrete and categorical explanatory variables

If the explanatory variables are discrete then the form of the functions have the same form as in the case of interval-valued explanatory variables, i.e. φ_j may be $\varphi_j(x) = x$, $\ln x$ or $1/x$.

If the jth explanatory variable is categorical and take k_j different values, then $x_j(\cdot)$ is understood as a $(k_j - 1)$-dimensional vector

$$x_j(\cdot) = (x_{j1}(\cdot), ..., x_{j,k_j-1}(\cdot))^T,$$

taking k_j different values

$$(0, 0, \cdots, 0)^T, (1, 0, \cdots, 0)^T, (0, 1, 0, \cdots, 0)^T, \cdots, (0, 0, \cdots, 0, 1)^T,$$

and β_j is $(k_j - 1)$-dimensional:

$$\beta_j = (\beta_{j1}, ..., \beta_{j,k_j-1})^T.$$

So, if the jth explanatory variable is categorical, and others are interval-valued or discrete then

$$\beta^T x = \beta_0 + \beta_1 x_1 + \cdots + \beta_{j-1} x_{j-1} + \sum_{l=1}^{k_j-1} \beta_{jl} x_{jl} + \beta_{j+1} x_{j+1} + \cdots + \beta_m x_m. \quad (7.18)$$

The obtained model is equivalent to the model (5.5) with $m + k_j - 2$ univariate explanatory variables. If $k_j = 2$, the explanatory variable x_j is dichotomous, taking two values 0 or 1.

Let us consider the interpretation of the parameters β_j under the model(7.3).

7.3.3 Models without interactions

a) *Interval-valued or discrete explanatory variables*

Suppose that the jth explanatory variable x_j is interval-valued or discrete. Then

$$e^{\beta_j} = \frac{\alpha_{(x_1,...,x_j+1,...,x_m)}(t)}{\alpha_{(x_1,...,x_j,...,x_m)}(t)}, \qquad (7.19)$$

is the ratio of hazard rates corresponding to the change of x_j by the unity.

b) *Categorical explanatory variables*

Suppose that $x_j = (x_{j1}, \cdot, ..., x_{j,k_j-1})^T$ is categorical. Its first value is $(0, \cdots, 0)^T$

and the $(i+1)$th value is $(0, \cdots, 0, 1, 0, \cdots, 0)^T$, where the unity is the ith coordinate. Then

$$e^{\beta_{ji}} = \frac{\alpha_{(x_1,\ldots,x_{j-1},(0,0,\cdots,0,1,0,\cdots,0),x_{j+1},\ldots,x_m)}(t)}{\alpha_{(x_1,\ldots,x_{j-1},(0,0,\cdots,0),x_{j+1},\ldots,x_m)}(t)} \qquad (7.20)$$

is the ratio of hazard rates corresponding to the change of x_j from the first to the $(i+1)$th value.

7.3.4 Models with interactions

a) Interaction between interval-valued or discrete explanatory variables

If there are two interval-valued or discrete explanatory variables and there is interaction between them then

$$\beta^T x = \beta_0 + \beta_1 x_1 + \beta_2 x_2 + \beta_3 x_1 x_2. \qquad (7.21)$$

For three explanatory variables

$$\beta^T x = \beta_0 + \beta_1 x_1 + \beta_2 x_2 + \beta_3 x_3 + \beta_4 x_1 x_2 + \beta_5 x_1 x_3 + \beta_6 x_2 x_3 + \beta_7 x_1 x_2 x_3, \qquad (7.22)$$

and so on.

In the case of two explanatory variables the ratio of hazard rates is

$$\frac{\alpha_{(x_1,x_2+1)}(t)}{\alpha_{(x_1,x_2)}(t)} = e^{\beta_1 + \beta_3 x_1} \qquad (7.23)$$

depends on the value of x_1.

So

$$e^{\beta_1 + \beta_3 x_1} \qquad (7.24)$$

is the ratio of hazard rates corresponding to the change of x_2 by the unity, other explanatory variable being fixed and equal to x_1

b) Interaction between interval-valued or discrete and categorical explanatory variables

Suppose that there are two explanatory variables: x_1 is interval-valued or discrete and x_2 is categorical, with k_2 possible values. Then

$$\beta^T x = \beta_1 x_1 + \sum_{i=1}^{k_2-1} \beta_{2i} x_{2i} + \sum_{i=1}^{k_2-1} \beta_{12i} x_1 x_{2i}, \qquad (7.25)$$

and the mean ratio

$$\frac{\alpha_{(x_1,(0,\ldots,0,1,0,\cdots,0))}(t)}{\alpha_{(x_1,(0,\ldots,0))}(t)} = e^{\beta_{2i} + \beta_{12i} x_1} \qquad (7.26)$$

depends on the value of x_1.

So in this example

$$e^{\beta_{2i} + \beta_{12i} x_1} \qquad (7.27)$$

is the ratio of hazard rates corresponding to the change of x_2 from the first to the $(i+1)$th value, other explanatory variable being fixed and equal to x_1.

c) *Interaction between categorical explanatory variables*

Suppose that both x_1 and x_2 are categorical with three values for each. Then

$$x_1 = (x_{11}, x_{12})^T, \quad x_2 = (x_{21}, x_{22})^T,$$

and

$$\beta^T x = \beta_{11} x_{11} + \beta_{12} x_{12} + \beta_{21} x_{21} + \beta_{22} x_{22} +$$

$$\beta_{1121} x_{11} x_{21} + \beta_{1122} x_{11} x_{22} + \beta_{1221} x_{12} x_{21} + \beta_{1222} x_{12} x_{22}. \tag{7.28}$$

In this case the ratio

$$\frac{\alpha_{(x_1,(1,0))}}{\alpha_{(x_1,(0,0))}} = e^{\beta_{21} + \beta_{1121} x_{11} + \beta_{1221} x_{12}} \tag{7.29}$$

depends on the value of $x_1 = (x_{11}, x_{12})^T$.
So

$$e^{\beta_{21} + \beta_{1121} x_{11} + \beta_{1221} x_{12}} \tag{7.30}$$

is the ratio of hazard rates corresponding to the change of x_2 from the first to the second value, other explanatory variable being fixed and equal to $x_1 = (x_{11}, x_{12})^T$.

Generalization is evident if the explanatory variables take three or more values.

7.3.5 Time dependent regression coefficients

We shall consider the vector of regression coefficients $\beta(t)$ in the form (5.37) as was done for the AFT model, so the model (2.120) can be written as the usual PH model:

$$\alpha_{x(\cdot)}(t) = e^{\theta^T z(t)} \alpha_0(t),$$

where the unknown parameter and the explanatory variables are defined by the formula (5.37).

7.4 Semiparametric FTR data analysis for the PH model

7.4.1 Model and data

Let us consider the PH model:

$$\alpha_{x(\cdot)}(t) = e^{\beta^T x(t)} \alpha(t). \tag{7.31}$$

The baseline hazard rate function $\alpha(t)$ is supposed to be unknown.
In terms of the survival functions

$$S_{x(\cdot)}(t) = \exp\left\{ -\int_0^t e^{\beta^T x(u)} dA(u) \right\}, \tag{7.32}$$

where

$$A(t) = \int_0^t \alpha(u)du.$$

The parameter $\beta = (\beta_1, \cdots, \beta_m)^T$ and the vector of the explanatory variables $x(\cdot) = (x_1(\cdot), \cdots, x_m(\cdot))^T$ have m coordinates.

Data

n units are observed. The ith unit is tested under the value $x^{(i)}(\cdot) = (x_1^{(i)}(\cdot), ..., x_m^{(i)}(\cdot))^T$ of a possibly time-varying and multidimensional explanatory variable $x(\cdot)$.

The data are supposed to be right censored.

Let T_i and C_i be the failure and censoring times of the ith unit,

$$X_i = T_i \wedge C_i, \quad \delta_i = \mathbf{1}_{\{T_i \leq C_i\}}.$$

In Chapter 4 was noted that right censored data may be presented in the form (4.1) or (4.3).

Denote by S_i and α_i the survival and the hazard rate functions under $x^{(i)}(\cdot)$. Under the AFT model they have the forms:

$$S_i(t; \beta) = \exp\left\{ -\int_0^t e^{\beta^T x^{(i)}(u)} dA(u) \right\}, \quad \alpha_i(t; \beta) = e^{\beta^T x^{(i)}(t)} \alpha(t). \quad (7.33)$$

If $x^{(i)}$ is constant then

$$S_i(t, \beta) = \exp\left\{ -e^{\beta^T x^{(i)}} A(t) \right\} = S(t)^{e^{\beta^T x^{(i)}}}, \quad \alpha_i(t, \beta) = e^{\beta^T x^{(i)}} \alpha_0(t); \quad (7.34)$$

here $S(t) = \exp\{-A(t)\}$ is the baseline survival function.

7.4.2 Semiparametric estimation of the regression parameters

Two equivalent methods of semiparametric estimation of the regression parameters β are considered here. The idea of the first is the same as in the case of the AFT model: write the parametric score function U and obtain the modified score function from it by replacing the unknown baseline cumulative hazard function $A(t)$ by its estimator $\tilde{A}(t, \beta)$ (depending on β). Another method is to write the parametric likelihood function as a product of two factors: one depending only on β and other depending on both β and A. The first factor is called the *partial likelihood function*. The estimator of the parameter β is obtained by maximizing this function.

It is interesting to note that both methods give exactly the same estimators.

First method

If α is specified then under the model (3.2) the maximum likelihood estimator of the parameter β is obtained by solving the system of equations

$$U_j(\beta) = 0, \quad (j = 1, \cdots, m),$$

where (see (4.22))

$$U_j(\beta) = \sum_{i=1}^{n} \int_0^\infty \frac{\partial}{\partial \beta_j} \log\{\alpha_i(u,\beta)\}\{dN_i(u) - Y_i(u)\alpha_i(u,\beta)du\}.$$

In the particular case of the PH model

$$\log\{\alpha_i(t,\beta)\} = \beta^T x^{(i)}(t) + \ln \alpha(t), \qquad \frac{\partial}{\partial \beta_j} \log\{\alpha_i(t,\beta)\} = x_j^{(i)}(t).$$

So the functions $U_j(\beta)$ can be written in the form

$$U_j(\beta) = \sum_{i=1}^{n} \int_0^\infty x_j^{(i)}(u)\,\{dN_i(u) - Y_i(u)e^{\beta^T x^{(i)}(u)}\}\,dA(u). \qquad (7.35)$$

If α is unknown then the score functions U_j depend not only on β but also on the unknown function A.

The estimator $\tilde{A}_0(t,\beta)$ is implied by Theorem A.2:

$$\mathbf{E}\{N_i(t)\} = \mathbf{E}\{\int_0^t Y_i(u)\,\alpha_i(u,\beta)du\} = \mathbf{E}\{\int_0^t Y_i(u)\,e^{\beta^T x^{(i)}(u)}dA(u)\}. \qquad (7.36)$$

Set

$$S^{(0)}(v,\beta) = \sum_{i=1}^{n} Y_i(v)\,e^{\beta^T x^{(i)}(v)}. \qquad (7.37)$$

The equality (7.36) implies

$$\mathbf{E}\{N(t)\} = \mathbf{E}\{\int_0^t S^{(0)}(u,\beta)dA(u)\}. \qquad (7.38)$$

This induces an estimator $\tilde{A}(t,\beta)$ for $A(t)$ defined by the equation:

$$N(t) = \int_0^t S^{(0)}(u,\beta)d\tilde{A}(u,\beta).$$

So

$$\tilde{A}(t,\beta) = \int_0^t \frac{dN(u)}{S^{(0)}(u,\beta)}. \qquad (7.39)$$

Replacing the function A_0 by $\tilde{A}_0(t,\beta)$ in (7.38), the modified score functions $\tilde{U}_j(\beta)$, depending only on β, are obtained. The random vector

$$\tilde{U}(\beta) = (\tilde{U}_1(\beta), \cdots, \tilde{U}_m(\beta))^T$$

has the form:

$$\tilde{U}(\beta) = \sum_{i=1}^{n} \int_0^\infty x^{(i)}(u)\{dN_i(u) - Y_i(u)e^{\beta^T x^{(i)}(u)}\}\frac{dN(u)}{S^{(0)}(u,\beta)},$$

or, shortly,

$$\tilde{U}(\beta) = \sum_{i=1}^{n} \int_0^\infty \{x^{(i)}(u) - E(u,\beta)\}\,dN_i(u), \qquad (7.40)$$

where

$$E(v, \beta) = \frac{S^{(1)}(v, \beta)}{S^{(0)}(v, \beta)},$$

$$S^{(1)}(v, \beta) = \sum_{i=1}^{n} x^{(i)}(v) Y_i(v)\, e^{\beta^T x^{(i)}(v)}.$$

Note that for the PH model the optimal weights

$$x^{(i)}(t) = \frac{\partial}{\partial \beta} \log\{\alpha_i(t, \beta)\}$$

do not depend on β and A. It was not so in the case of the AFT model.

The score function is calculated very simply:

$$\tilde{U}(\beta) = \sum_{i:\,\delta_i=1} \left\{ x^{(i)}(X_i) - \frac{\sum\limits_{j:X_j\geq X_i} x_j(X_i)\, e^{\beta^T x_j(X_i)}}{\sum\limits_{l:X_l\geq X_i} e^{\beta^T x_l(X_i)}} \right\}. \qquad (7.41)$$

If $x^{(i)}$ are constant then

$$\tilde{U}(\beta) = \sum_{i:\,\delta_i=1} \left\{ x^{(i)} - \frac{\sum\limits_{j:X_j\geq X_i} x_j(X_i)\, e^{\beta^T x_j}}{\sum\limits_{l:X_l\geq X_i} e^{\beta^T x_l}} \right\}. \qquad (7.42)$$

Second method

The score function (7.40) may be obtained another way using the notion of the partial likelihood.

Indeed, write the full parametric likelihood function as a product of several terms:

$$L = \prod_{i=1}^{n} \alpha_i^{\delta_i}(X_i, \beta)\, S_i(X_i, \beta)$$

$$= \prod_{i=1}^{n} \left(\int_0^{\infty} \alpha_i(u, \beta)\, dN_i(u) \right)^{\delta_i} \exp\left\{ -\int_0^{\infty} Y_i(u)\alpha_i(u, \beta)\, du \right\}$$

$$= \prod_{i=1}^{n} \left(\int_0^{\infty} \frac{e^{\beta^T x_i(u)} dN_i(u)}{S^{(0)}(u, \beta)} \right)^{\delta_i} \prod_{i=1}^{n} \left(\int_0^{\infty} \alpha_0(u) S^{(0)}(u, \beta) dN_i(u) \right)^{\delta_i} \times$$

$$\exp\left\{ -\int_0^{\infty} \alpha_0(u) S^{(0)}(u, \beta)\, du \right\}.$$

We set $0^0 = 1$.

The first term in the full likelihood depends on β and does not depend on the unknown baseline hazard rate function $\alpha_0(t)$. It is called the *partial likelihood* function:

$$\tilde{L}(\beta) = \prod_{i=1}^{n} \left(\int_0^{\infty} \frac{e^{\beta^T x_i(u)} dN_i(u)}{S^{(0)}(u, \beta)} \right)^{\delta_i}. \qquad (7.43)$$

In terms of (X_i, δ_i) it is written simply:

$$\tilde{L}(\beta) = \prod_{i:\delta_i=1} \frac{e^{\beta^T x_i(X_i)}}{S^{(0)}(X_i, \beta)} = \prod_{i:\delta_i=1} \frac{e^{\beta^T x_i(X_i)}}{\sum_{j:X_j \geq X_i} e^{\beta^T x_j(X_i)}}. \tag{7.44}$$

So

$$\ln \tilde{L}(\beta) = \sum_{i=1}^{n} \delta_i \{\beta^T x_i(X_i) - \ln S^{(0)}(X_i, \beta)\},$$

and

$$\tilde{U}(\beta) = \sum_{i=1}^{n} \delta_i(x_i(X_i) - E(X_i, \beta)) = \sum_{i=1}^{n} \int_0^\infty \{x^{(i)}(u) - E(u, \beta)\} \, dN_i(u).$$

It is the score function (7.40).

7.4.3 Estimators of the main reliability characteristics

Suppose that $x(\cdot)$ is an arbitrary explanatory variable which may be different from $x^{(i)}(\cdot), (i = 1, \cdots, n)$.

Estimator of the survival function $S_{x(\cdot)}(t)$:

The formula (7.39) implies that the baseline cumulative hazard and $A(t)$ and the cumulative hazard $A_{x(\cdot)}(t)$ are estimated by the statistics

$$\hat{A}(t) = \tilde{A}(t, \hat{\beta}), \quad \hat{A}_{x(\cdot)}(t) = \int_0^t e^{\hat{\beta}^T x(u)} d\hat{A}(u) = \int_0^t e^{\hat{\beta}^T x(u)} \frac{dN(u)}{S^{(0)}(u, \hat{\beta})}. \tag{7.45}$$

The estimator of the survival function $S_{x(\cdot)}(t)$ is the product integral

$$\hat{S}_{x(\cdot)}(t) = \Pi_{0 \leq u \leq t}(1 - d\hat{A}_{x(\cdot)}(u)) = \prod_{s:s \leq t} \left(1 - \frac{e^{\hat{\beta}^T x(u)} \Delta N(s)}{S^{(0)}(s, \hat{\beta})}\right). \tag{7.46}$$

The estimator (7.46) is calculated simply:

$$\hat{S}_{x(\cdot)}(t) = \prod_{j:\delta_j=1} \left(1 - \frac{e^{\hat{\beta}^T x(X_j)}}{S^{(0)}(X_j, \hat{\beta})}\right) = \prod_{j:\delta_j=1} \left(1 - \frac{e^{\hat{\beta}^T x(X_j)}}{\sum_{l:X_l \geq X_j} e^{\hat{\beta}^T x_l(X_j)}}\right). \tag{7.47}$$

If x is constant then

$$\hat{S}_x(t) = \prod_{j:\delta_j=1} \left(1 - \frac{e^{\hat{\beta}^T x}}{\sum_{l:X_l \geq X_j} e^{\hat{\beta}^T x_l}}\right). \tag{7.48}$$

Estimator of the p-quantile $t_p(x(\cdot))$

$$\hat{t}_p(x(\cdot)) = \sup\{t : \hat{S}_{x(\cdot)}(t) \geq 1 - p\}. \tag{7.49}$$

Estimator of the mean failure time $m(x(\cdot))$:

$$\hat{m}(x) = \sum_{j:\delta_j=1} X_j\{\hat{S}_{x(\cdot)}(X_j-) - \hat{S}_{x(\cdot)}(X_j)\} \tag{7.50}$$

The estimator of the mean may be underestimated if the last X_i is the censoring time.

Estimators of the hazard ratios

The hazard ratio $HR(x, y)$ (see (7.4)) is estimated by

$$\hat{H}R(x, y) = e^{\hat{\beta}^T (y-x)}. \tag{7.51}$$

7.4.4 Asymptotic distribution of the regression parameters estimators

The score function (7.40) can be written in the form

$$\tilde{U}(\beta) = \tilde{U}(\tau, \beta),$$

where $\tau = \sup\{t : \sum_{i=1}^n Y_i(t) > 0\}$,

$$\tilde{U}(t, \beta) = \sum_{i=1}^n \int_0^t \{x^{(i)}(v) - E(v, \beta)\}dM_i(v, \beta), \tag{7.52}$$

and

$$M_i(t, \beta) = N_i(t) - \int_0^t \alpha_i(v, \beta)dv = N_i(t) - \int_0^t Y_i(v)\, e^{\beta^T x^{(i)}(v)}dA(v).$$

The components of the random vector $\tilde{U}(\beta) = (\tilde{U}_1(\beta), \cdots, \tilde{U}_m(\beta))^T$ have the form

$$\tilde{U}_j(\beta) = \tilde{U}_j(\tau, \beta), \tag{7.53}$$

where

$$\tilde{U}_j(t, \beta) = \sum_{i=1}^n \int_0^t \{x_j^{(i)}(v) - E_j(v, \beta)\}dM_i(v, \beta), \tag{7.54}$$

and $E_j(v, \beta)$ is the jth component of $E(v, \beta)$:

$$E_j(v, \beta) = S_j^{(1)}(v, \beta)/S^{(0)}(v, \beta), \quad S_j^{(1)}(v, \beta) = \sum_{i=1}^n x_j^{(i)}(v)Y_i(v)\, e^{\beta^T x^{(i)}(v)}.$$

$$\tag{7.55}$$

Asymptotic properties of the regression parameter β are investigated similarly as in the case of parametric regression models because the structure of the score function (7.53) is the same as the structure of the score function (4.22). The score functions (7.53) (semiparametric estimation under the PH model) and the score functions (4.22) (parametric estimation in the regression models) have the form (4.27) with only difference that in the first case

$$H_{ij}(t, \theta) = \frac{\partial}{\partial \theta_j} \ln\{\alpha_i(t, \theta)\},$$

and in the second:

$$H_{ij}(t, \beta) = x_j^{(i)}(t) - E_j(t, \beta).$$

Denote by β_0 the true value of the parameter β. Similarly as in Chapter 4 (formula (4.19))

$$n^{1/2}(\hat{\beta} - \beta_0) = \left(\frac{1}{n}\mathbf{I}(\beta_0)\right)_{m \times m}^{-1} n^{-1/2} U(\beta_0) + \Delta, \qquad (7.56)$$

where (cf. (4.18))

$$\mathbf{I}(\beta_0) = \left(-\frac{\partial}{\partial \beta_{j'}} U_j(\beta)\right)_{m \times m} = \int_0^\tau V(u, \beta) dN(u), \qquad (7.57)$$

and $V(u, \beta) = (V_{jj'}(\beta))_{m \times m}$ is the matrix with the elements

$$V_{jj'}(t, \beta) = \frac{S_{jj'}^{(2)}(t, \beta)}{S^{(0)}(t, \beta)} - \frac{S_j^{(1)}(t, \beta)S_{j'}^{(1)}(t, \beta)}{(S^{(0)}(t, \beta))^2},$$

$$S_{jj'}^{(2)}(t, \beta) = \sum_{i=1}^n x_j^{(i)}(t)x_{j'}^{(i)}(t)Y_i(t)\, e^{\beta^T x^{(i)}(t)}. \qquad (7.58)$$

If $\Delta \xrightarrow{\mathbf{P}} 0$ then the asymptotic distributions of the random variables

$$n^{1/2}(\hat{\beta} - \beta_0) \quad \text{and} \quad \left(\frac{1}{n}\mathbf{I}(\beta_0)\right)^{-1} n^{-1/2}\tilde{U}(\beta_0) \qquad (7.59)$$

are the same and

$$n^{1/2}(\hat{\beta} - \beta_0) = \left(\frac{1}{n}\mathbf{I}(\beta_0)\right)^{-1} \sum_{i=1}^n \int_0^\tau \{x^{(i)}(v) - E(v, \beta)\} dM_i(v, \beta) + o_p(1). \qquad (7.60)$$

So if the random matrix $n^{-1}\mathbf{I}(\beta_0)$ converges in probability to a nondegenerated nonrandom matrix then the asymptotic properties of the maximum likelihood estimator $\hat{\beta}$ can be obtained from the asymptotic properties of the score function $\tilde{U}(\beta)$.

Theorem A.7 implies that if the processes $x_j^{(i)}(v) - E_j(v, \beta_0)$ are caglad adapted on $[0, \tau]$, for all $t \in [0, \tau]$

$$< \frac{1}{\sqrt{n}}U_j, \frac{1}{\sqrt{n}}U_{j'} > (t, \beta_0) \xrightarrow{\mathbf{P}} \sigma_{jj'}(t, \beta_0), \qquad (7.61)$$

$$< \frac{1}{\sqrt{n}}U_{\varepsilon j}(\tau) > \xrightarrow{\mathbf{P}} 0, \tag{7.62}$$

and for any $t \in [0, \tau]$ the matrix $\Sigma_0(t) = (\sigma_{jj'}(t, \beta_0))$ is positively definite then the score function $\tilde{U}(\theta_0)$ is asymptotically normal:

$$n^{-1/2}U(\beta_0) \xrightarrow{\mathcal{D}} N(0, \Sigma_0(\tau)) \text{ as } n \to \infty, \tag{7.63}$$

moreover,

$$n^{-1/2}U(\cdot, \beta_0) \xrightarrow{\mathcal{D}} Z(\cdot, \beta_0) \text{ on } (D[0, \tau])^m, \tag{7.64}$$

where Z is a m-variate Gaussian process having components with independent increments, $Z_j(0) = 0$ a.s. and for all $0 \le s \le t \le \tau$:

$$\mathbf{cov}(Z_j(s), Z_{j'}(t)) = \sigma_{jj'}(s).$$

The predictable covariations (7.61) and (7.62) are

$$< \frac{1}{\sqrt{n}}U_j, \frac{1}{\sqrt{n}}U_{j'} > (t, \beta_0) =$$

$$\frac{1}{n}\sum_{i=1}^{n}\int_0^t \{x_j^{(i)}(v) - E_j(v, \beta)\}\{x_{j'}^{(i)}(v) - E_{j'}(v, \beta)\}Y_i(v)\, e^{\beta^T x^{(i)}(v)}dA(v)$$

$$= \frac{1}{n}\int_0^\tau V_{jj'}(u, \beta_0)S^{(0)}(u, \beta_0)dA(u), \tag{7.65}$$

$$< \frac{1}{n}U_{\varepsilon j} > (t, \beta_0) =$$

$$\frac{1}{n}\sum_{i=1}^{n}\int_0^t \{x_j^{(i)}(v) - E_j(v, \beta_0)\}^2 \mathbf{1}_{\{|x_j^{(i)}(v)-E_j(v,\beta)| \ge \sqrt{n}\varepsilon\}}Y_i(v)\, e^{\beta^T x^{(i)}(v)}dA(v).$$
$$\tag{7.66}$$

Which sufficient conditions are needed for (7.61) and (7.62) to hold?

The convergence (7.61) holds if the stochastic process $n^{-1}V_{jj'}S^{(0)}$ (cf.(7.65)) stabilizes when n is large. This stabilization holds if the stochastic processes $n^{-1}S^{(i)}$ stabilize and $S^{(0)}$ (its square is in the denominator of $V_{jj'}$ expression) approaches a positive function on $[0, \tau]$ when n is going to infinity.

For the convergence (7.62) to hold, a complementary condition on boundedness of the explanatory variables is needed (the indicator should go to zero). If even the function under the integral (7.66) is small when n is large, the integral may be large if $A(\tau) = \infty$. So a natural condition is $A(\tau) < \infty$.

Conditions on differentiability of the limit functions of the stochastic processes $n^{-1}S^{(i)}$ are needed when writing the Taylor expansion (7.56) and differentiating the score function by interchanging the order of integration and differentiation.

Denote by $\|A\| = \sup_{i,j}|a_{ij}|$ the norm of any matrix A,

$$S^{(2)}(t, \beta) = (S_{jj'}^{(2)}(t, \beta))_{m \times m}.$$

So let us consider the following conditions.

Assumptions A:

a) *There exists a neighborhood B of β_0 and the scalar, m-vector, and $m \times m$ matrix functions $s^{(0)}(t,\beta)$, $s^{(1)}(t,\beta)$, and $s^{(2)}(t,\beta)$ such that for $k = 0, 1, 2$*

$$\sup_{\beta \in B,\, t \in [0,\tau]} \left\| \frac{1}{n} S^{(k)}(t,\beta) - s^{(k)}(t,\beta) \right\| \xrightarrow{\mathbf{P}} 0.$$

b) *$s^{(k)}(t,\beta)$ are continuous functions of $\beta \in B$ uniformly in $t \in [0,\tau]$ and bounded on $B \times [0,\tau]$.*

c) *$\sup_{t \in [0,\tau]} s^{(0)}(t,\beta) > 0$.*

d) *$\frac{\partial}{\partial \beta} s^{(0)}(t,\beta) = s^{(1)}(t,\beta)$ and $\frac{\partial^2}{\partial \beta^2} s^{(0)}(t,\beta) = s^{(2)}(t,\beta)$.*

e) *For all i, j $(i = 1, \cdots, n; j = 1, \cdots, m)$*

$$\sup_{t \in [0,\tau]} |x_j^{(i)}(t)| < \infty.$$

f) *$A(\tau) < \infty$.*

g) *The matrix*

$$\Sigma_0(\tau) = \int_0^\tau v(u,\beta) s^{(0)}(u,\beta) dA(u)$$

is positively definite; here

$$v = s^{(2)}/s^{(0)} - ee^T, \quad e = s^{(1)}/s^{(0)}.$$

Note that Assumptions A are sufficient for the matrix $n^{-1}\, \mathbf{I}(\beta_0)$ to converge in probability to the matrix $\Sigma(\beta_0)$ because

$$\frac{1}{n}\, \mathbf{I}(\beta_0) - < \frac{1}{\sqrt{n}} U_j, \frac{1}{\sqrt{n}} U_{j'} > (\tau, \beta_0) = \frac{1}{n} \int_0^\tau V_{jj'}(u, \beta_0) dM(u, \beta_0) \xrightarrow{P} 0;$$

here $dM(u, \beta_0) = dN(u) - S^{(0)}(u, \beta_0) dA(u)$. The last convergence is implied by the Corollary A6.

Assumptions A are sufficient not only for the asymptotic normality of $\tilde{U}$ and for convergence in probability of $n^{-1}\mathbf{I}(\beta_0)$ but also for the convergence in probability to zero of the term Δ.

The following result is implied by Andersen and Gill (1982).

Theorem 7.1. *Assume conditions A hold. Then*

1) *there exists a neighborhood of β_0 within which, with probability tending to 1 as $n \to \infty$, the root $\hat{\beta}$ of the system of equations $U(\beta) = 0$ is uniquely defined;*

2) *$\hat{\beta} \xrightarrow{\mathbf{P}} \beta_0$;*

3) *$n^{-1/2}U(\cdot, \beta_0) \xrightarrow{D} Z(\cdot, \beta_0)$ on $(D[0,\tau])^m$;*

4) *$n^{1/2}(\hat{\beta} - \beta_0) \xrightarrow{D} N(0, \Sigma^{-1}(\beta_0))$;*

5) *$\|\frac{1}{n}\mathbf{I}(\hat{\beta}) - \Sigma(\beta_0)\| \xrightarrow{P} 0$.*

7.4.5 Asymptotic distribution of the main reliability characteristics

In the books on survival analysis, the limit distribution of the baseline cumulative hazard estimator is usually done. In the case of FTR data it is more important to give the asymptotic distribution of the estimator $\hat{S}_x(t)$. Therefore we give the proof of the following theorem.

Theorem 7.2. *Under the assumptions of Theorem 7.1 for all $t \in [0, \tau]$*

$$n^{1/2}\{\hat{S}_x(t) - S_x(t)\} \xrightarrow{D} N(0, S_x^2(t)\sigma_x^2(t, \beta_0)) \qquad as \qquad n \to \infty,$$

where

$$\sigma_x^2(t, \beta_0) = b^T(t, \beta_0) \Sigma^{-1}(\beta_0) \, b(t, \beta_0) + e^{2\beta_0^T x} \int_0^t \frac{dA(v)}{s^{(0)}(v, \beta_0)}$$

and

$$b(t, \beta_0) = x A_x(t) - e^{\beta_0^T x} \int_0^t e(v, \beta_0) dA(v).$$

Proof. Let us consider the difference

$$\sqrt{n}(\hat{A}_x(t) - A_x(t)) = \sqrt{n}\left(e^{\hat{\beta}^T x_0} \int_0^t \frac{dN(v)}{S^{(0)}(v, \hat{\beta})} - A_x(t)\right) =$$

$$\sqrt{n}\left\{\left(e^{\hat{\beta}^T x} - e^{\beta_0^T x}\right) \int_0^t \frac{dN(v)}{S^{(0)}(v, \beta_0)}\right.$$

$$+ e^{\beta_0^T x} \int_0^t \left(\frac{1}{S^{(0)}(v, \hat{\beta})} - \frac{1}{S^{(0)}(v, \beta_0)}\right) dN(v) +$$

$$\left(e^{\hat{\beta}^T x} - e^{\beta_0^T x}\right) \int_0^t \left(\frac{1}{S^{(0)}(v, \hat{\beta})} - \frac{1}{S^{(0)}(v, \beta_0)}\right) dN(v) +$$

$$\left. e^{\beta_0^T x} \left(\int_0^t \frac{dN(v)}{S^{(0)}(v, \beta_0)} - A(t)\right)\right\}.$$

Write the finite increments formula for differences in first three terms:

$$\sqrt{n}(\hat{A}_x(t) - A_x(t)) =$$

$$\sqrt{n}(\hat{\beta} - \beta_0)^T \left\{ x \, e^{\beta^{*T}x} \int_0^t \frac{dN(v)}{S^{(0)}(v, \beta_0)} - e^{\beta_0^T x} \int_0^t \frac{E(v, \beta^{**})}{S^{(0)}(v, \beta^{**})} dN(v)\right\}$$

$$+ e^{\beta_0^T x} \sqrt{n}(\hat{\beta} - \beta_0)^T \sqrt{n}(\hat{\beta} - \beta_0) \frac{1}{\sqrt{n}} x^T \int_0^t \frac{E(v, \beta^{**})}{S^{(0)}(v, \beta^{**})} dN(v) +$$

$$e^{\beta_0^T x} \sqrt{n} \int_0^t \frac{dM(v)}{S^{(0)}(v, \beta_0)},$$

where β^*, β^{**} are points on the line segment between β_0 and $\hat{\beta}$. Note that

$$x e^{\beta^{*T}x_0} \int_0^t \frac{dN(v)}{S^{(0)}(v, \beta_0)} - e^{\beta_0^T x} \int_0^t \frac{E(v, \beta^*)}{S^{(0)}(v, \beta^*)} dN(v) \xrightarrow{P} b(t, \beta_0).$$

and

$$\frac{1}{\sqrt{n}} x^T \int_0^t \frac{E(v, \beta^*)}{S^{(0)}(v, \beta^*)} dN(v) \xrightarrow{\mathbf{P}} 0 \quad \text{uniformly on} \quad [0, \tau].$$

The formula (7.60) and the result 5) from Theorem 7.1 imply

$$\sqrt{n}(\hat{A}_x(t) - A_x(t)) = b^T(t, \beta_0) \Sigma^{-1}(\beta_0) n^{-1/2} \tilde{U}(\tau, \beta_0) +$$

$$e^{\beta_0^T x} \sqrt{n} \int_0^t \frac{dM(v)}{S^{(0)}(v, \beta_0)} + \Delta(t, \beta_0) = A_1(t) + A_2(t) + \Delta(t),$$

where

$$\sup_{[0, \tau]} |\Delta(t)| \xrightarrow{\mathbf{P}} 0.$$

The formulas (7.65) and (7.66) imply that for all $0 \le t \le \tau$

$$< A_1, A_1 > (t) \xrightarrow{\mathbf{P}} b^T(t, \beta_0) \Sigma^{-1}(\beta_0) b(t, \beta_0)$$

and

$$< A_{1\epsilon}, A_{1\epsilon} > (t) \xrightarrow{\mathbf{P}} 0.$$

For all $0 \le t \le \tau$ the predictable variation

$$< A_2, A_2 > (t) = e^{2\beta_0^T x} n \int_0^t \frac{dA(v)}{S^{(0)}(v, \beta_0)} \xrightarrow{\mathbf{P}} e^{2\beta_0^T x} \int_0^t \frac{dA(v)}{s^{(0)}(v, \beta_0)}$$

and

$$< A_{2\epsilon}, A_{2\epsilon} > (t) = e^{2\beta_0^T x} n \int_0^t \mathbf{1}_{\{ | \frac{e^{\beta_0^T x} \sqrt{n}}{S^{(0)}(v, \beta_0)} | \ge \epsilon \}} \frac{dA(v)}{S^{(0)}(v, \beta_0)} \xrightarrow{\mathbf{P}} 0.$$

The predictable covariation

$$< A_1(t), A_2 > (t) =$$

$$2 b^T(t, \beta_0) \Sigma^{-1}(\beta_0) e^{\beta_0^T x} \sum_{i=1}^n \int_0^t \{x_i(v) - E(v, \beta_0)\} \frac{e^{\beta_0 x} Y_i(v)}{S^{(0)}(v, \beta_0)} dA(v) = 0.$$

Corollary A.5 implies that

$$n^{1/2} \{ \hat{A}_x(\cdot) - A_x(\cdot) \} \xrightarrow{D} Z^*(\cdot, \beta_0) \quad \text{on } (D[0, \tau]), \tag{7.67}$$

where Z^* is the mean zero Gaussian process such that for all $0 \le s \le t \le \tau$:

$$\mathbf{cov}(Z^*(s), Z^*(t)) = \sigma_x^2(s \wedge t, \beta_0).$$

In particular, for any $t \in [0, \tau]$

$$n^{1/2} \{ \hat{A}_x(t) - A_x(t) \} \xrightarrow{D} N(0, \sigma_x^2(t, \beta_0)) \quad \text{as} \quad n \to \infty.$$

The delta method implies the result of the theorem.

Corollary. If $t_p(x) < \tau$ and $\alpha_x(t_p(x)) > 0$ then under Assumptions A

$$n^{1/2} \{ \hat{t}_p(x) - t_p(x) \} \xrightarrow{D} N(0, \sigma^2(t_p(x), \beta_0) / \alpha_x^2(t_p(x))) \quad \text{as} \quad n \to \infty. \tag{7.68}$$

Proof. Note that

$$\hat{t}_p(x) = \hat{A}_x^{-1}(-\ln(1 - p)).$$

The convergence (7.67) and Theorem A.10 imply that

$$n^{1/2}\{\hat{t}_p(x) - t_p(x)\} \xrightarrow{D} -\frac{Z^*(t_p(x))}{\alpha_x(t_p(x))},$$

which implies (7.68).

7.4.6 Tests for nullity of the regression coefficients

Let us consider the hypothesis

$$H_{k_1,k_2,\cdots,k_l} : \beta_{k_1} = \cdots = \beta_{k_l} = 0, \quad (1 \le k_1 \le k_2 \le \cdots \le k_l). \qquad (7.69)$$

If $H_{k_1,k_2,\cdots,k_l}$ is verified, the variables $x_{k_1}, \cdots, x_{k_l}$ are excluded from the model. In this particular case, the hypothesis

$$H_{1,2,\cdots,m} : \beta_1 = \cdots = \beta_m = 0$$

means that none of the explanatory variables improve the prediction, i.e. there is no regression.

The hypothesis

$$H_k : \beta_k = 0, \quad (k = 1, ..., m).$$

means that the model with and without the explanatory variable x_k gives the same prediction.

Likelihood ratio test

Let

$$\hat{L} = L(\hat{\beta}) = \max_{\beta} L(\beta) \qquad (7.70)$$

be the maximum of the partial likelihood function under the full model with m explanatory variables and

$$\hat{L}_{k_1...k_l} = \max_{\beta:\beta_{k_1}=...=\beta_{k_l}=0} \tilde{L}(\beta) \qquad (7.71)$$

be the maximum of the likelihood function under the hypothesis $H_{k_1...k_l}$ which is the maximum of the likelihood function, corresponding to the model with $(m - l)$ explanatory variables $\{x_1, ..., x_m\} \setminus \{x_{k_1}, ..., x_{k_l}\}$.

If n is large then the distribution of the likelihood ratio statistic (see (4.40)):

$$LR^{(*)} = -2\ln\frac{\tilde{L}(\beta)}{\hat{L}} \qquad (7.72)$$

is approximately chi-square with m degrees of freedom. Under $H_{k_1...k_l}$ the coefficients $\beta_{k_1}, \cdots, \beta_{k_l}$ are equal to zero in (7.72).

Similarly under $H_{k_1...k_l}$ the distribution of the likelihood ratio statistic

$$LR^{(**)} = -2\ln\frac{L(\beta)}{\hat{L}_{k_1...k_l}} \qquad (7.73)$$

is approximately chi-square with $m - l$ degrees of freedom. The coefficients $\beta_{k_1}, \cdots, \beta_{k_l}$ are equal to zero in (7.73).

It can be shown that the statistics

$$LR^{(**)} \quad \text{and} \quad LR^{(*)} - LR^{(**)}$$

are asymptotically independent.

Thus, under $H_{k_1 \ldots k_l}$ the distribution of the likelihood ratio test statistic

$$LR_{k_1, \ldots, k_l} = -2 \ln \frac{\hat{L}_{k_1 \ldots k_l}}{\hat{L}} \tag{7.74}$$

is approximated by the chi-square distribution with $l = m - (m - l)$ degrees of freedom when n is large.

The hypothesis $H_{k_1 \ldots k_l} : \beta_{k_1} = \ldots = \beta_{k_l} = 0$ is rejected with the significance level α if

$$LR_{k_1, \ldots, k_l} > \chi^2_{1-\alpha}(l); \tag{7.75}$$

the hypothesis $H_{12 \ldots m} : \beta_1 = \ldots = \beta_m = 0$ is rejected if

$$LR_{1, \ldots, m} > \chi^2_{1-\alpha}(m);$$

the hypothesis $H_k : \beta_k = 0$ is rejected if

$$LR_k > \chi^2_{1-\alpha}(1).$$

The statistic LR_k is often used in stepwise regression procedures when the problem of including or rejecting the explanatory variable x_k is considered.

In the particular case the likelihood ratio test statistics may be used when testing hypotheses about absence of interactions. For example, for the model with

$$\beta^T x = \beta_1 x_1 + \beta_2 x_2 + \beta_3 x_3 + \beta_4 x_1 x_2 + \beta_5 x_1 x_3 + \beta_6 x_2 x_3$$

the hypothesis $H_{456} : \beta_4 = \beta_5 = \beta_6 = 0$ or $H_5 : \beta_5 = 0$ may be tested.

Wald's tests

Let $A_{k_1 \ldots k_l}(\hat{\beta})$ be the submatrix of $I^{-1}(\hat{\beta})$ which is in the intersection of the $k_1, \ldots, k_l$ rows and $k_1, \ldots, k_l$ columns. Under $H_{k_1, \ldots, k_l}$ and large n the distribution of the statistic (see (4.39))

$$W_{k_1 \ldots k_l} = (\hat{\beta}_{k_1}, \ldots, \hat{\beta}_{k_l})^T A^{-1}_{k_1 \ldots k_l}(\hat{\beta})(\hat{\beta}_{k_1}, \ldots, \hat{\beta}_{k_l})^T$$

is approximated by the chi-square distribution with l degrees of freedom.

The hypothesis $H_{k_1 \ldots k_l}$ is rejected with the significance level α if

$$W_{k_1 \ldots k_l} > \chi^2_{1-\alpha}(l),$$

the hypothesis $H_{12 \ldots m}$ is rejected if

$$W_{1 \ldots m} > \chi^2_{1-\alpha}(m),$$

and the hypothesis H_k is rejected if

$$W_k = \frac{\hat{\beta}_k^2}{I^{kk}(\hat{\beta})} = \frac{\hat{\beta}_k^2}{\hat{\mathrm{Var}}\,(\hat{\beta}_k)} > \chi_{1-\alpha}^2(1).$$

Score tests

If n is large then the distribution of the statistic (see (4.38)):

$$(U_{k_1}(\beta), \cdots, U_{k_l}(\beta))^T\, A_{k_1 \ldots k_l}(\hat{\beta})\, (U_{k_1}(\beta), \cdots, U_{k_l}(\beta))$$

is approximately chi-square with l degrees of freedom. Under $H_{k_1 \ldots k_l}$ the limit distribution of this statistic does not change if β (with the components $\beta_i = 0$ for $i = k_1, \cdots, k_l$) is replaced by $\tilde{\beta}$ verifying the condition

$$L(\tilde{\beta}) = \max_{\beta:\beta_{k_1} = \ldots = \beta_{k_l} = 0} L(\beta).$$

So the score statistic is

$$U_{k_1 \ldots k_l} = (U_{k_1}(\tilde{\beta}), \cdots, U_{k_l}(\tilde{\beta}))^T\, A_{k_1 \ldots k_l}(\hat{\beta})\, (U_{k_1}(\tilde{\beta}), \cdots, U_{k_l}(\tilde{\beta})).$$

The hypothesis $H_{k_1 \ldots k_l}$ is rejected with the significance level α if

$$U_{k_1 \ldots k_l} > \chi_{1-\alpha}^2(l).$$

7.4.7 Graphical test for the PH model

If the explanatory variables are constant then under the PH model

$$A_x(t) = e^{\beta^T x}\, A(t).$$

Hence

$$\ln A_x(t) = \beta^T x + \ln A(t).$$

Under different values of x the graphs of the time functions $\ln A_x(t)$ are parallel. So, if x is discrete or categorical with s possible values and the data are stratified into s groups according to these values, then the Nelson-Aalen estimators $\hat{A}_j(t)$ $(j = 1, \cdots, s)$ may be constructed. Then the graphs of $\ln \hat{A}_j(t)$ should be approximately parallel under the PH model.

7.4.8 Stratified PH model

In some situations when the PH model is not verified under the vector $x = (x_1, \cdots, x_m)$ of the explanatory variables, the units can be divided into disjoint groups or strata such that in each stratum the PH model

$$\alpha_x(t) = e^{\beta^T x}\, A_j(t) \quad (j = 1, \cdots, s). \tag{7.76}$$

The baseline hazard α_j is distinct for each stratum but the regression parameters β are common.

The model (7.76) is called the stratified PH model. All data are used for estimation of the parameter β.

Generally the stratified PH model is used when the influence of, say, the first p components $x_1, \cdots, x_p$ of the vector of the explanatory variables $x = (x_1, \cdots, x_m)^T$ on the survival is studied and the components $x_{p+1}, \cdots, x_m$ are treated as discrete or categorical confounding variables. The strata are defined by the different values of the confounding variables.

The partial likelihood for the stratified PH model is defined as the product of the partial likelihood functions corresponding the distinct strata:

$$L(\beta) = \prod_{j=1}^{s} L_j(\beta). \tag{7.77}$$

One can read more about the Cox model, for example, in Altman and Andersen (1986), Andersen and Gill (1982), Andersen, Borgan, Gill and Keiding (1993), Aranda-Ordaz (1983), Arjas (1988), Bagdonavičius and Nikulin (1997d), Breslow (1975a,b), Breslow and Crowley (1982), Chappelle (1992), Cox and Oakes (1984), Fleming and Harrington (1991), Huber (2000), Therneau and Grambsch (2000).

CHAPTER 8

GPH models: FTR analysis

8.1 Introduction

We shall consider only semiparametric estimation, because under any specified GPH1 model a unique parametric family of survival distributions such that for any constant explanatory variable the survival distribution belongs to this family can be found. For example, for the GPH_{GW} model this family is the family of generalized Weibull distributions, for the GPH_{GLL} model this family is the family of generalized loglogistic distributions, etc. But under a specified GPH1 model and the corresponding class of survival distributions this model coincides with the AFT model. Parametric estimation for the AFT model is considered in Chapter 5.

8.2 Semiparametric FTR data analysis for the GPH1 models

8.2.1 Models

Let us consider the GPH1 model:

$$\alpha_{x(\cdot)}(t) = e^{\beta^T x(t)} \, q\{A_{x(\cdot)}(t), \gamma\} \, \alpha(t). \tag{8.1}$$

The baseline hazard rate function $\alpha(t)$ is supposed to be unknown. The function q belongs to a specified class of functions:

1) $q(u; \gamma) = (1 + u)^{-\gamma+1}$, ($GPH_{GW}$ model);
2) $q(u; \gamma) = e^{\gamma u}$, ($GPH_{GLL}$ model);
3) $q(u; \gamma) = (1 + \gamma u)^{-1}$, ($IGF$ model).

Other parametrizations can be considered.

It was shown in Chapter 2 that if under the GPH_{GW} model the ratio of the hazard rates under any two different constant explanatory variables is superior to 1 at the beginning of functioning then these ratios are increasing $(0 < \gamma < 1)$ or decreasing $(\gamma > 1)$ but remain superior to 1. In the case of the PH model the ratio of hazard rates is constant. Under the GPH_{GLL} model this ratio goes to 1, i.e. the hazard rates meet at infinity. Under the IGF model the hazard rates are equal at the beginning of functioning and the ratio of the hazard rates are monotone time functions.

Under the GPH1 model the survival function $S_{x(\cdot)}(t)$ has the form

$$S_{x(\cdot)}(t) = G\left\{\int_0^t e^{\beta^T x(u)} dA(u); \gamma\right\}, \tag{8.2}$$

where

$$A(t) = \int_0^t \alpha(u)du$$

is the baseline cumulative hazard and $G(t; \gamma)$ is the survival function of the resource (see Chapter 2.5.3). The function $G(t; \gamma)$ is the inverse of the function

$$H(u; \gamma) = \int_0^{-\ln u} \frac{dv}{q(v; \gamma)}$$

with respect to the first argument. The cumulative hazard function is:

$$A_{x(\cdot)}(t) = -\ln G \left\{ \int_0^t e^{\beta^T x(u)} dA(u); \gamma \right\}. \tag{8.3}$$

For the GPH_{GW} model:

$$A_{x(\cdot)}(t) = \left\{ 1 + \gamma \int_0^t e^{\beta^T x(u)} dA(u) \right\}^{\frac{1}{\gamma}} - 1, \quad (\gamma \neq 0),$$

$$A_{x(\cdot)}(t) = \exp \left\{ \int_0^t e^{\beta^T x(u)} dA(u) \right\} - 1, \quad (\gamma = 0). \tag{8.4}$$

If $\gamma < 0$, the support is finite (see Section 2.5.7). For the GPH_{GLL} model:

$$A_{x(\cdot)}(t) = -\frac{1}{\gamma} \ln \left\{ 1 - \gamma \int_0^t e^{\beta^T x(u)} dA(u) \right\}, \quad (\gamma \neq 0),$$

$$A_{x(\cdot)}(t) = \int_0^t e^{\beta^T x(u)} dA(u), \quad (\gamma = 0). \tag{8.5}$$

If $\gamma > 0$, the support is finite (see Section 2.5.7). For the IGF model:

$$A_{x(\cdot)}(t) = \frac{1}{\gamma} \left\{ \left(1 + 2\gamma \int_0^t e^{\beta^T x(u)} dA(u) \right)^{1/2} - 1 \right\}, \quad (\gamma \neq 0),$$

$$A_{x(\cdot)}(t) = \int_0^t e^{\beta^T x(u)} dA(u), \quad (\gamma = 0). \tag{8.6}$$

8.2.2 Data

n units are observed. The ith unit is tested under the value $x^{(i)}(\cdot)$ of a possibly time-varying and multidimensional explanatory variable $x(\cdot)$.

The data are supposed to be independently right censored.

Let T_i and C_i be the failure and censoring times of the ith unit,

$$X_i = T_i \wedge C_i, \quad \delta_i = \mathbf{1}_{\{T_i \leq C_i\}}.$$

In Chapter 4 it was noted that right censored data may be presented in the form (4.1) or (4.3).

Denote by S_i, α_i and A_i the survival, hazard rate, and cumulative hazard functions under $x^{(i)}(\cdot)$.

8.2.3 Semiparametric estimation of the parameters θ

Procedures for semiparametric estimation in GPH1 models for any specified q with time-varying covariates and properties of these estimators were given by Bagdonavičius & Nikulin (1994, 1995, 1997a,b). An interesting application of this model was considered by Ceci, Delattre, Hoffmann and Mazliak (2000). In the case of constant covariates, semiparametric estimation for linear transformation models was given by Dabrowska & Doksum (1988b) and Cheng, Wei & Ying (1995). Dabrowska & Doksum considered a resampling scheme for computing maximum rank likelihood estimation. Cheng, Wei & Ying (1995) considered methods of estimation based on generalised estimating equations. Murphy, Rossini & Van der Vaart (1997) considered estimation for the proportional odds model (the GPH1 model with the standard loglogistic distribution of the resource) and constant covariates. Andersen, Borgan, Gill & Keiding (1993) discuss estimation in the gamma frailty model with covariates by using the EM algorithm. Parner (1998) considered estimation in the correlated gamma-frailty model. Bagdonavičius & Nikulin (2001) give estimation for the GPH1 models using a modified partial likelihood method.

Let us consider estimation by the general method used in this book: construction of modified score functions by replacing the unknown infinite-dimensional parameter in the parametric score functions by its efficient estimator depending on finite-dimensional parameters of the model.

Denote by

$$\theta = (\beta_1, \cdots, \beta_m, \gamma)^T = (\theta_1, \cdots, \theta_{m+1})^T$$

the finite-dimensional parameters of the model.

If α is completely known then under the model (8.1) the parametric maximum likelihood estimator of the parameter θ is obtained by solving the system of equations

$$U_j(\theta) = 0, \quad (j = 1, \cdots, m+1),$$

where

$$U_j(\theta) = \sum_{i=1}^{n} \int_0^\infty w_j^{(i)}(u, \theta, A)\{dN_i(u) - Y_i(u)\alpha_i(u, \theta)du\},$$

where

$$w_j^{(i)}(t, \theta, A) = \frac{\partial}{\partial \beta_j} \log\{\alpha_i(t, \theta)\} = x_j^{(i)}(t) + (\ln q)_1'\{A_i(t, \theta), \gamma\}\frac{\partial}{\partial \beta_j} A_i(t, \theta),$$

$$(j = 1, \cdots, m),$$

$$w_{m+1}^{(i)}(t, \theta, A) = \frac{\partial}{\partial \gamma} \log\{\alpha_i(t, \theta)\} =$$

$$(\ln q)_2'\{A_i(t, \theta), \gamma\} + (\ln q)_1'\{A_i(t, \theta), \gamma\}\frac{\partial}{\partial \gamma} A_i(t, \theta); \qquad (8.7)$$

here $(\ln q)_l'(t, \gamma)$ denotes the partial derivative of $\ln q(t, \gamma)$ with respect to the

lth argument ($l = 1, 2$). Note that

$$\frac{\partial}{\partial \beta_j} A_i(t, \theta) = q(A_i(t, \theta), \gamma) \int_0^t x_j^{(i)}(u) e^{\beta^T x^{(i)}(u)} dA(u),$$

$$\frac{\partial}{\partial \gamma} A_i(t, \theta) = (- \ln G)_2'(H(e^{-A_i(t, \theta)}, \gamma), \gamma). \qquad (8.8)$$

So the functions $U_j(\theta)$ can be written in the form

$$U_j(\theta) = \sum_{i=1}^n \int_0^\infty w_j^{(i)}(u, \theta, A) \{dN_i(u) - Y_i(u) e^{\beta^T x^{(i)}(u)} q(A_i(u, \theta), \gamma) dA(u)\}.$$
$$(8.9)$$

If α is unknown then the score functions U_j depend not only on θ but also on the unknown function A.

The estimator $\tilde{A}(t, \theta)$ of $A(t)$ is implied by Theorem A.1:

$$\mathbf{E}\{N_i(t)\} = \mathbf{E}\left\{ \int_0^t Y_i(u) e^{\beta^T x^{(i)}(u)} q\{A_i(u, \theta), \gamma\} dA(u) \right\}. \qquad (8.10)$$

Set

$$S^{(0)}(v, \theta, A) = \sum_{i=1}^n Y_i(v) e^{\beta^T x^{(i)}(v)} q\{A_i(v, \theta), \gamma\}. \qquad (8.11)$$

The equality (8.10) implies:

$$\mathbf{E}\{N(t)\} = \mathbf{E}\left\{ \int_0^t S^{(0)}(u, \theta, A) d\, A(u) \right\}. \qquad (8.12)$$

This induces the estimator $\tilde{A}(t, \theta)$ of $A(t)$ defined by the equation:

$$N(t) = \int_0^t S^{(0)}(u-, \theta, \tilde{A}) \, d\tilde{A}(u, \theta).$$

The estimator $\tilde{A}$ can be found recurrently:

$$\tilde{A}(t, \theta) = \int_0^t \frac{dN(u)}{S^{(0)}(u-, \theta, \tilde{A})}. \qquad (8.13)$$

Replacing the function A by $\tilde{A}(t, \theta)$ in (8.9), the modified score functions $\tilde{U}_j(\theta)$, depending only on θ, are obtained. The random vector

$$\tilde{U}(\theta) = (\tilde{U}_1(\theta), \cdots, \tilde{U}_{m+1}(\theta))^T$$

has the form:

$$\tilde{U}(\theta) = \sum_{i=1}^n \int_0^\infty w^{(i)}(u, \theta) \left\{ dN_i(u) - Y_i(u) e^{\beta^T x^{(i)}(u)} q(\tilde{A}_i(u, \theta), \gamma) \frac{dN(u)}{S^{(0)}(u, \theta)} \right\},$$

where

$$\tilde{A}_i(v, \theta) = - \ln G \left\{ \int_0^t e^{\beta^T x^{(i)}(u)} d\tilde{A}(u, \theta); \gamma \right\}, \qquad (8.14)$$

$$S^{(0)}(v,\theta) = S^{(0)}(v,\theta,\tilde{A}), \quad w^{(i)}(u,\theta) = (w_1^{(i)}(u,\theta), \cdots, w_{m+1}^{(i)}(u,\theta))^T,$$

$$w_j^{(i)}(u,\theta) = w_j^{(i)}(u,\theta,\tilde{A}).$$

Shortly,

$$\tilde{U}(\theta) = \sum_{i=1}^{n} \int_0^\infty \{w^{(i)}(u,\theta) - E(u,\theta)\} \, dN_i(u), \tag{8.15}$$

where

$$E(v,\theta) = \frac{S^{(1)}(v,\theta)}{S^{(0)}(v,\theta)}, \quad S^{(1)}(v,\theta) = \sum_{i=1}^{n} w^{(i)}(v,\theta) Y_i(v) \, e^{\beta^T x^{(i)}(v)} q\{\tilde{A}_i(v,\theta),\gamma\}.$$

Note that as in the case of the PH model we consider the optimal weights.

8.2.4 Algorithm for computing the score functions

Suppose at first that there is no *ex aequo*. Let $T_{(1)} < ... < T_{(r)}$ be observed and ordered failure times, $r \le n$. Here (i) notes the index of the unit which failure is observed the ith.

The formulas (8.11), (8.13), and (8.14) imply that for fixed θ the estimator $\tilde{A}(t;\theta)$ is calculated by the following recurrent formulas:

$$\tilde{A}(0;\theta) = 0, \quad \tilde{A}(T_{(1)};\theta) = \left(\sum_{i=1}^{n} Y_i(T_{(1)}) e^{\beta^T x^{(i)}(T_{(1)})} \right)^{-1},$$

and for $j = 1, ..., r-1$

$$\tilde{A}(T_{(j+1)};\theta) =$$

$$\tilde{A}(T_{(j)};\theta) + \left(\sum_{i=1}^{n} Y_i(T_{(j+1)}) e^{\beta^T x^{(i)}(T_{(j+1)})} q^* \left(\sum_{l=1}^{j} e^{\beta^T x^{(i)}(T_{(l)})} \Delta\tilde{A}(T_{(l)};\theta), \gamma \right) \right)^{-1},$$

$$\tag{8.16}$$

where

$$T_{(0)} = 0, \quad \Delta\tilde{A}(T_{(l)};\theta) = \tilde{A}(T_{(l)};\theta) - \tilde{A}(T_{(l-1)};\theta), \quad q^*(t,\gamma) = q(-\ln G(t,\gamma),\gamma)$$

So the score function is calculated simply:

$$\tilde{U}_j(\theta) = \sum_{i=1}^{r} \{w_j^{((i))}(T_{(i)},\theta) - E_j(T_{(i)},\theta)\}, \tag{8.17}$$

where

$$E_j(T_{(i)},\theta) = \frac{S_j^{(1)}(T_{(i)},\theta)}{S^{(0)}(T_{(i)},\theta)},$$

$$S^{(0)}(T_{(i)},\theta) = \sum_{s=1}^{n} Y_s(T_{(i)}) \, e^{\beta^T x^{(s)}(T_{(i)})} q\{\tilde{A}_s(T_{(i)},\theta),\gamma\},$$

$$S_j^{(1)}(T_{(i)},\theta) = \sum_{s=1}^{n} w_j^{(s)}(T_{(i)},\theta) Y_s(T_{(i)}) \, e^{\beta^T x^{(s)}(T_{(i)})} q\{\tilde{A}_s(T_{(i)},\theta),\gamma\},$$

$$w_j^{(s)}(T_{(i)}, \theta) = x_j^{(s)}(T_{(i)}) + q_1' \{ \tilde{A}_s(T_{(i)}, \theta), \gamma \} \sum_{l=1}^{i} x_j^{(s)}(T_{(l)}) e^{\beta^T x^{(s)}(T_{(l)})} \Delta \tilde{A}(T_{(l)}; \theta),$$

$$(j = 1, \cdots, m)$$

$$\tilde{A}_s(T_{(i)}, \theta) = - \ln G(\sum_{l=1}^{i} e^{\beta^T x^{(s)}(T_{(l)})} \Delta \tilde{A}(T_{(l)}; \theta); \gamma),$$

$$w_{m+1}^{(s)}(T_{(i)}, \theta) = (\ln q)_2' \{ \tilde{A}_s(T_{(i)}, \theta), \gamma \} +$$

$$(\ln q)_1'(\tilde{A}_s(T_{(i)}, \theta), \gamma) \, (- \ln G)_2' \{ \sum_{l=1}^{i} e^{\beta^T x^{(s)}(T_{(l)})} \Delta \tilde{A}(T_{(l)}; \theta); \gamma \}. \qquad (8.18)$$

If $x^{(s)}$ are constant then

$$x^s(T_{(i)}) = x^s, \quad \sum_{l=1}^{i} e^{\beta^T x^{(s)}(T_{(l)})} \Delta \tilde{A}(T_{(l)}; \theta) = e^{\beta^T x^{(s)}} \tilde{A}(T_{(i)}; \theta),$$

$$\sum_{l=1}^{i} x_j^{(s)}(T_{(l)}) e^{\beta^T x^{(s)}(T_{(l)})} \Delta \tilde{A}(T_{(l)}; \theta) = x_j^{(s)} e^{\beta^T x^{(s)}} \tilde{A}(T_{(i)}; \theta)$$

in (8.18).

Note that to find an initial estimator $\hat{\theta}^{(0)}$ verifying the equations (8.17) you need an initial estimator $\hat{A}^{(0)}(t)$ and vice versa: to find an initial estimator $\hat{A}^{(0)}(t) = \tilde{A}(t, \hat{\theta}^{(0)})$ you need an initial estimator $\hat{\theta}^{(0)}$. This magic circle can be entered in the following way.

The initial estimator $\hat{\theta}^{(0)}$ can be obtained (see Vonta (2000) for the case of constant explanatory variables and uncensored data) as follows: the baseline cumulative hazard function A is approximated by a piecewise linear function

$$A(t, \mu) = \int_0^t \sum_{i=1}^{r} \mathbf{1}_{(a_{i-1}, a_i]}(s) e^{\mu_i} ds,$$

where $0 = a_0 < a_1 < \cdots < a_r = \infty$ are given constants and $\mu = (\mu_1, \cdots, \mu_r)^T$ are unknown parameters.

Then the parametric model

$$S_{x(\cdot)}(t; \mu, \beta, \gamma) = G \left\{ \sum_{i=1}^{r} e^{\mu_i} \int_0^t e^{\beta^T x(u)} \mathbf{1}_{(a_{i-1}, a_i]}(u) du, \gamma \right\}$$

can be considered. Denote by $\hat{\mu}^{(0)}, \hat{\beta}^{(0)}, \hat{\gamma}^{(0)}$ the parametric maximum likelihood estimators of the parameters μ, β, γ. So $\hat{\theta}^{(0)} = (\hat{\beta}^{(0)}, \hat{\gamma}^{(0)})$ can be considered as an initial estimator for θ.

If the initial estimator $\hat{\theta}^{(0)}$ is obtained then the initial estimator $\tilde{A}^{(0)}(t, \hat{\theta}^{(0)})$ is obtained recurrently by (8.16) using $\theta = \hat{\theta}^{(0)} = (\hat{\beta}^{(0)}, \hat{\gamma}^{(0)})$. The estimator $\hat{\theta}^{(1)}$ is obtained by solving the equations $\tilde{U}_j(\theta) = 0$, where $\tilde{U}_j(\theta)$ are given by (8.17), and using $\tilde{A}^{(0)}(t, \hat{\theta}^{(0)})$ instead of $\tilde{A}(t, \theta)$. And so on: the estimator $\tilde{A}^{(1)}(t, \hat{\theta}^{(1)})$ is obtained recurrently by (8.16) using $\theta = \hat{\theta}^{(1)}$, etc.

8.2.5 Expressions of the score functions for specified GPH1 models

For any specified GPH1 model the score functions are computed using the formulas (8.16)-(8.18). The functions q, $-\ln G$, q^*, q'_1, $(\ln q)'_1$, $(-\ln G)'_2$ are different for different models.

GPH$_{GW}$ model

$$q(u,\gamma) = (1+u)^{-\gamma+1}; \quad -\ln G(u,\gamma) = \{1+\gamma u\}^{\frac{1}{\gamma}} - 1, \quad (\gamma \neq 0),$$

$$-\ln G(u,0) = e^u - 1; q^*(u,\gamma) = \{1+\gamma u\}^{\frac{1-\gamma}{\gamma}} \quad (\gamma \neq 0), \quad q^*(u,0) = e^u;$$

$$q'_1(u,\gamma) = (1-\gamma)(1+u)^{-\gamma}; \quad (\ln q)'_1(u,\gamma) = \frac{1-\gamma}{1+u};$$

$$(-\ln G)'_2(u,\gamma) = \{1+\gamma u\}^{\frac{1-\gamma}{\gamma}} \frac{(1+\gamma u)\ln(1+\gamma u) - \gamma u}{\gamma^2}, \quad (\gamma \neq 0)$$

$$(-\ln G)'_2(u,0) = \frac{u^2}{2} e^u.$$

GPH$_{GLL}$ model

$$q(u,\gamma) = e^{\gamma u}; \quad -\ln G(u,\gamma) = -\frac{1}{\gamma} \ln(1-\gamma u), \quad (\gamma \neq 0); \quad -\ln G(u,\gamma) = u;$$

$$q^*(u,\gamma) = (1-\gamma u)^{-1}; \quad q'_1(u,\gamma) = \gamma e^{\gamma u}; \quad (\ln q)'_1(u,\gamma) = \gamma;$$

$$(-\ln G)'_2(u,\gamma) = \frac{(1-\gamma u)\ln(1-\gamma u) + \gamma u}{\gamma^2(1-\gamma u)}, \quad (\gamma \neq 0), \quad (-\ln G)'_2(u,0) = \frac{u^2}{2}.$$

IGF model

$$q(u,\gamma) = (1+\gamma u)^{-1}; \quad -\ln G(u,\gamma) = \frac{1}{\gamma}\{(1+2\gamma u)^{\frac{1}{2}} - 1\}, \quad (\gamma \neq 0),$$

$$(-\ln G)'_2(u,0) = u; q^*(u,\gamma) = (1+2\gamma u)^{-\frac{1}{2}};$$

$$q'_1(u,\gamma) = -\gamma(1+\gamma u)^{-2}; \quad (\ln q)'_1(u,\gamma) = -\gamma(1+\gamma u)^{-1};$$

$$(-\ln G)'_2(u,\gamma) = \frac{(1+2\gamma u)^{1/2} - (1+\gamma u)}{\gamma^2(1+2\gamma u)^{1/2}}, \quad (\gamma \neq 0), \quad (-\ln G)'_2(u,0) = -\frac{u^2}{2}.$$

Similarly as in the case of the PH model the score function (8.15) may be obtained by other way-using the notion of the modified partial likelihood (see Bagdonavicius and Nikulin (1999)).

8.2.6 Case of ex aequo

Suppose that *ex aequo* are possible. Let $T_1^* < ... < T_r^*$ be observed and ordered distinct failure times, $r \leq n$. Note by d_i the number of failures at the moment T_i^*. Let $(i_1), \cdots, (i_{d_i})$ be the indices of the units failed at T_i^*.

The estimator $\tilde{A}(t; \theta)$ is calculated by the following recurrent formulas :

$$\tilde{A}(0; \theta) = 0, \quad \tilde{A}(T_1^*; \theta) = d_1 \left(\sum_{i=1}^{n} Y_i(T_1^*) e^{\beta^T x^{(i)}(T_1^*)} \right)^{-1} ;$$

for $j = 1, ..., r - 1$

$$\tilde{A}(T_{j+1}^*; \theta) =$$

$$\tilde{A}(T_j^*; \theta) + d_{j+1} \left(\sum_{i=1}^{n} Y_i(T_{j+1}^*) e^{\beta^T x^{(i)}(T_{j+1}^*)} q^* (\sum_{l=1}^{j} e^{\beta^T x^{(i)}(T_l^*)} \Delta \tilde{A}(T_l^*; \theta), \gamma) \right)^{-1}.$$

$$(8.19)$$

So the score function is modified:

$$\tilde{U}_j(\theta) = \sum_{i=1}^{r} \sum_{s=1}^{k_i} \{w_j^{((i_s))}(T_i^*, \theta) - E_j(T_i^*, \theta)\}. \qquad (8.20)$$

8.2.7 Estimators of the main reliability characteristics

Suppose that $x(\cdot)$ is an arbitrary explanatory variable which may be different from $x^{(i)}(\cdot), (i = 1, \cdots, n)$.

Estimator of the survival function $S_{x(\cdot)}(t)$:

The estimator of the survival function $S_{x(\cdot)}(t)$ is

$$\hat{S}_{x(\cdot)}(t) = G\left\{ \int_O^t e^{\hat{\beta}^T x(u)} d\tilde{A}(u, \hat{\theta}), \hat{\gamma} \right\} = G\left\{ \int_O^t e^{\hat{\beta}^T x(u)} \frac{dN(u)}{S^{(0)}(u, \hat{\theta})}, \hat{\gamma} \right\}.$$

$$(8.21)$$

This estimator is calculated simply:

$$\hat{S}_{x(\cdot)}(t) = G\left\{ \sum_{i:T_{(i)} \leq t} \frac{e^{\hat{\beta}^T x(T_{(i)})}}{S^{(0)}(T_{(i)}, \hat{\theta})}, \hat{\gamma} \right\}, \qquad (8.22)$$

where $S^{(0)}(T_{(i)}, \hat{\theta})$ is calculated using the formula (8.18). If x is constant then

$$\hat{S}_x(t) = G\left\{ \sum_{i:T_{(i)} \leq t} \frac{e^{\hat{\beta}^T x}}{S^{(0)}(T_{(i)}, \hat{\theta})}, \hat{\gamma} \right\}. \qquad (8.23)$$

Estimator of the p-quantile $t_p(x(\cdot))$

$$\hat{t}_p(x(\cdot)) = \sup\{t : \hat{S}_{x(\cdot)}(t) \geq 1 - p\}. \tag{8.24}$$

Estimator of the mean failure time $m(x(\cdot))$:

$$\hat{m}(x) = \sum_{j:\delta_j=1} X_j\{\hat{S}_{x(\cdot)}(X_j-) - \hat{S}_{x(\cdot)}(X_j)\} \tag{8.25}$$

The estimator of the mean may be underestimated if the last X_i is the censoring time.

8.2.8 Some remarks on asymptotic distribution estimators

Investigations by simulation (see Hafdi (2000), Hafdi, El Himdi and al (2001)) show good properties of estimators (even in the case of finite supports of failure time distributions) for finite samples.

For models with specified specified G (or q) the asymptotic properties of the estimators of the regression parameters and reliability characteristics are given by Bagdonavičius and Nikulin (1997a, grouped data, 1997b, general right censored data), for models with parametrized via γ - by Bagdonavičius and Nikulin (2001). The proofs of asymptotic properties are rather technical and are not given here.

As an example, we give the asymptotic properties of estimators for the GPH_{LL} model with constant explanatory variables. The properties are analogous for other GPH1 models.

Denote by θ_0 the true value of the parameter θ and set

$$z_j(u;\theta) = \left(\begin{array}{c} x_j(u) \\ \tilde{A}_j(u;\theta) \end{array} \right), \quad S^{(2)}(u;\theta) = \sum_{i=1}^{n} \frac{\partial w^{(i)}(u;\theta)}{\partial\theta} Y_j(u)e^{\theta^T z_j(u;\theta)},$$

$$S_*^{(0)}(u;\theta) = \sum_{i=1}^{n} Y_i(u)e^{2\theta^T z_i(u;\theta)}, \quad S_*^{(1)}(u;\theta) = \sum_{i=1}^{n} w^{(i)}(u;\theta)Y_i(u)e^{2\theta^T z_i(u;\theta)}.$$

Conditions A:
a) There exists a neighborhood Θ of θ_0 and continuous on Θ uniformly in $t \in [0,\tau]$ and bounded on $\Theta \times [0,\tau]$ scalar functions $s^{(0)}(u;\theta)$, $s_*^{(0)}(u;\theta)$, vector functions $s^{(1)}(u;\theta)$, $s_*^{(1)}(u;\theta)$ and $(k+1) \times (k+1)$ matrix $s^{(2)}(u;\theta)$ such that $s^{(0)}(u;\theta_0) > 0$ on $[0,\tau]$,

$$\sup_{\theta\in\Theta,\, u\in[0,\tau]} \|\frac{1}{n}S^{(i)}(u;\theta) - s^{(i)}(u;\theta)\| \to 0$$

$$\sup_{\theta\in\Theta,\, u\in[0,\tau]} \|\frac{1}{n}S_*^{(i)}(u;\theta) - s_*^{(i)}(u;\theta)\| \to 0$$

$$\sup_{\theta \in \Theta,\, u \in [0,\tau]} \left\| \frac{\partial E(u;\theta)}{\partial \theta} - \frac{\partial e(u;\theta)}{\partial \theta} \right\| \xrightarrow{P} 0, \quad \text{as} \quad n \to \infty,$$

where $e(u;\theta) = s^{(1)}(u;\theta)/s^{(0)}(u;\theta)$.

b) $A(\tau) < \infty$.

Set

$$e_*(u;\theta) = s_*^{(0)}(u;\theta)/s^{(0)}(u;\theta), \quad h(t;\theta) = exp\{-\gamma \int_0^t e_*(u;\theta)dA(u)\},$$

$$g(u;\theta) = \frac{s^{(1)}(u;\theta)s_*^{(0)}(u;\theta) - s^{(0)}(u;\theta)s_*^{(1)}(u;\theta)}{s^{(0)}(u;\theta)},$$

$$w(u;\theta) = e(u;\theta) - \frac{\gamma}{h(u;\theta)s^{(0)}(u;\theta)} \int_u^\tau g(v;\theta)h(v;\theta)dA(v).$$

Denote by $A^{\otimes 2}$ the product AA^T, $J(u) = \mathbf{1}_{\{Y(u)>0\}}$.

c)

$$\frac{1}{n} \sum_{i=1}^n \int_0^\tau J(u)\{w^{(i)}(u;\theta_0) - w(u;\theta_0)\}^{\otimes 2} e^{\theta_0^T z_i(u;\theta_0)} Y_i(u)dA(u) \xrightarrow{P} \Sigma(\theta_0).$$

d) *The matrix*

$$\Sigma_1(\theta_0) = -\int_0^\tau \left\{ s^{(2)}(u;\theta_0) - \frac{\partial e(u;\theta_0)}{\partial \theta} s^{(0)}(u;\theta_0) \right\} dA_0(u)$$

is positive definite.

Theorem 8.1. *Under Conditions A*

$$n^{1/2}(\hat{\theta} - \theta_0) \xrightarrow{D} N(0, \Sigma_1^{-1}(\theta_0)\Sigma(\theta_0)\left(\Sigma_1^{-1}(\theta_0)\right)^T).$$

Set

$$C(\theta_0) = -\int_0^t \frac{\partial}{\partial \theta} \ln S^{(0)}(u;\theta_0)dA(u),$$

$$H_i(u;\theta_0) = J(u) \left[C(\theta_0)\Sigma_1^{-1}(\theta_0)\{w^{(i)}(u;\theta_0) - w(u;\theta_0)\} + \frac{1}{s^{(0)}(u;\theta_0)} \right],$$

$$v_i(u;\theta_0) = \Sigma_1(\theta_0)\{w_i(u;\theta_0) - w(u;\theta_0)\} = \begin{pmatrix} v_i^{(1)}(u;\theta_0) \\ v_i^{(2)}(u;\theta_0) \end{pmatrix},$$

$$c_{ix}(u;\theta_0) = \gamma_0 e^{\beta_0^T x} H_i(u;\theta_0) + e^{\beta_0^T x} A(t)v_i^{(2)}(u;\theta_0) + \gamma_0 A(t)e^{\beta_0^T x}\{v_i^{(1)}(u;\theta_0)\}^T x,$$

where $v_i^{(1)}(u;\theta)$ has the dimension k.

Condition B

$$\frac{1}{n} \sum_{i=1}^n \int_0^t c_{ix}^2(u;\theta_0)Y_i(u)e^{\theta_0^T z_i(u;\theta)}dA(u) \xrightarrow{P} \sigma_{11x}(t),$$

$$\frac{1}{n} \sum_{i=1}^n \int_0^t c_{ix}(u;\theta_0)v_i^{(2)}(u;\theta_0)Y_i(u)e^{\theta_0^T z_i(u;\theta_0)}dA(u) \xrightarrow{P} \sigma_{12x}(t),$$

$$\frac{1}{n}\sum_{i=1}^{n}\int_0^t (v_i^{(2)}(u;\theta_0))^2 Y_i(u)e^{\theta_0^T z_i(u;\theta_0)} dA(u) \xrightarrow{\mathbf{P}} \sigma_{22x}(t).$$

Set

$$\sigma_{21x} = \sigma_{12x}, \quad g_{1x}(t) = \frac{1}{\gamma_0} S_x^{1-\gamma_0}(t), \quad g_{2x}(t) = -\frac{1}{\gamma_0} S_x(t) \ln S_x(t).$$

Theorem 8.2. *Under Conditions A and B for all* $t \in [0, \tau]$

$$\sqrt{n}\{\hat{S}_x(t) - S_x(t)\} \xrightarrow{D} N(0, \sigma_x^2(t)),$$

where $\sigma_x^2(t) = \sum_{i=1}^2 \sum_{j=1}^2 g_{ix}(t)\sigma_{ijx}(t)g_{jx}(t).$

8.2.9 Graphical tests for the GPH1 models

If the explanatory variables x_1, x_2 are constant then under the GPH1 model (cf. (2.73)):
$$H(e^{-A_{x_2}(t)}, \gamma) = e^{\beta^T(x_2 - x_1)} H(e^{-A_{x_1}(t)}, \gamma).$$
Hence
$$\ln H(e^{-A_{x_2}(t)}, \gamma) - \ln H(e^{-A_{x_1}(t)}, \gamma) = \beta^T(x_2 - x_1).$$
Under different values of x the graphs of the time functions $\ln H(e^{-A_x(t)}, \gamma)$ are parallel. So, if x is discrete or categorical with s possible values and the data are stratified into s groups according to these values, then the Nelson-Aalen estimators $\hat{A}_j(t)$ $(j = 1, \cdots, s)$ may be constructed and the parameter γ estimated. Then the graphs of $\ln H(e^{-\hat{A}_x(t)}, \hat{\gamma})$ should be approximately parallel under the PH1 model.

So under the GPH_{GW} model the graphs of

$$\ln |(1 + \hat{A}_j(t))^{\hat{\gamma}} - 1|, \quad (\hat{\gamma} \neq 0),$$

$$\ln \ln(1 + \hat{A}_j(t)), \quad (\hat{\gamma} = 0);$$

under the GPH_{GLL} model the graphs of

$$\ln |1 - e^{-\hat{\gamma}\hat{A}_j(t)}|, \quad (\hat{\gamma} \neq 0), \ln \hat{A}_j(t), \quad (\hat{\gamma} = 0);$$

and for the IGF model the graphs of

$$\ln\{\hat{A}_j(t) + \frac{\hat{\gamma}}{2}\hat{A}_j^2(t)\}, \quad (\hat{\gamma} \neq 0), \ln \hat{A}_j(t), \quad (\hat{\gamma} = 0)$$

are approximately parallel.

8.3 Semiparametric FTR data analysis: intersecting hazards

8.3.1 Model and data

Let us consider the GPH2 model with constant explanatory variables:

$$\alpha_x(t) = u\{x, A_x(t)\} \alpha(t). \tag{8.26}$$

The baseline hazard rate function $\alpha(t)$ is supposed to be unknown. A cross-effect of hazard rates is obtained under the following parametrizations:

1) CRE1 model:

$$\alpha_x(t) = e^{\beta^T x}(1 + A_x(t))^{\gamma^T x + 1}\,\alpha(t) \tag{8.27}$$

or in terms of the baseline cumulative hazard

$$\alpha_x(t) = e^{\beta^T x}(1 - \gamma^T x e^{\beta^T x} A(t))^{-1-1/\gamma^T x}\,\alpha(t),$$

$$A_x(t) = (1 - \gamma^T x e^{\beta^T x} A(t))^{-1/\gamma^T x} - 1, \tag{8.28}$$

where $x = (x_1, \cdots, x_m)^T$, $\beta = (\beta_1, \cdots, \beta_m)^T$, $\gamma = (\gamma_1, \cdots, \gamma_m)^T$.

2) CRE2 model:

$$\alpha_x(t) = e^{(\beta+\gamma)^T x}(A(t))^{e^{\gamma^T x}-1}\,\alpha(t), \quad A_x(t) = e^{(\beta+\gamma)^T x}\gamma^T x^{-1}(A(t))^{e^{\gamma^T x}}. \tag{8.29}$$

The data are the same as in Chapter 8.2.1.

Denote by S_i, α_i and A_i the survival, hazard rate, and cumulative hazard functions under $x^{(i)}(\cdot)$.

8.3.2 Semiparametric estimation of the parameters θ

Semiparametric estimation for the GPH2 models using is given in Bagdon-avičius and Nikulin (2000d), for the second model by Hsieh (2000), Ebrahimi (1998).

If α is completely known then similarly as in the case of the GPH1 models we obtain that the parametric maximum likelihood estimator of the parameter $\theta = (\beta_1, \cdots, \beta_m, \gamma_1, \cdots, \gamma_m)^T$ is obtained by solving the system of equations

$$U_j(\theta) = 0, \quad (j = 1, \cdots, 2m),$$

where

$$U_j(\theta) = \sum_{i=1}^{n} \int_0^\infty w_j^{(i)}(t, \theta, A)\{dN_i(u) - Y_i(u)\alpha_i(u, \theta)du\}; \tag{8.30}$$

for the CRE1 model:

$$w_j^{(i)}(t, \theta, A) = \frac{\partial}{\partial \beta_j} \log\{\alpha_i(t, \theta)\} = \frac{x_j^{(i)} e^{\beta^T x^{(i)}} A(t)}{1 - \gamma^T x^{(i)} e^{\beta^T x^{(i)}} A(t)},$$

$$w_{m+j}^{(i)}(t, \theta, A) = \frac{\partial}{\partial \gamma_j} \log\{\alpha_i(t, \theta)\} =$$

$$\frac{1}{\gamma^T x^{(i)}} \left\{ \frac{x_j^{(i)}}{\gamma^T x^{(i)}} \ln\{1 - \gamma^T x^{(i)} e^{\beta^T x^{(i)}} A(t)\} + w_j^{(i)}(t, \theta, A) - x_j^{(i)} \right\}$$

$$(j = 1, \cdots, m); \tag{8.31}$$

for the CRE2 model:

$$w_j^{(i)}(t, \theta, A) = x_j^{(i)},$$

$$w^{(i)}_{m+j}(t, \theta, A) = x^{(i)}_j \{1 + e^{\gamma^T x^{(i)}} \ln A(t)\} \quad (j = 1, \cdots, m). \tag{8.32}$$

For the first model set

$$B(x, \theta, A(t)) = (1 - \gamma^T x \, e^{\beta^T x} A(t))^{-1-1/(\gamma^T x)}, \tag{8.33}$$

and for the second

$$B(x, \theta, A(t)) = e^{\gamma^T x} \{A(t)\}^{e^{\gamma^T x} - 1}. \tag{8.34}$$

If α is unknown then the score functions U_j depend not only on θ but also on the unknown function A.

Similarly as for the GPH1 models the estimator $\tilde{A}(t, \theta)$ of $A(t)$ can be found recurrently:

$$\tilde{A}(t, \theta) = \int_0^t \frac{dN(u)}{S^{(0)}(u-, \theta, \tilde{A})}, \tag{8.35}$$

where

$$S^{(0)}(v, \theta, A) = \sum_{i=1}^n Y_i(v) \, e^{\beta^T x^{(i)}(v)} B(x^{(i)}, \theta, A(v)). \tag{8.36}$$

Replacing the function A by $\tilde{A}(t, \theta)$ in (8.30), the modified score functions $\tilde{U}_j(\theta)$, depending only on θ, are obtained:

$$\tilde{U}_j(\theta) = \sum_{i=1}^n \int_0^\infty \{w^{(i)}_j(u, \theta, \tilde{A}) - E_j(u, \theta, \tilde{A})\} \, dN_i(u), \tag{8.37}$$

where

$$E_j(v, \theta, A) = \frac{S_j^{(1)}(v, \theta, A)}{S^{(0)}(v, \theta, A)},$$

$$S_j^{(1)}(v, \theta, A) = \sum_{i=1}^n w^{(i)}_j(v, \theta, A) Y_i(v) \, e^{\beta^T x^{(i)}(v)} B(x^{(i)}, \theta, A(v)).$$

8.3.3 Algorithm for computing the score functions

Suppose at first that there is no *ex aequo*. Let $T_{(1)} < \ldots < T_{(r)}$ be observed and ordered failure times, $r \leq n$. Here (i) notes the index of the unit which failure is observed the ith.

For fixed θ the estimator $\tilde{A}(t; \theta)$ is calculated by the following recurrent formulas:

$$\tilde{A}(0; \theta) = 0, \quad \tilde{A}(T_{(1)}; \theta) = \left(\sum_{i=1}^n Y_i(T_{(1)}) e^{\beta^T x^{(i)}} \right)^{-1};$$

for $j = 1, \ldots, r - 1$

$$\tilde{A}(T_{(j+1)}; \theta) = \tilde{A}(T_{(j)}; \theta) + \left(\sum_{i=1}^n Y_i(T_{(j+1)}) e^{\beta^T x^{(i)}} B(x^{(i)}, \theta, \tilde{A}(T_{(j)}, \theta)) \right)^{-1}, \tag{8.38}$$

where $T_{(0)} = 0$. So the score function is calculated simply:

$$\tilde{U}_j(\theta) = \sum_{i=1}^{r} \{w_j^{((i))}(T_{(i)}, \theta, \tilde{A}) - E_j(T_{(i)}, \theta, \tilde{A})\}. \qquad (8.39)$$

The initial estimator for θ is obtained approximating the baseline hazard rate by a piecewise constant function

$$\alpha(t, \mu) = \sum_{i=1}^{r} e^{\mu_i} \mathbf{1}_{(a_{i-1}, a_i]}(t),$$

where $0 = a_0 < a_1 < \cdots < a_r = \infty$ are given constants and $\mu = (\mu_1, \cdots, \mu_r)^T$ are unknown parameters.

Then the parametric model (corresponding to the CRE1 model)

$$\alpha_x(t) = e^{\beta^T x} \left(1 - \gamma^T x e^{\beta^T x} \int_0^t \sum_{i=1}^{r} \mathbf{1}_{(a_{i-1}, a_i]}(s) e^{\mu_i} ds\right)^{-1-1/\gamma^T x} \times$$

$$\sum_{i=1}^{r} e^{\mu_i} \mathbf{1}_{(a_{i-1}, a_i]}(t), \qquad (8.40)$$

or (corresponding to the CRE2 model)

$$\alpha_x(t) = e^{(\beta+\gamma)^T x} \left(\int_0^t \sum_{i=1}^{r} \mathbf{1}_{(a_{i-1}, a_i]}(s) e^{\mu_i} ds\right)^{e^{\gamma^T x}-1} \sum_{i=1}^{r} e^{\mu_i} \mathbf{1}_{(a_{i-1}, a_i]}(t), \qquad (8.41)$$

can be used to obtain the initial estimator of θ.

The initial estimator $\tilde{A}^{(0)}(t, \theta^{(0)})$ is obtained recurrently using (8.38). The estimator $\hat{\theta}^{(1)}$ is obtained by solving the equations $\tilde{U}_j(\theta) = 0$, where $\tilde{U}_j(\theta)$ are given by (8.39) with $\tilde{A}^{(0)}(t, \theta^{(0)})$ replacing $\tilde{A}(t, \theta)$. And so on.

8.3.4 Case of ex aequo

Suppose that *ex aequo* are possible. Let $T_1^* < \ldots < T_r^*$ be observed and ordered distinct failure times, $r \leq n$. Note by d_i the number of failures at the moment T_i^*. Let $(i_1), \cdots, (i_{d_i})$ be the indices of the units failed at T_i^*.

The estimator $\tilde{A}(t; \theta)$ is calculated by the following recurrent formulas :

$$\tilde{A}(0; \theta) = 0, \quad \tilde{A}(T_1^*; \theta) = d_1 \left(\sum_{i=1}^{n} Y_i(T_1^*) e^{\beta^T x^{(i)}}\right)^{-1} ;$$

for $j = 1, ..., r - 1$

$$\tilde{A}(T_{j+1}^*; \theta) = \tilde{A}(T_j^*; \theta) + d_{j+1} \left(\sum_{i=1}^{n} Y_i(T_{j+1}^*) e^{\beta^T x^{(i)}} B(x^{(i)}, \theta, \tilde{A}(T_j^*; \theta))\right)^{-1}. \qquad (8.42)$$

So the score function is modified:

$$\tilde{U}_j(\theta) = \sum_{i=1}^{r} \sum_{s=1}^{k_i} \{w_j^{((i_s))}(T_i^*, \theta, \tilde{A}) - E_j(T_i^*, \theta, \tilde{A})\}. \qquad (8.43)$$

8.3.5 Estimators of the main reliability characteristics

Suppose that $x(\cdot)$ is an arbitrary explanatory variable which may be different from $x^{(i)}(\cdot), (i = 1, \cdots, n)$.

Estimator of the survival function $S_{x(\cdot)}(t)$:

The estimator of the survival function $S_{x(\cdot)}(t)$ for the first model is

$$\hat{S}_{x(\cdot)}(t) = \exp\left\{1 - (1 - \hat{\gamma}^T x \, e^{\hat{\beta}^T x} \, \tilde{A}(t, \hat{\theta}))^{-1/(\hat{\gamma}^T x)}\right\}; \qquad (8.44)$$

for the second model

$$\hat{S}_{x(\cdot)}(t) = \exp\left\{-e^{\hat{\beta}^T x} (\tilde{A}(t, \hat{\theta}))^{e^{\hat{\gamma}^T x}}\right\}. \qquad (8.45)$$

Estimator of the p-quantile $t_p(x(\cdot))$

$$\hat{t}_p(x(\cdot)) = \sup\{t : \hat{S}_{x(\cdot)}(t) \geq 1 - p\}. \qquad (8.46)$$

Estimator of the mean failure time $m(x(\cdot))$:

$$\hat{m}(x) = \sum_{j:\delta_j=1} X_j\{\hat{S}_{x(\cdot)}(X_j-) - \hat{S}_{x(\cdot)}(X_j)\} \qquad (8.47)$$

The estimator of the mean may be underestimated if the last X_i is the censoring time.

   Asymptotic properties of the estimators are given by Bagdonavičius and Nikulin (2000d). See also, Ciampi and Etezadi-Amoli (1985), Hyde (1977), Nielsen, Gill, Andersen and Sorensen (1992), Vonta (2000).

CHAPTER 9

Changing scale and shape model

9.1 Parametric FTR data analysis

Let us consider the CHSS model in the form:

$$S_{x(\cdot)}(t) = S_0 \left(\int_0^t e^{\beta^T x(u)} u^{e^{\gamma^T x(u)}-1} du \right), \qquad (9.1)$$

where $\beta = (\beta_0, \cdots, \beta_m)^T$, $\gamma = (\gamma_0, \cdots, \gamma_m)^T$, $x(\cdot) = (x_0(\cdot), \cdots, x_m(\cdot))^T$, $x_0(t) \equiv 1$, and S_0 belongs to a specified scale-shape class of survival functions:

$$S_0(t) = G_0 \left\{ (t/\eta)^\nu \right\} \quad (\eta, \nu > 0).$$

The parameter η can be included in the coefficient β_0, so suppose that

$$S_0(t; \sigma) = G_0(t^{1/\sigma}), \quad \sigma = 1/\nu.$$

Suppose that n units are observed. The ith unit is tested under the explanatory variable $x^{(i)}(\cdot)$.

The data are supposed to be right censored.

Set $S_i = S_{x^{(i)}}$. Under the model (9.1)

$$S_i(t; \beta, \gamma, \sigma) = G_0 \left\{ \left(\int_0^t e^{\beta^T x^{(i)}(u)} u^{e^{\gamma^T x^{(i)}(u)}-1} du \right)^{1/\sigma} \right\}. \qquad (9.2)$$

Set

$$G(u) = G_0(e^u), \quad u \in \mathbf{R}, \quad g(u) = -G'(u), \quad h(u) = g(u)/G(u),$$

$$\theta = (\beta^T, \gamma^T, \sigma)^T, \quad f_i(t, \theta) = \int_0^t e^{\beta^T x^{(i)}(u)} u^{e^{\gamma^T x^{(i)}(u)}-1} du.$$

The likelihood function is

$$L(\theta) = \prod_{i=1}^n \left\{ \frac{1}{\sigma} e^{\beta^T x^{(i)}(X_i)} X_i^{e^{\gamma^T x^{(i)}(X_i)}-1} \left(f_i(X_i, \theta) \right)^{-1} \times \right.$$

$$\left. h \left(\frac{1}{\sigma} \ln(f_i(X_i, \theta)) \right) \right\}^{\delta_i} G \left(\frac{1}{\sigma} \ln(f_i(X_i, \theta)) \right). \qquad (9.3)$$

Denote by $\hat{\theta}$ the maximum likelihood estimator of the parameter θ.

Estimator of the survival function $S_{x(\cdot)}(t)$:

$$\hat{S}_{x(\cdot)}(t) = G_0 \left\{ \left(\int_0^t e^{\hat{\beta}^T x(u)} u^{e^{\hat{\gamma}^T x(u)} - 1} du \right)^{1/\hat{\sigma}} \right\}. \tag{9.4}$$

Full FTR data analysis can be done similarly as for the AFT model.

We shall follow from here Bagdonavičius and Nikulin (1999a, 1999b, 2000)

9.2 Semiparametric FTR data analysis

Let us consider the CHSS model in the form:

$$S_{x(\cdot)}(t) = S_0 \left(\int_0^t e^{\beta^T x(u)} u^{e^{\gamma^T x(u)} - 1} du \right), \tag{9.5}$$

where $\beta = (\beta_1, \cdots, \beta_m)^T$, $\gamma = (\gamma_1, \cdots, \gamma_m)^T$, $x(\cdot) = (x_1(\cdot), \cdots, x_m(\cdot))^T$, and suppose that the baseline survival function S_0 is completely unknown. If $\gamma = 0$, we have the AFT model.

Suppose that the data is as in the previous section. As in Chapter 6, it can be presented in the form

$$(X_1, \delta_1), \cdots, (X_n, \delta_n).$$

or

$$(N_1(t), Y_1(t), t \geq 0), \cdots, (N_n(t), Y_n(t), t \geq 0).$$

Denote by S_i and α_i the survival and the hazard rate functions under $x^{(i)}(\cdot)$. The hazard rates are:

$$\alpha_i(t; \theta) = \alpha_0 \left\{ \int_0^t e^{\beta^T x^{(i)}(u)} u^{e^{\gamma^T x^{(i)}(u)} - 1} du \right\} e^{\beta^T x^{(i)}(t)} t^{e^{\gamma^T x^{(i)}(t)} - 1};$$

here $\alpha_0 = -S_0'/S_0$ is the baseline hazard function and $\theta = (\beta^T, \gamma^T)^T$. If $x^{(i)}$ is constant then

$$\alpha_i(t; \theta) = \alpha_0 \left\{ e^{(\beta - \gamma)^T x^{(i)}} t^{e^{\gamma^T x^{(i)}}} \right\} e^{\beta^T x^{(i)}} t^{e^{\gamma^T x^{(i)}} - 1};$$

9.2.1 Semiparametric estimation of the regression parameters

If S_0 is specified then the maximum likelihood estimator of the parameter θ is obtained by solving the system of equations

$$U(\theta) = 0,$$

where

$$U(\theta) = \sum_{i=1}^n \int_0^\infty \frac{\partial}{\partial \theta} \log\{\alpha_i(u, \theta)\}\{dN_i(u) - Y_i(u)\,dA_0\{f_i(u, \theta)\}\},$$

$$f_i(t, \theta) = \int_0^t e^{\beta^T x^{(i)}(u)} u^{e^{\gamma^T x^{(i)}(u)} - 1} du, \quad A_0(t) = -\ln\{S_0(t)\}.$$

Note that

$$\frac{\partial}{\partial \beta} \log\{\alpha_i(t, \theta)\} =$$

$$x^{(i)}(t) + \frac{\alpha_0'(f_i(t, \theta))}{\alpha_0(f_i(t, \theta))} \int_0^t x^{(i)}(u) e^{\beta^T x^{(i)}(u)} u^{e^{\gamma^T x^{(i)}(u)} - 1} du,$$

$$\frac{\partial}{\partial \gamma} \log\{\alpha_i(t, \theta)\} =$$

$$x^{(i)}(t) e^{\gamma^T x^{(i)}(t)} \ln t + \frac{\alpha_0'(f_i(t, \theta))}{\alpha_0(f_i(t, \theta))} \int_0^t x^{(i)}(u) e^{(\beta+\gamma)^T x^{(i)}(u)} u^{e^{\gamma^T x^{(i)}(u)} - 1} du.$$

If S_0 is unknown then the score function U depends not only on θ but also on unknown functions A_0, α_0 and α_0'.

The idea of semiparametric estimation of θ is the same as in the case of the AFT model: to replace in the expression of $U(\theta)$ the unknown baseline cumulative hazard function A_0 by its good estimator (still depending on θ), and the weight functions by suitable functions which do not depend on the unknown baseline functions α_0 and α_0'. In such a way the obtained modified score function does not contain unknown infinite-dimensional parameters A_0, α_0, and α_0' and contains only the finite-dimensional parameter θ.

The optimal weights depend on the derivative of the baseline hazard rate which estimation is complicated when the law is unknown. So simplest weights $x_j^{(i)}(t)$ (in the first m score functions) and $x_j^{(i)}(t) e^{\gamma^T x^{(i)}(t)} \ln t$ (in the last m score functions) may be chosen. They are optimal when the baseline distribution is exponential.

The idea of the estimator $\tilde{A}_0(t, \beta)$ construction is the same as in the case of the AFT model: A_0 is estimated by the Nelson-Aalen estimator

$$\tilde{A}_0(t, \theta) = \sum_{i=1}^n \int_0^t \frac{dN_i(g_i(u, \theta))}{\sum_{l=1}^n Y_l(g_l(u, \theta))},$$

where $g_i(u, \theta)$ is the function inverse to $f_i(u, \theta)$ with respect to the first argument.

Replacing the function A_0 by $\tilde{A}_0(t, \beta)$ and using the above mentioned weights, modified score function $\tilde{U}(\theta)$, depending only on θ, is obtained:

$$\tilde{U}(\theta) = (\tilde{U}_1^T(\theta), \tilde{U}_2^T(\theta))^T, \tag{9.6}$$

where

$$\tilde{U}_1(\theta) = \sum_{i=1}^n \int_0^\infty \{x^{(i)}(u) - \bar{x}_1(f_i(u, \theta), \theta)\} dN_i(u),$$

$$\tilde{U}_2(\theta) = \sum_{i=1}^n \int_0^\infty \{x^{(i)}(u) e^{\gamma^T x^{(i)}(u)} \ln u - \bar{x}_2(f_i(u, \theta), \theta)\} dN_i(u),$$

$$\bar{x}_1(v, \theta) = \frac{\sum_{j=1}^n x^{(j)}(g_j(v, \theta)) Y_j(g_j(v, \theta))}{\sum_{j=1}^n Y_j(g_j(v, \theta))},$$

$$\bar{x}_2(v,\theta) = \frac{\sum_{j=1}^n x^{(j)}(g_j(v,\theta))e^{\gamma^T x^{(j)}(g_j(v,\theta))}\ln g_j(v,\theta)\,Y_j(g_j(v,\theta))}{\sum_{j=1}^n Y_j(g_j(v,\theta))}.$$

The values of the random vector $\tilde{U}(\theta)$ are dispersed around zero. An estimator of the parameter θ is obtained by minimizing the distance of $\tilde{U}(\theta)$ from zero.

9.2.2 Estimators of the main reliability characteristics

Suppose that $x(\cdot)$ is an arbitrary explanatory variable which may be different from $x^{(i)}(\cdot), (i=1,\cdots,n)$. Set

$$f_{x(\cdot)}(t,\theta) = \int_0^t e^{\beta^T x(u)} u^{e^{\gamma^T x(u)}-1}du.$$

Estimator of the survival function $S_{x(\cdot)}(t)$:

Similarly as in the case of AFT model:

$$\hat{S}_{x(\cdot)}(t) = \prod_{j:\delta_j=1,f_j(X_j,\hat\theta)\le f_{x(\cdot)}(t,\hat\theta)}\left(1 - \frac{1}{\sum_{l=1}^n 1_{\{f_l(X_l,\hat\theta)\ge f_j(X_j,\hat\theta)\}}}\right). \tag{9.7}$$

Estimator of the p-quantile $t_p(x(\cdot))$

$$\hat{t}_p(x(\cdot)) = \sup\{t: \hat{S}_{x(\cdot)}(t) \ge 1-p\}. \tag{9.8}$$

Estimator of the mean failure time $m(x(\cdot))$:

$$\hat{m}(x(\cdot)) = -\int_0^\infty u\,d\hat{S}_{x(\cdot)}(u) = \sum_{j=1}^n \delta_j\, g_{x(\cdot)}(f_j(X_j,\hat\theta),\hat\theta)\times$$
$$\{\tilde{S}_0(f_j(X_j,\hat\theta)-,\hat\theta) - \tilde{S}_0(f_j(X_j,\hat\theta),\hat\theta)\}, \tag{9.9}$$

where $g_{x(\cdot)}(t,\theta)$ is the function inverse to $f_{x(\cdot)}(t,\theta)$ with respect to the first argument, $X_0=0$.

9.3 Semiparametric estimation in ALT

9.3.1 First plan of experiments

The first plan is the particular case of the plan of experiments considered in 9.2.1., when $x^{(i)}$ have the form (5.135). So the score function has the form (9.4).

Estimators of the main reliability characteristics under the usual stress x_0 are defined by (9.5)-(9.7), taking x_0 instead of $x(\cdot)$ in all formulas.

9.3.2 Case of unspecified functions r and α

Suppose that the CHSS model holds on a set of stresses E:

$$S_{x(\cdot)}(t) = S_0 \left(\int_0^t r\{x(u)\} u^{\alpha\{x(u)\}-1} du \right),$$

and the functions r, α, and $S_0(t)$ are completely unknown. Then the third plan of experiments considered in 6.2.2 may be considered.

Set

$$S_0 = S_{x_0}, \quad S_1 = S_{x_1}, \quad S_2 = S_{x_2(\cdot)}, \quad A_i = -\ln S_i \quad (i = 1, 2, 0),$$

$$\theta = r(x_1)/\alpha(x_1), \quad \alpha = \alpha(x_1), \quad S_0^*(t) = S_{x_0}(\theta t).$$

Under the CHSS model

$$S_{x_1}(t) = S_0^*(t^\alpha),$$

and

$$S_{x_2(\cdot)}(t) = \begin{cases} S_0^*(t^\alpha), & 0 \le t < t_1, \\ S_0^*(t_1^\alpha + \frac{t-t_1}{\theta}), & t \ge t_1, \end{cases}$$

which implies that

$$\ln \alpha_{x_1}(t) = \ln \alpha_0^*(t^\alpha) + \ln \alpha + (\alpha - 1) \ln t,$$

and

$$\ln \alpha_{x_2}(t) = \begin{cases} \ln \alpha_0^*(t^\alpha) + \ln \alpha + (\alpha - 1)\ln t, & 0 \le t < t_1, \\ \ln \alpha_0^*(t_1^\alpha + \frac{t-t_1}{\theta}) - \ln \theta, & t \ge t_1, \end{cases}$$

Suppose at first that S^* is known. Then

$$\frac{\partial \ln \alpha_{x_1}(t)}{\partial \alpha} = (\ln \alpha_0^*)'(t^\alpha)t^\alpha \ln t + \frac{1}{\alpha} + \ln t,$$

$$\frac{\partial \ln \alpha_{x_1}(t)}{\partial \theta} = 0,$$

$$\frac{\partial \ln \alpha_{x_2}(t)}{\partial \alpha} = \begin{cases} (\ln \alpha_0^*)'(t^\alpha)t^\alpha \ln t + \frac{1}{\alpha} + \ln t, & 0 \le t < t_1, \\ (\ln \alpha_0^*)'(t_1^\alpha + \frac{t-t_1}{\theta})t_1^\alpha \ln t_1, & t \ge t_1, \end{cases}$$

$$\frac{\partial \ln \alpha_{x_2}(t)}{\partial \theta} = \begin{cases} 0, & 0 \le t < t_1, \\ (\ln \alpha_0^*)'(t^\alpha)(-(t-t_1)/\theta^2) - \theta^{-1}, & t \ge t_1, \end{cases}$$

Similarly as in 6.2.2 for the AFT model we obtain score functions (with optimal weights for $S^*(t) = \exp\{-t\}$):

$$\tilde{U}_1(\alpha, \theta) =$$

$$\int_0^\infty (\frac{1}{\alpha} + \ln u) \left\{ dN_1(u) - Y_1(u) \frac{dN_1(u) + dN_2(g_2(f_1(u; \alpha, \theta); \alpha, \theta))}{Y_1(u) + Y_2(g_2(f_1(u; \alpha, \theta); \alpha, \theta))} \right\} +$$

$$\int_0^{t_1} (\frac{1}{\alpha} + \ln u) \left\{ dN_2(u) - Y_2(u) \frac{dN_1(g_1(f_2(u; \alpha, \theta); \alpha, \theta)) + dN_2(u)}{Y_1(g_1(f_2(u; \alpha, \theta); \alpha, \theta)) + Y_2(u)} \right\} =$$

$$\frac{1}{\alpha} \int_{t_1}^{t_2} \left\{ 1 + \ln(t_1^\alpha + \frac{v - t_1}{\theta}) \right\} \times$$

$$\frac{Y_2(v)dN_1((t_1^\alpha + \frac{v-t_1}{\theta})^{1/\alpha}) - Y_1((t_1^\alpha + \frac{v-t_1}{\theta})^{1/\alpha}))dN_2(v)}{Y_1((t_1^\alpha + \frac{v-t_1}{\theta})^{1/\alpha})) + Y_2(v)}.$$

By the same way we obtain

$$\tilde{U}_2(\alpha,\theta) = -\frac{1}{\theta}\int_{t_1}^{t_2} \frac{Y_1(g_1 \circ f_2(u))dN_2(u) - Y_2(u)dN_1(g_1 \circ f_2(u))}{Y_1(g_1 \circ f_2(u)) + Y_2(u)}.$$

Hence the two modified score functions may be considered:

$$\hat{U}_1(\alpha,\theta) =$$

$$\int_{t_1}^{t_2} \ln(t_1^\alpha + \frac{u-t_1}{\theta})\frac{Y_2(u)dN_1((t_1^\alpha + \frac{u-t_1}{\theta})^{1/\alpha}) - Y_1((t_1^\alpha + \frac{u-t_1}{\theta})^{1/\alpha}))dN_2(u)}{Y_1((t_1^\alpha + \frac{u-t_1}{\theta})^{1/\alpha})) + Y_2(u)}$$

and

$$\hat{U}_2(\alpha,\theta) = \int_{t_1}^{t_2} \frac{Y_1(t_1^\alpha + \frac{v-t_1}{\theta})dN_2(u) - Y_2(u)dN_1(t_1^\alpha + \frac{v-t_1}{\theta})}{Y_1(t_1^\alpha + \frac{v-t_1}{\theta}) + Y_2(u)}.$$

The function $\hat{U}_2(\alpha,\theta)$ is increasing with respect to θ under any fixed α and $\lim_{\theta\to+0}\hat{U}_2(\alpha,\theta) < 0$, $\lim_{\theta\to+\infty}\hat{U}_2(\alpha,\theta) > 0$ a.s. Define

$$\tilde{\theta}(\alpha) = \sup\{\theta : \hat{U}(\alpha,\theta) \le 0\}. \tag{9.10}$$

The estimator α is defined as the minimizer of $|\hat{U}_1(\alpha,\tilde{\theta}(\alpha))|$.

The survival function $S_{x_0}(t)$ is estimated by:

$$\hat{S}_{x_0}(t) = \prod_{(i,j)\in B(t)} \left(1 - \frac{1}{Y_1(T_{1i}) + Y_2(t_1 \wedge T_{1i} + \hat{\theta}((T_{1i}^{\hat\alpha} - t_1^{\hat\alpha}) \vee 0))}\right) \times$$

$$\left(1 - \frac{1}{Y_2(T_{2j}) + Y_1(((t_1 \wedge T_{2j})^{\hat\alpha} + \frac{T_{2j}-t_1}{\theta} \vee 0)^{1/\hat\alpha})}\right), \tag{9.11}$$

where

$$B(t) = \{(i,j)|\ \hat{\theta}T_{1i}^{\hat\alpha} \le t,\ \hat{\theta}(T_{2j} \wedge t_1)^{\hat\alpha} + (T_{2j} - t_1) \vee 0 \le t\}.$$

The quantile $t_p(x_0)$ is estimated by

$$\hat{t}_p(x_0) = \sup\{t : \hat{S}_{x_0}(t) \ge 1 - p\}$$

and the mean time-to-failure $m(x_0)$ by

$$\hat{m}(x_0) = \int_0^\infty \hat{S}_{x_0}(u)\,du.$$

CHAPTER 10

GAH and GAMH models

10.1 GAH model

Let us consider the generalized additive hazards model:

$$S_{x(\cdot)}(t) = G\{H(S_0(t)) + \int_0^t \gamma^T x(\tau)d\tau\}.$$

with a specified G, $H = G^{-1}$ and an unknown baseline survival function S_0. In the particular case $G(t) = e^{-t}, t \geq 0$, we have the additive hazards model.

Suppose that n units are observed and the ith unit is tested under the explanatory variable $x^{(i)}(\cdot) = (x_1^{(i)}(\cdot), ..., x_m^{(i)}(\cdot))^T$.

Assume that the data are right censored and the multiplicative intensity model holds. The data can be presented in the form

$$(X_1, \delta_1, x^{(1)}(\cdot)), \cdots, (X_n, \delta_n, x^{(n)}(\cdot)).$$

or

$$(N_1(t), Y_1(t), x^{(i)}(t), t \geq 0), \cdots, (N_n(t), Y_n(t), x^{(i)}(t), t \geq 0).$$

Denote by S_i and α_i the survival and the hazard rate functions under $x^{(i)}(\cdot)$ and set $\alpha = -G'/G$, $\psi = \alpha \circ H$.

The Doob-Meyer decomposition of $N = \sum_i N_i$ implies that

$$dN(t) = dM(t) + \sum_{i=1}^n \psi(S_i(t))Y_i(t)\{dH(S_0(t)) + \gamma^T x^{(i)}(t)dt\}$$

and

$$\int_0^t \frac{J(u)(dN(u) - S_*^{(0)}(u,\gamma)du)}{S^{(0)}(u)} = \int_0^t J(u)dH(S_0(u)) + \int_0^t \frac{J(u)dM(u)}{S^{(0)}(u)}, \tag{10.1}$$

where

$$J(u) = \mathbf{1}_{\{Y(u)>0\}}, \quad S^{(0)}(u,\gamma) = \sum_{i=1}^n Y_i(u)\psi(S_i(u)),$$

$$S_*^{(0)}(u,\gamma) = \sum_{i=1}^n Y_i(u)\psi(S_i(u))\gamma^T x^{(i)}(u).$$

If $Y(t) > 0$, then

$$\int_0^t J(u)dH(S_0(u)) = H(S_0(t)). \tag{10.2}$$

The equalities (10.1) and (10.2) imply that a reasonable estimator $\hat{H}_0(t, \gamma)$ for $H_0(t) = H(S_0(t))$ (still depending on γ) is determined by the recurrent equation

$$\hat{H}_0(t, \gamma) = \int_0^t J(u) \frac{dN(u) - \tilde{S}_*^{(0)}(u, \gamma) du}{\tilde{S}^{(0)}(u, \gamma)},$$

where

$$\tilde{S}^{(0)}(u, \gamma) = \sum_{i=1}^n Y_i(u) \alpha(\hat{H}_i(u, \gamma)), \quad \tilde{S}_*^{(0)}(u, \gamma) = \sum_{i=1}^n Y_i(u) \alpha(\hat{H}_i(u, \gamma)) \gamma^T x^{(i)}(u),$$

$$\hat{H}_i(u, \gamma) = \hat{H}_0(u-, \gamma) + \int_0^u \gamma^T x^{(i)}(v) dv.$$

Set $\tau^* = \sup\{t : Y(t) > 0\}$. Using the fact that

$$dM_i(u) = dN_i(u) - Y_i(u) \alpha(H_i(u, \gamma)) dH_i(u, \gamma)$$

with

$$H_i(u, \gamma) = H_0(u) + \int_0^u \gamma^T x^{(i)}(v) dv,$$

Bagdonavičius and Nikulin (1997a) proposed to estimate the parameter γ by solving the estimating equations

$$U(\gamma, \tau) = 0,$$

where the estimating function is given by the next formula

$$U(t, \gamma) = \sum_{i=1}^n \int_0^t x^{(i)}(u) \{dN_i(u) - Y_i(u) \alpha(\hat{H}_i(u, \gamma)) d\hat{H}_i(u, \gamma)\} =$$

$$\sum_{i=1}^n \int_0^t \{x^{(i)}(u) - \tilde{E}(u, \gamma)\} \left\{ dN_i(u) - Y_i(u) \alpha(\hat{H}_i(u, \gamma)) \gamma^T x^{(i)}(u) du \right\},$$

$$\tilde{E}(u, \gamma) = \frac{\tilde{S}^{(1)}(u, \gamma)}{\tilde{S}^{(0)}(u, \gamma)}, \quad \tilde{S}^{(1)}(\gamma, u) = \sum_{i=1}^n x^{(i)}(u) Y_i(u) \alpha(\hat{H}_i(u, \gamma)).$$

These equations generalize the estimating equations of Lin and Ying (1994) for the additive hazards model (taking $\alpha(p) \equiv 1$).

If we denote by $\hat{\gamma}$ the estimator of γ then the estimator of the survival function $S_{x(\cdot)}(t)$ is

$$\hat{S}_{x(\cdot)}(t) = G \left\{ \tilde{H}_0(t, \hat{\gamma}) + \int_0^t \hat{\gamma}^T x(u) du \right\}.$$

Asymptotic properties of the estimators are given in Bagdonavičius and Nikulin (1997b).

If several groups of units (as in ALT) are tested, then implicit estimator of the parameters γ (see Bagdonavičius and Nikulin (1995)) can be obtained. Let us consider them.

Suppose that n_i units are tested under the stress $x^{(i)}(\cdot)$ $(i = 1, 2, ..., k)$, $n = \sum_{i=1}^{k} n_i$. The data can be presented in the form

$$(X_{ij}, \delta_{ij}, x^{(i)}(\cdot))$$

or

$$(N_{ij}(t), Y_{ij}(t), x^{(i)}(t), t \geq 0) \quad (i = 1, \cdots, k; j = 1, \cdots, n_i).$$

Set

$$N_i(t) = \sum_{j=1}^{n_i} N_{ij}(t), \quad N(t) = \sum_{i=1}^{k} N_i(t), \quad Y_i(t) = \sum_{j=1}^{n_i} Y_{ij}(t), \quad Y(t) = \sum_{i=1}^{k} Y_i(t).$$

The functions $S^{(0)}$ and $S_*^{(0)}$ in the equality (10.1) can be written

$$S^{(0)}(u) = \sum_{i=1}^{k} Y_i(u)\psi(S_i(u)),$$

$$S_*^{(0)}(u, \gamma) = \sum_{i=1}^{k} Y_i(u)\gamma^T x^{(i)}(u)\psi(S_i(u)),$$

and the survival functions S_i can be estimated by the Kaplan-Meier estimators

$$\hat{S}_i(t) = \prod_{u \leq t} \left(1 - \frac{\Delta N_i(u)}{Y_i(u)}\right)$$

where $\Delta N_i(u) = N_i(u) - N_i(u - 0)$. So the equalities (10.1) and (10.2) imply that a reasonable estimator for $H_0(t))$ (still depending on γ) is

$$\hat{H}_0(t)) = \int_0^t J(u)\frac{dN(u) - \tilde{S}_*^{(0)}(u, \gamma)du}{\tilde{S}^{(0)}(u)},$$

where

$$\tilde{S}^{(0)}(u) = \sum_{i=1}^{k} Y_i(u)\psi(\hat{S}_i(u)),$$

$$\tilde{S}_*^{(0)}(u, \gamma) = \sum_{i=1}^{k} Y_i(u)\gamma^T x^{(i)}(u)\psi(\hat{S}_i(u)).$$

The parameter γ is estimated by solving the estimating equations

$$U(\tau, \gamma) = 0,$$

where

$$U(t, \gamma) =$$

$$\sum_{i=1}^{k} \int_0^t x_i(u) \left\{dN_i(u) - Y_i(u)\psi(\hat{S}_i(u-)) \left(dH(\hat{S}_0(u)) + \gamma^T x^{(i)}(u)du\right)\right\} =$$

$$\sum_{i=1}^{k} \int_0^t \{x_i(u) - \tilde{E}(u)\} \left\{dN_i(u) - Y_i(u)\psi(\hat{S}_i(u-))\gamma^T x^{(i)}(u)du\right\},$$

where

$$\tilde{E}(u) = \frac{\tilde{S}^{(1)}(u)}{\tilde{S}^{(0)}(u)}, \quad \tilde{S}^{(1)}(u) = \sum_{i=1}^{k} x^{(i)}(u) Y_i(u) \psi(\hat{S}_i(u-)).$$

Denote by $\hat{\gamma}$ the estimator satisfying this equation. Note that it has an explicit form:

$$\hat{\gamma} = \left\{ \sum_{i=1}^{k} \int_0^{\tau} (x^{(i)}(u) - \tilde{E}(u))^{\otimes 2} Y_i(u) \psi(\hat{S}_i(u-)) du \right\}^{-1} \times$$

$$\left\{ \sum_{i=1}^{k} \int_0^{\tau} (x^{(i)}(u) - \tilde{E}(u)) dN_i(u) \right\}. \tag{10.3}$$

The estimator of the survival function under any $x(\cdot)$ is

$$\hat{S}_{x(\cdot)}(t) = G \left\{ \int_0^t J(u) \frac{dN(u) - \tilde{S}_*^{(0)}(u, \hat{\gamma}) du}{\tilde{S}^{(0)}(u)} + \int_0^t \hat{\gamma}^T x(u) du \right\}. \tag{10.4}$$

Asymptotic properties of the simplified estimators (10.3) and (10.4) were studied also by Bordes (1996).

10.2 GAMH model

Let us consider the generalized additive-multiplicative hazards model:

$$S_{x(\cdot)}(t) = G \left\{ \int_0^t e^{\beta^T z(u)} dH(S_0(u)) + \int_0^t \gamma^T w(u) du \right\}.$$

with specified G, $H = G^{-1}$ and an unknown baseline survival function S_0, here $\beta = (\beta_1, ..., \beta_p)^T$, $\gamma = (\gamma_1, ..., \gamma_s)^T$ $x^T(t) = (z^T(t), w^T(t))$. Set $\theta = (\beta^T, \gamma^T)^T$.

If $p = m$ or $s = m$, we have the GPH or the GAH model, respectively.

Suppose that n units are observed and the ith unit is tested under the explanatory variable $x^{(i)}(\cdot)$.

Assume that the data are right censored and the multiplicative intensity model holds.

Denote by S_i and α_i the survival and the hazard rate functions under $x^{(i)}(\cdot)$ and set $\alpha = -G'/G$, $\psi = \alpha \circ H$.

Similarly as in the case of the GAH model, Bagdonavičius and Nikulin (1997a) proposed to estimate the parameter θ by solving the estimating equations

$$U(\tau, \theta) = 0,$$

where

$$U(t, \theta) = \sum_{i=1}^{n} \int_0^t J(u) \{ x^{(i)}(u) - \tilde{E}(u, \theta) \} \left\{ dN_i(u) - Y_i(u) \alpha(\hat{H}_i(u, \theta)) \gamma^T w^{(i)}(u) du \right\},$$

$$\hat{H}_i(u, \theta) = \int_0^{u-} e^{\beta^T z^{(i)}(v)} d\hat{H}_0(v, \theta) + \int_0^u \gamma^T w^{(i)}(v) dv.$$

$$\hat{H}_0(t,\theta) = \int_0^t J(u)\frac{dN(u) - \tilde{S}_*^{(0)}(u,\theta)du}{\tilde{S}^{(0)}(u,\theta)},$$

$$\tilde{S}^{(0)}(u,\theta) = \sum_{i=1}^n Y_i(u)e^{\beta^T z^{(i)}(u)}\alpha(\hat{H}_i(u,\theta)),$$

$$\tilde{S}_*^{(0)}(u,\theta) = \sum_{i=1}^n Y_i(u)\alpha(\hat{H}_i(u,\theta))\gamma^T w^{(i)}(u),$$

$$\tilde{E}(u,\theta) = \frac{\tilde{S}^{(1)}(u,\theta)}{\tilde{S}^{(0)}(u,\theta)}, \quad \tilde{S}^{(1)}(u,\theta) = \sum_{i=1}^n x^{(i)}(u)Y_i(u)e^{\beta^T z^{(i)}(u)}\alpha(\hat{H}_i(u,\theta)).$$

These equations generalize the estimating equations Lin and Ying (1996) for the additive-multiplicative model (taking $\alpha(p) \equiv 1$).

The estimator of the survival function under any $x(\cdot) = (z^T(\cdot), w^T(\cdot))^T \in E$ is

$$\hat{S}_{x(\cdot)}(t) = G\left\{\int_0^t J(u)e^{\hat{\beta}^T z(u)}\frac{dN(u) - \tilde{S}_*^{(0)}(u,\hat{\theta})du}{\tilde{S}^{(0)}(u,\hat{\theta})} + \int_0^t \hat{\gamma}^T w(u)du\right\}.$$

Asymptotic properties of estimators are given by Bagdonavičius and Nikulin (1997a).

If several groups of units (as in ALT) are tested, then simpler estimators can be obtained (Bagdonavičius and Nikulin (1995)), solving the estimating equations $U(\tau,\theta) = 0$, where

$$U(t,\theta) = \sum_{i=1}^k \int_0^t J(u)\{x_i(u) - \tilde{E}(u,\beta)\}\left\{dN_i(u) - Y_i(u)\psi(\hat{S}_i(u-))\gamma^T w_i(u)du\right\}.$$

$$\tilde{E}(u,\beta) = \frac{\tilde{S}^{(1)}(u,\beta)}{\tilde{S}^{(0)}(u,\beta)}, \quad \tilde{S}^{(0)}(u,\beta) = \sum_{i=1}^k Y_i(u)e^{\beta^T z_i(u)}\psi(\hat{S}_i(u)),$$

$$\tilde{S}^{(1)}(u,\beta) = \sum_{i=1}^k x_i(u)Y_i(u)e^{\beta^T z_i(u)}\psi(\hat{S}_i(u-)).$$

The estimator of the survival function under $x(\cdot) \in E$ is

$$\hat{S}_{x(\cdot)}(t) = G\left\{\int_0^t J(u)e^{\hat{\beta}^T z(u)}\frac{dN(u) - \tilde{S}_*^{(0)}(u,\hat{\gamma})du}{\tilde{S}^{(0)}(u,\hat{\beta})} + \int_0^t \hat{\gamma}^T w(u)du\right\}.$$

The estimator $\hat{\gamma}$ can be written as explicit function of $\hat{\beta}$.

Asymptotic properties of estimators are given by Bagdonavičius and Nikulin (1998).

10.3 AAR model

Let us consider Aalens additive risk model:

$$\alpha_{x(\cdot)}(t) = x^T(t)\,\alpha(t)$$

with unknown baseline function $\alpha(\cdot) = (\alpha_1(\cdot), \cdots, \alpha_m(\cdot))^T$.

Suppose that n units are observed and the ith unit is tested under the explanatory variable $x^{(i)}(\cdot)$.

Assume that the data are right censored and the multiplicative intensity model holds.

Denote by S_i and α_i the survival and the hazard rate functions under $x^{(i)}(\cdot)$.

Let us consider a submodel

$$\alpha(t) = \alpha_0(t) + \eta\varphi(t), \qquad (10.5)$$

in which η is a one-dimensional parameter and φ, α_0 is a given m-vector of functions.

The score function obtained from the parametric likelihood function for the parameter η is

$$U(\eta) = \sum_{i=1}^{n} \int_0^\infty \frac{\partial}{\partial\eta} \log\alpha_i(t)(dN_i(t) - Y_i(t)\alpha_i(t)dt) =$$

$$\sum_{i=1}^{n} \int_0^\infty \frac{\varphi^T(t)x^{(i)}(t)}{\alpha_i(t)}(dN_i(t) - Y_i(t)(x^{(i)}(t))^T dA(t)),$$

where

$$A(t) = \int_0^t \alpha(u)du.$$

The weights $\varphi^T(t)x^{(i)}(t)(\alpha_i(t))^{-1}$ depend on the hazard rates $\alpha_i(u)$. If A is unknown and we want to estimate it, the estimator should be the same for all φ. Setting $U(\eta) = 0$ for all functions φ implies that for all t

$$\frac{x^{(i)}(t)}{\alpha_i(t)}(dN_i(t) - Y_i(t)(x^{(i)}(t))^T dA(t)) = 0,$$

which implies the estimator (still depending on α_i):

$$\tilde{A}(t) = \sum_{j=1}^{n} \int_0^t \left(\sum_{i=1}^{n} x^{(i)}(u)(x^{(i)}(t))^T(\alpha_i(u))^{-1}\right)^{-1} x^{(j)}(u)(\alpha_j(u))^{-1}dN_j(u).$$

$$(10.6)$$

Replacing $\alpha_i(u)$ by 1 in the expression of the estimator, we obtain the Aalen's ordinary least squares (OLS) estimator $\hat{A}_{ls}$.

To obtain efficient estimators of A the hazard rates α_i should be replaced in (10.6) by their estimators

$$\hat{\alpha}_i(t) = (x^{(i)}(t))^T\hat{\alpha}(t),$$

where

$$\hat{\alpha}(t) = \frac{1}{b}\int_O^\infty K\left(\frac{t-u}{b}\right)d\hat{A}_{ls}(u);$$

K is a left-continuous kernel function of bounded variation, having integral 1, support $(\varepsilon, 1]$ for some $0 < \varepsilon < 1$, and $b > 0$ is a bandwidth parameter.

Thus, we have the weighted least squares (WLS) estimator given by Huffer and McKeague (1991):

$$\hat{A}(t) = \sum_{j=1}^{n} \int_0^t \hat{H}_j(u) dN_j(u), \qquad (10.7)$$

where

$$\hat{H}_j(u) = (\sum_{i=1}^{n} x^{(i)}(u)(x^{(i)}(t))^T (\hat{\alpha}_i(u))^{-1})^{-1} x^{(j)}(u)(\hat{\alpha}_j(u))^{-1}. \qquad (10.8)$$

In practice the gain of efficiency when using the WHS estimator instead of OWS estimator is small.

The WHS estimator is not very sensitive to the choice of kernel function. Taking the kernel to be constant on $(\varepsilon, 1]$, the estimator $\hat{a}(t)$ is proportional to the sum of increments of $\hat{A}_{ls}$. The estimators of the cumulative hazard $A_{x(\cdot)}$ and the survival function $S_{x(\cdot)}$ are

$$\hat{A}_{x(\cdot)}(t) = \sum_{j=1}^{n} \int_0^t x^T(u) \hat{H}_j(u) dN_j(u) = \sum_{j:\delta_i=1, X_j \leq t} x^T(X_j) \hat{H}_j(X_j),$$

$$\hat{S}_{x(\cdot)}(t) = \pi_{0 \leq s \leq t}(1 - d\hat{A}_{x(\cdot)}(s)) = \prod_{j:\delta_i=1, X_j \leq t} \left(1 - x^T(X_j) \hat{H}_j(X_j)\right).$$

Asymptotic properties of the estimator $\hat{A}$ are given in Huffer and McKeague (1991), Andersen et al (1993), Section VII.4.2. See also McKeague and Utical (1991).

10.4 PPAR model

Let us consider the partly parametric additive risk model:

$$\alpha_x(t) = x_1^T \alpha(t) + \beta^T x_2, \qquad (10.9)$$

where x_1 and x_2 are q and p dimensional components of the explanatory variable x, $\alpha(t) = (\alpha_1, \cdots, \alpha_q)$, $\beta = (\beta_1, \cdots, \beta_p)^T$ are unknown, with unknown baseline function $\alpha(\cdot) = (\alpha_1(\cdot), \cdots, \alpha_q(\cdot))^T$.

Suppose that the data is as in the previous section. The estimation procedure given by McKeague and Sasieni (1994), is analogous as in the case of AAR model: under the submodel of (10.9) with α defined by (10.5) the maximum likelihood estimators of the parameters η and β verify the system of equations

$$U_1(\eta, \beta) = \sum_{i=1}^{n} \int_0^\infty \frac{\partial}{\partial \eta} \log \alpha_i(t)(dN_i(t) - Y_i(t)\alpha_i(t)dt) =$$

$$\sum_{i=1}^{n} \int_0^\infty \frac{\varphi^T(t)x_1^{(i)}}{\alpha_i(t)}(dN_i(t) - Y_i(t)(x_1^{(i)})^T dA(t) - \beta^T x_2 Y_i(t)dt) = 0,$$

$$U_2(\eta, \beta) = \sum_{i=1}^{n} \int_0^\infty \frac{\partial}{\partial \beta} \log \alpha_i(t)(dN_i(t) - Y_i(t)\alpha_i(t)dt =$$

$$\sum_{i=1}^{n} \int_0^\infty \frac{x_2^{(i)}}{\alpha_i(t)}(dN_i(t) - Y_i(t)(x_1^{(i)})^T dA(t) - \beta^T x_2^{(i)} Y_i(t)dt) = 0. \qquad (10.10)$$

If A is unknown and we want to estimate it, the estimator should be the same for all φ. Setting $U_1(\eta, \beta) = 0$ for all functions φ implies that for all t

$$\frac{x^{(i)}}{\alpha_i(t)}(dN_i(t) - Y_i(t)(x_1^{(i)})^T dA(t) - \beta^T x_2^{(i)} Y_i(t)dt) = 0,$$

which implies the estimator (still depending on α_i and β):

$$\tilde{A}(t) = \sum_{j=1}^{n} \int_0^t (\sum_{i=1}^{n} x_1^{(i)}(x_1^{(i)})^T Y_i(u)(\alpha_i(u))^{-1})^{-1} x_1^{(j)}(\alpha_j(u))^{-1} \times$$

$$(dN_j(u) - \beta^T x_2^{(j)} Y_j(u)du). \qquad (10.11)$$

Shortly,

$$\tilde{A}(t) = \int_0^t (A_1^T(u)W(u)A_1(u))^{-1}(A_1^T(u)W(u)dN(u) - A_1^T(u)W(u)A_2(u)\beta du),$$

$$(10.12)$$

where $W(t) = diag(\alpha_j(t))$ (diagonal matrix with elements $\alpha_j(t)$ on the diagonal), $A_1(t) = (x_1^{(1)}Y_1(t), \cdots, x_1^{(n)}Y_n(t))^T$, $A_2(t) = (x_2^{(1)}Y_1(t), \cdots, x_2^{(n)}Y_n(t))^T$, $N(t) = (N_1(t), \cdots, N_n(t))^T$.

Replacing A by $\tilde{A}$ in the equation (10.10) and solving with respect to β, we obtain the estimator (still depending on α_i) for β:

$$\tilde{\beta} = \left(\int_0^\infty A_2^T(u)H(u)A_2(u)du \right)^{-1} \int_0^\infty A_2^T(u)H(u)dN(u), \qquad (10.13)$$

where $H = W - WA_1(A_1^T W A_1)^{-1}A_1^T W$.

Replacing W by the identity matrix I yields an estimator $\hat{\beta}$. Replacing β in (10.12) by $\hat{\beta}$ and using I in place of W gives unbiased, consistent, and asymptotically Gaussian estimator of A.

As in the previous section, properties of estimators can be slightly improved defining estimators of β and A, taking consistent estimators of α_i in the expression of W (see McKeague and Sasieni (1994)). See also Gasser and Muller (1979).

The estimators of the cumulative hazard $A_{x(\cdot)}$ and the survival function $S_{x(\cdot)}$ are

$$\hat{A}_{x(\cdot)}(t) = \sum_{j=1}^{n} \int_0^t x_1^T d\hat{A}(u) + \hat{\beta}^T x_2 t,$$

$$\hat{S}_{x(\cdot)}(t) = \pi_{0 \leq s \leq t}(1 - d\hat{A}_{x(\cdot)}(s)).$$

Asymptotic properties of the estimator $\hat{A}$ are given in McKeague and Sasieni (1994).

CHAPTER 11

Estimation when a process of production is unstable

Suppose that a process of production is *unstable*, i.e. the reliability of units produced in nonintersecting time intervals $I_1 = (t_0, t_1], \cdots, I_q(t_{m-1}, t_m]$ is different: under the same stress conditions the survival functions of units, produced in the intervals I_i and I_j $(i \neq j)$, are different.

If the heredity hypothesis (Definition 2.13) is satisfied on E and sufficiently large usual and accelerated data are accumulated during a long period of observations, then good estimators of the functions $\rho(x^{(0)}, x^{(1)})$ (or $b(x^{(0)}, x^{(1)})$) can be obtained. The reliability of newly produced units under the usual stress x_0 can be estimated from accelerated life data under the stress $x^{(1)} > x^{(0)}$, using the estimators $\hat{\rho}(x^{(0)}, x^{(1)})$ or $\hat{b}(x^{(0)}, x^{(1)})$ and *without using the experiment under the normal stress*. We shall follow here Bagdonavičius and Nikulin (1997f).

11.1 Application of the AFT model

Suppose that for units produced in each of fixed time intervals I_i the AFT model holds.

If the function $r(x)$ and the survival function $S_{x^{(0)}}$ are completely unknown, the third plan of experiments (Chapter 5.6.3) may be used for units produced in the interval I_i.

Two groups of units are tested:

a) The first group of $n_1^{(i)}$ units under a constant accelerated stress $x^{(1)}$;

b) The second group of $n_2^{(i)}$ units under a step-stress: time $t_1^{(i)}$ under $x^{(1)}$, and after this moment under the usual stress $x^{(0)}$ until the moment $t_2^{(i)}$, i.e. under the stress:

$$x^{(2i)}(\tau) = \begin{cases} x^{(1)}, & 0 \leq \tau \leq t_1^{(i)}, \\ x^{(0)}, & t_1^{(i)} < \tau \leq t_2^{(i)}. \end{cases}$$

In the particular case when some failures can be obtained in the interval $[0, t_2^{(i)}]$ under the usual stress, the value $t_1^{(i)} = 0$ may be taken in the second experiment, i.e. the units may be tested only under $x^{(0)}$.

Set $r = \rho(x^{(0)}, x^{(1)})$. The parameter r is the same for units produced in different time intervals under the heredity hypothesis. The estimator $\hat{r}^{(i)}$ of the parameter r is defined by the formula (6.24)(adding the upper index (i) in all formulas). The asymptotic distribution of this estimator is given by (6.37). So

$$a_n^{(i)}(\hat{r}^{(i)} - r_0) \xrightarrow{D} N(0, (\sigma_r^{(i)})^2),$$ (11.1)

where

$$(\sigma_r^{(i)})^2 = \frac{S_2^{(i)}(t_1) - S_2^{(i)}(t_2)}{\{(U^{(i)})'(r_0)\}^2},$$ (11.2)

and $(U^{(i)})'(r_0)$ is given by (6.38).

The estimator obtained from all data is

$$\hat{r} = \frac{\sum_{i=1}^m a^{(i)} \hat{r}^{(i)}}{\sum_{i=1}^m a^{(i)}}.$$ (11.3)

If sufficiently large preliminary data is accumulated, the estimator $\hat{r}$ has small variance. The reliability of newly produced units can be estimated using the data from experiments under the accelerated stress $x^{(1)}$. If $\hat{S}_{x^{(1)}}$, $\hat{t}_p(x^{(1)})$ and $\hat{m}(x^{(1)})$ are estimators of the survival function, p-quantile and mean under $x^{(1)}$ obtained from such experiments, then the estimator of the survival function under the usual stress $x^{(0)}$ are:

$$\hat{S}_{x^{(0)}}(t) = \hat{S}_{x^{(1)}}(t/\hat{r}), \quad \hat{t}_p(x^{(0)}) = \hat{r}\,\hat{t}_p(x^{(1)}), \quad \hat{m}(x^{(0)}) = \hat{r}\,\hat{m}(x^{(1)}).$$ (11.4)

Suppose that the models are parametric, for example,

$$S_{x^{(0)}}^{(i)}(t) = S_0\left(\left(t/\theta^{(i)}\right)^{\alpha^{(i)}}\right)$$

for the units produced in the ith time interval. If the AFT model

$$S_{x(\cdot)}^{(i)}(t) = S_0\left\{\left(\int_0^t r\{x(\tau)\}d\tau/\theta^{(i)}\right)^{\alpha^{(i)}}\right\},$$

holds then the heredity hypothesis means that the function $r(\cdot)$ does not depend on i.

Suppose that for a group of units produced in some time interval the above considered experiment is used. Then the estimator of $r := r(x^{(1)})/r(x^{(0)})$ is $\hat{r}^{(i)} = e^{\hat{\rho}^{(i)}}$, where the estimator $\hat{\rho}^{(i)}$ is obtained from the system of equations (5.143). The unified estimator from q groups of units is obtained using the formula (11.3). The reliability of newly produced units can be estimated using the data from an experiment under the accelerated stress $x^{(1)}$. The estimators of the reliability characteristics have the form (11.4), expressions of the estimators $\hat{S}_{x^{(1)}}$, $\hat{t}_p(x^{(1)})$ and $\hat{m}(x^{(1)})$ being evidently different then in the nonparametric case.

11.2 Application of the GPH1 model

Suppose that for the units produced in a particular time interval (we skip the upper index (i)) the GPH1 model with specified q (or, equivalently, G) holds.

Suppose that n_j units are tested under the stress $x^{(j)}$ (j=0,1), the data are right censored and the multiplicative intensity model holds.

Denote $N_j(t)$ the numbers of observed failures and $Y_j(t)$ - the numbers of units at risk just prior to t for the jth group on test.

We suppose that all failure and censoring processes are censored at the moment τ.

The parameter $\rho = \rho(x^{(0)}, x^{(1)})$ (which is common for units produced in different intervals) is estimated by solving the estimating equations

$$U(\rho, \tau) = 0,$$

where

$$U(\rho, t) = N_2(t) - \rho(x_1, x_2) \int_0^t \psi(\hat{S}_2(u-))Y_2(u)dH \circ \hat{S}_1(u),$$

and $\hat{S}_1, \hat{S}_2$ are the Kaplan-Meier estimators of the reliability functions S_1 and S_2. So we obtain the estimator

$$\hat{\rho} = \hat{\rho}(x_1, x_2) = \frac{N_2(\tau)}{\int_0^\tau \psi(\hat{S}_2(u-))Y_2(u)dH \circ \hat{S}_1(u)}.$$

Suppose that the estimator $\hat{\rho}$ is obtained from the preliminary experiments and the heredity principle is satisfied. If the newly produced units are tested only under the accelerated stress $x^{(1)}$ and the Kaplan-Meier estimator $\tilde{S}_{x^{(1)}}$ is obtained, then the reliability function $S_{x^{(0)}}$ can be estimated as follows:

$$\tilde{S}_{x^{(0)}}(t) = G\left\{\hat{\rho}H \circ \tilde{S}_{x^{(1)}}(t)\right\}.$$

Let us consider the asymptotic properties of the estimators $\hat{\rho}$ and $\hat{b}$; Set $n = n_0 + n_1$ and suppose that

$$n_j/n \to l_j \in]0, 1[, \quad \sup_{u \in [0,\tau]} |Y_j(u)/n_i - y_j(u)| \xrightarrow{P} 0 \quad \text{as} \quad n \to \infty.$$

Then

$$\sqrt{n}(\hat{\rho} - \rho) = \sqrt{n}\frac{A_n}{B_n} = \sqrt{n}\times$$

$$\frac{M_2(\tau) + \rho\left\{\int_0^{\tau^n} \psi \circ S_2(u)Y_2(u)dH \circ S_1(u) - \int_0^{\tau^n} \psi \circ \hat{S}_2(u-)Y_2(u)dH \circ \hat{S}_1(u)\right\}}{\int_0^{\tau^n} \psi \circ \hat{S}_2(u)Y_2(u)dH \circ \hat{S}_1(u)}.$$

We have

$$\frac{B_n}{n} \xrightarrow{P} \int_0^\tau \psi \circ S_2(u)l_2 y_2(u)dH \circ S_1(u) = B.$$

By Theorem A7

$$\frac{M_j}{\sqrt{n}} \xrightarrow{D} V_j \quad \text{in} \quad \mathcal{D}[0, \tau],$$

where V_j is a Gaussian martingale with $V_j(0) = 0$ and for all $0 \le s \le t \le \tau$

$$\mathbf{Cov}\left(V_j(s), V_j(t)\right) = \sigma_j^2(s),$$

where

$$\sigma_j^2(s) = l_j \int_0^s y_j(u)\alpha_j(u)du.$$

The Kaplan-Meier estimators are asymptotically Gaussian (see Theorem A9):

$$\sqrt{n}(\hat{S}_j - S_j)(\cdot) \xrightarrow{\mathcal{D}} -\frac{S_j(\cdot)}{l_j} \int_0^{\cdot} \frac{dV_j(u)}{y_j(u)} \quad \text{in} \quad \mathcal{D}[0,\tau].$$

So we have:

$$\frac{1}{\sqrt{n}} A_n = \frac{1}{\sqrt{n}} \left\{ M_2(\tau^n) - \rho \left[\int_0^{\tau^n} \left(\psi \circ \hat{S}_2(u) - \psi \circ S_2(u) \right) Y_2(u)dH \circ S_1(u) + \right. \right.$$

$$\left. \left. \int_0^{\tau^n} \psi \circ S_2(u)Y_2(u)d(H \circ \hat{S}_1(u) - H \circ S_1(u)) \right] \right\} + o_p(1) \xrightarrow{\mathcal{D}}$$

$$\int_0^{\tau} (a_1(u)dV_1(u) + a_2(u)dV_2(u)) = Z_M.$$

Note that

$$\mathbf{E}(Z_M) = 0, \quad \mathbf{Var}(Z_M) = \int_0^{\tau} a_1^2(u)d\sigma_1^2(u) + a_2^2(u)d\sigma_2^2(u),$$

$$\sqrt{n}(\hat{\rho} - \rho) \xrightarrow{\mathcal{D}} N(0, \sigma_\rho^2),$$

where $\sigma_\rho^2 = \mathbf{Var}(Z_M)/B^2$. The variance σ_ρ^2 can be consistently estimated by the statistic

$$\hat{\sigma}_\rho^2 = \hat{\mathbf{Var}}(Z_M)/\hat{B}^2,$$

where

$$\hat{B} = B_n/n, \quad \hat{\mathbf{Var}}(Z_M) = \frac{1}{n} \int_0^{\tau^n} (\hat{a}_1^2(u)dN_1(u) + \hat{a}_2^2(u)dN_2(u)),$$

$$\hat{a}_1(u) = \frac{\hat{\rho}}{Y_1(u)} \left\{ \psi \circ \hat{S}_2(\tau^n)Y_2(\tau^n)H' \circ \hat{S}_1(\tau^n)\hat{S}_1(\tau^n) - \right.$$

$$\left. \int_u^{\tau^n} H' \circ \hat{S}_1(v)\hat{S}_1(v)d(\psi \circ \hat{S}_2(v)Y_2(v)) \right\},$$

$$\hat{a}_2(u) = \left\{ 1 + \frac{\hat{\rho}}{Y_2(u)} \int_u^{\tau^n} Y_2(v)\psi' \circ \hat{S}_2(v)\hat{S}_2(v)dH \circ \hat{S}_1(v) \right\}.$$

Taking into account that ρ is positive, the rate of convergence to the normal distribution can be increased considering the estimator $\hat{\rho}^* = \ln \hat{\rho}$. If we denote $\rho^* = \ln \rho$, then

$$\sqrt{n}(\hat{\rho}^* - \rho^*) \xrightarrow{\mathcal{D}} N(0, \sigma_{\rho^*}^2), \quad \text{where} \quad \sigma_{\rho^*}^2 = \frac{\sigma_\rho^2}{\rho^2}.$$

The variance $\sigma_{\rho^*}^2$ can be consistently estimated by $\hat{\sigma}_{\rho^*}^2 = \hat{\sigma}_\rho^2/\hat{\rho}^2$.

If during m different periods the estimators

$$\hat{\rho}^{*(1)} = \ln \hat{\rho}^{(1)}, \dots, \hat{\rho}^{*(m)} = \ln \hat{\rho}^{(m)} \quad \text{and} \quad \hat{b}^{(1)}, \dots, \hat{b}^{(m)}$$

are obtained and the heredity principle is satisfied, we can determine

$$\hat{\rho}^* = \sum_{i=1}^{m} \frac{\hat{\rho}^{*(i)}}{\hat{\sigma}_{\rho^*}^{(i)}} \Big/ \sum_{i=1}^{m} \frac{1}{\hat{\sigma}_{\rho^*}^{(i)}}, \quad \hat{\rho} = e^{\hat{\rho}^*},$$

where $(\hat{\sigma}_{\rho^*}^{(i)})^2$ is the estimator of the variance of $\hat{\rho}^{*(i)}$.

The estimator $\hat{\rho}$ is used to estimate the newly produced units from the data from testing of these units under the accelerated stress $x^{(1)}$. The estimator of the survival function under the usual stress is

$$S_{x^{(0)}}(t) = G\left\{ H \circ \hat{S}_{x^{(1)}}(t)/\hat{\rho} \right\}.$$

CHAPTER 12

Goodness-of-fit for accelerated life models

12.1 Goodness-of-fit for the GS model

Let E_m be a set of step-stresses of the form

$$x(\tau) = \begin{cases} x_1, & 0 \le \tau < t_1, \\ x_2, & t_1 \le \tau < t_2, \\ \dots & \dots \\ x_m, & t_{m-1} \le \tau < t_m. \end{cases} \quad (12.1)$$

Set $t_0 = 0$.

If the GS model holds on E_m then the survival function $S_{x(\cdot)}(t)$ verifies the equality:

$$S_{x(\cdot)}(t) = S_{x_i}(t - t_{i-1} + t^*_{i-1}), \quad \text{if} \quad t \in [t_{i-1}, t_i) \, (i = 1, 2, \dots, m), \quad (12.2)$$

where t^*_i can be found by solving the equations

$$S_{x_1}(t_1) = S_{x_2}(t^*_1), \dots, S_{x_i}(t_i - t_{i-1} + t^*_{i-1}) = S_{x_{i+1}}(t^*_i) \, (i = 1, \dots, m-1). \quad (12.3)$$

Note also that

$$S_{x(\cdot)}(t_i) = S_{x_{i+1}}(t^*_i).$$

The moment t_i under the stress $x(\cdot)$ is equivalent to the moment t^*_i under the stress x_{i+1}.

We considered several alternatives to the GS model. For example, under the PH model the time-shift rule does not take place if failure times under constant stresses are not exponentially distributed.

Another alternative was formulated by taking into account the influence of switch-ups of stresses on reliability of units: after the switch-up at the moment t_i from the stress x_i to the stress x_{i+1} the survival function has a jump:

$$S_{x(\cdot)}(t_i) = S_{x(\cdot)}(t_i-)\,\delta_i;$$

here δ_i is the probability for an unit not to fail because of the switch-up at the moment t_i. In this case the GS model for step-stresses can be modified as follows:

$$S_{x(\cdot)}(t) = S_{x_i}(t - t_{i-1} + t^{**}_{i-1}), \quad (12.4)$$

where

$$t^{**}_1 = S^{-1}_{x_2}\{S_{x_1}(t_1)\,\delta_1\}, \quad t^{**}_i = S^{-1}_{x_{i+1}}\{S_{x_i}(t_i - t_{i-1} + t^{**}_{i-1})\,\delta_i\}. \quad (12.5)$$

12.1.1 Test statistic for the GS model

Suppose that a group of n_0 units is tested under the step-stress (12.1) and m groups of $n_1, \cdots, n_m$ units are tested under constant in time stresses $x_1 \cdots, x_m$ ($x_1 < \cdots < x_m$), respectively. The units are observed time t_m given for the experiment.

We write $x(\cdot) < y(\cdot)$ if $S_{x(\cdot)}(t) > S_{y(\cdot)}(t)$ for all $t > 0$.

The idea of goodness-of-fit is based on comparing two estimators $\hat{A}^{(1)}_{x(\cdot)}$ and $\hat{A}^{(2)}_{x(\cdot)}$ of the cumulative hazard rate $A_{x(\cdot)}$. One estimator can be obtained from the experiment under the step-stress (12.1) and another from the experiments under the stresses $x_1, \cdots, x_m$ by using the equalities (12.2) and (12.3).

Denote by $N_i(t)$ and $Y_i(t)$ the numbers of observed failures in the interval $[0, t]$ and the number of units at risk just prior the moment t, respectively, for the group of units tested under the stress x_i and $N(t)$, $Y(t)$ the analogous numbers for the group of units tested under the stress $x(\cdot)$.

Set

$$\alpha_i = \alpha_{x_i}, \quad \alpha = \alpha_{x(\cdot)}, \quad A_i = A_{x_i}, \quad A = A_{x(\cdot)} \quad (i = 1, ..., m).$$

If the GS model holds on $E = \{x_1, \cdots, x_m, x(\cdot)\}$, then the cumulative hazard A can be written in terms of cumulative hazards A_i (cf. (12.2) and (12.3)):

$$A(t) = A_i(t - t_{i-1} + t^*_{i-1}), \ t \in [t_{i-1}, t_i) \ (i = 1, ..., m), \qquad (12.6)$$

where

$$t^*_0 = 0, \quad t^*_1 = A_2^{-1}(A_1(t_1)), \cdots,$$
$$t^*_i = A_{i+1}^{-1}(A_i(t_i - t_{i-1} + t^*_{i-1})) \quad (i = 1, ..., m-1). \qquad (12.7)$$

The first estimator $\hat{A}^{(1)}$ of the cumulative hazard A is the Nelson-Aalen estimator obtained from the experiment under the step-stress (12.1):

$$\hat{A}^{(1)}(t) = \int_0^t \frac{dN(v)}{Y(v)}.$$

The second is suggested by the GS model (formulas (12.6) and (12.7)) and is obtained from the experiments under constant stresses:

$$\hat{A}^{(2)}(t) = \hat{A}_i(t - t_{i-1} + \hat{t}^*_{i-1}), \ t \in [t_{i-1}, t_i), \ (i = 1, ..., m),$$

where

$$\hat{t}^*_0 = 0, \quad \hat{t}^*_1 = \hat{A}_2^{-1}(\hat{A}_1(t_1)), \ \cdots, \ \hat{t}^*_i = \hat{A}_{i+1}^{-1}(\hat{A}_i(t_i - t_{i-1} + \hat{t}^*_{i-1})),$$

$$\hat{A}_i^{-1}(s) = \inf\{u : \hat{A}_i(u) \geq s\}, \quad \hat{A}_i(t) = \int_0^t \frac{dN_i(v)}{Y_i(v)} \quad (i = 1, ..., m).$$

The test is based on the statistic

$$T_n = \int_0^{t_m} K(v) \, d\{\hat{A}^{(1)}(t) - \hat{A}^{(2)}(t)\}, \qquad (12.8)$$

where K is the weight function.

We shall consider the weight functions of the type: for $v \in [t_i, t_i + \Delta t_i)$

$$K(v) = \frac{1}{\sqrt{n}} \frac{Y(v)Y_{i+1}(v - t_i + \hat{t}_i^*)}{Y(v) + Y_{i+1}(v - t_i + \hat{t}_i^*)} \, g\left(\frac{Y(v) + Y_{i+1}(v - t_i + \hat{t}_i^*)}{n}\right),$$

where g is a nonnegative bounded continuous function with bounded variation on $[0,1]$ and $n = \sum_{i=0}^{m} n_i$.

Take notice that the properties of this statistic are different from the properties of the logrank-type statistics (see, for example, Moreau, Maccario, Lelouch and Huber (1992)) used testing the hypothesis of the equality of survival functions. The properties of T_n would be similar if t_i^* would be known. The problem is that the points t_i^* are unknown and are estimated. Thus when seeking the limit distribution of the statistic T_n we must keep in mind that the estimators $\hat{A}_i$ are approaching A_i with the same rate as the estimators $\hat{t}_i^*$ are approaching t_i^*.

The condition $x_1 < \cdots < x_m$ implies that

$$\mathbf{P}\{T_n \text{ is defined}\} \to 1 \quad \text{as} \quad n_i \to \infty.$$

12.1.2 Asymptotic distribution of the test statistic

To find the asymptotic distribution of the test statistic, consider at first the asymptotic distribution of the estimators $\hat{t}_i^*$.

Assumptions A.

a) The hazard rates α_i are positive and continuous on $(0, \infty)$;

b) $A_i(t) < \infty$ for all $t < 0$;

c) $n \to \infty$, $n_i/n \to l_i$, $l_i \in (0,1)$.

Lemma 12.1. *Suppose that Assumptions A hold. Then*

$$\sqrt{n}(\hat{t}_j^* - t_j^*) \xrightarrow{\mathcal{D}} a_j \sum_{l=1}^{j} d_{jl}\{U_l(t_{l-1}^* + \Delta t_{l-1}) - U_{l+1}(t_l^*)\}, \tag{12.9}$$

where

$$d_{jl} = \prod_{s=l}^{j-1} c_s, \quad l = 1, ..., j-1; \quad d_{jj} = 1,$$

$$a_j = \frac{1}{\alpha_{j+1}(t_j^*)}, \quad c_s = \frac{\alpha_{s+1}(t_s^* + \Delta t_s)}{\alpha_{s+1}(t_s^*)},$$

$U_1, \cdots, U_m$ *and* U *are independent Gaussian martingales with* $U_i(0) = U(0) = 0$ *and*

$$\mathbf{Cov}\left(U_i(s_1), U_i(s_2)\right) = \frac{1}{l_i} \frac{1 - S_i(s_1 \wedge s_2)}{S_i(s_1 \wedge s_2)} := \sigma_i^2(s_1 \wedge s_2),$$

$$\mathbf{Cov}\left(U(s_1), U(s_2)\right) = \frac{1}{l_0} \frac{1 - S(s_1 \wedge s_2)}{S(s_1 \wedge s_2)} := \sigma^2(s_1 \wedge s_2)$$

with $S_i = \exp\{-A_i\}$, $S = \exp\{-A\}$.

Proof. Under Assumptions A for any $t \in (0, t_m)$ the estimators $\hat{A}_i$ and $\hat{A}^{(1)}$ are uniformly consistent on $[0, t]$, and

$$\sqrt{n}(\hat{A}_i - A_i) \overset{D}{\to} U_i, \quad \sqrt{n}(\hat{A}^{(1)} - A) \overset{D}{\to} U \tag{12.10}$$

on $D[0, t]$.

We prove (12.9) by recurrence. If $i = 1$ then

$$\sqrt{n}(\hat{t}_1^* - t_1^*) = \sqrt{n}(\hat{A}_2^{-1}(\hat{A}_1(t_1)) - A_2^{-1}(\hat{A}_1(t_1))) + \sqrt{n}(A_2^{-1}(\hat{A}_1(t_1)) - A_2^{-1}(A_1(t_1))). \tag{12.11}$$

For any $0 < s_1 < s_2 < \infty$ Theorem A12 implies

$$\sqrt{n}(\hat{A}_2^{-1} - A_2^{-1}) \overset{D}{\to} U_2^* \tag{12.12}$$

on $D[s_1, s_2]$, where

$$U_2^*(s) = -\frac{e^{-s}U_2(A_2^{-1}(s))}{p_2(A_2^{-1}(s))}$$

and p_i is the density of T_{x_i}. Note that

$$U_2^*(A_1(t_1)) = -\frac{U_2(t_1^*)}{\alpha_2(t_1^*)}. \tag{12.13}$$

Consistency of the estimator $\hat{A}_1(t_1)$, the convergence (12.12), and the formula (12.13) imply that

$$\sqrt{n}\{\hat{A}_2^{-1}(\hat{A}_1(t_1)) - A_2^{-1}(\hat{A}_1(t_1))\} \overset{D}{\to} -\frac{U_2(t_1^*)}{\alpha_2(t_1^*)} = -a_1 U_2(t_1^*). \tag{12.14}$$

Using the delta method (Theorem A10) and the convergence (12.10), we obtain

$$\sqrt{n}\{A_2^{-1}(\hat{A}_1(t_1)) - A_2^{-1}(A_1(t_1))\} \overset{D}{\to} \frac{1}{\alpha_2(t_1^*)}U_1(t_1) = a_1 U_1(t_0^* + \Delta t_0). \tag{12.15}$$

Thus (12.11), (12.14) and (12.15) imply that

$$\sqrt{n}(\hat{t}_1^* - t_1^*) \overset{D}{\to} a_1 d_{11}\{U_1(t_0^* + \Delta t_0) - U_2(t_1^*)\}.$$

Suppose that (12.9) holds for $i = j$. Then similarly as in the case $i = 1$ we have

$$\sqrt{n}(\hat{t}_{j+1}^* - t_{j+1}^*) = \sqrt{n}\{\hat{A}_{j+2}^{-1}(\hat{A}_{j+1}(\hat{t}_j^* + \Delta t_j)) - A_{j+2}^{-1}(A_{j+1}(t_j^* + \Delta t_j))\} =$$

$$\sqrt{n}\{\hat{A}_{j+2}^{-1}(\hat{A}_{j+1}(\hat{t}_j^* + \Delta t_j)) - A_{j+2}^{-1}(\hat{A}_{j+1}(\hat{t}_j^* + \Delta t_j))\} +$$

$$\sqrt{n}\{A_{j+2}^{-1}(\hat{A}_{j+1}(\hat{t}_j^* + \Delta t_j)) - A_{j+2}^{-1}(A_{j+1}(\hat{t}_j^* + \Delta t_j))\} =$$

$$\sqrt{n}\{A_{j+2}^{-1}(A_{j+1}(\hat{t}_j^* + \Delta t_j)) - A_{j+2}^{-1}(A_{j+1}(t_j^* + \Delta t_j))\} =$$

$$a_{j+1}\{U_{j+1}(t_j^* + \Delta t_j) - U_{j+2}(t_{j+1}^*)\} + a_{j+1}\frac{c_j}{a_j}\sqrt{n}(\hat{t}_j^* - t_j^*) + \Delta_n,$$

where $\Delta_n \xrightarrow{P} 0$ as $n \to 0$. The last formula and the assumption of recurrency imply that

$$\sqrt{n}(\hat{t}^*_{j+1} - t^*_{j+1}) \xrightarrow{D} a_{j+1}\{U_{j+1}(t^*_j + \Delta t_j) - U_{j+2}(t^*_{j+1})\} + a_{j+1}\frac{c_j}{a_j}a_j \times$$

$$\left\{ \sum_{l=1}^{j-1}\prod_{s=l}^{j-1} c_s\{U_l(t^*_{l-1} + \Delta t_{l-1}) - U_{l+1}(t^*_l)\} + U_j(t^*_{j-1} + \Delta t_{j-1}) - U_{j+1}(t^*_j) \right\} =$$

$$a_{j+1}\left\{ U_{j+1}(t^*_j + \Delta t_j) - U_{j+2}(t^*_{j+1}) + \sum_{l=1}^{j}\prod_{s=l}^{j} c_s\{U_l(t^*_{l-1} + \Delta t_{l-1}) - U_{l+1}(t^*_l)\} \right\} =$$

$$a_{j+1}\sum_{l=1}^{j+1} d_{j+1,l}\{U_l(t^*_{l-1} + \Delta t_{l-1}) - U_{l+1}(t^*_l)\}.$$

Let us consider the limit distribution of the statistic T_n. Note that uniformly on $[0, t_m]$

$$\frac{K(v)}{\sqrt{n}} \xrightarrow{P} k(v) = \frac{l_0 l_{i+1}}{l_0 + l_{i+1}}S(v)\, g\left((l_0 + l_{i+1})S(v)\right).$$

Set

$$e_j = a_j\{k(t_j)\,\alpha_{j+1}(t^*_j) - k(t_j + \Delta t_j)\,\alpha_{j+1}(t^*_j + \Delta t_j)$$

$$+ \int_{t^*_j}^{t^*_j + \Delta t_j} \alpha_{i+1}(t^*_i + \Delta t_i)(v)d\,k(v + t_j - t^*_j)\}, (j = 1, ..., m-1),$$

and $f_0 = f_m = 0,\quad f_i = \sum_{j=i}^{m-1} e_j d_{ji}, (i = 1, ..., m-1)$.

Theorem 12.1. *Under Assumptions A*

$$T_n \xrightarrow{D} \int_0^{t_m} k(v)dU(v) + \sum_{i=0}^{m-1}\{f_{i+1}U_{i+1}(t^*_i + \Delta t_i) - f_i U_{i+1}(t^*_i)$$

$$- \int_{t^*_i}^{t^*_i + \Delta t_i} k(v + t_i - t^*_i)dU_{i+1}(v)\}. \tag{12.16}$$

Proof. Write the statistic (12.8) in the form

$$T_n = \int_0^{t_m} K(v)\,d\{\hat{A}^{(1)}(t) - A(t)\}$$

$$+ \sum_{i=1}^{m-1}\int_{t^*_i}^{\hat{t}^*_i} K(v + t_i - \hat{t}^*_i)\,d\hat{A}_{i+1}(v) - \sum_{i=1}^{m-1}\int_{t^*_i + \Delta t_i}^{\hat{t}^*_i + \Delta t_i} K(v + t_i - \hat{t}^*_i)\,d\hat{A}_{i+1}(v)$$

$$- \sum_{i=1}^{m-1}\int_{t^*_i}^{t^*_i + \Delta t_i}\{K(v + t_i - \hat{t}^*_i) - K(v + t_i - t^*_i)\}\,d\hat{A}_{i+1}(v)$$

$$- \sum_{i=0}^{m-1}\int_{t^*_i}^{t^*_i + \Delta t_i} K(v + t_i - t^*_i)\,d\{\hat{A}_{i+1}(v) - A_{i+1}(v)\}.$$

Under Assumptions A

$$\int_0^{t_m} K(v) \, d\{\hat{A}^{(1)}(t) - A(t)\} = \int_0^{t_m} k(v) \, dU(v) + o_p(1),$$

$$\int_{t_i^*}^{\hat{t}_i^*} K(v + t_i - \hat{t}_i^*) \, d\hat{A}_{i+1}(v) = k(t_i) \, \alpha_{i+1}(t_i^*) \sqrt{n}(\hat{t}_i^* - t_i^*) + o_p(1),$$

$$\int_{t_i^*+\Delta t_i}^{\hat{t}_i^*+\Delta t_i} K(v+t_i-\hat{t}_i^*) \, d\,\hat{A}_{i+1}(v) = k(t_i+\Delta t_i) \, \alpha_{i+1}(t_i^*+\Delta t_i) \sqrt{n}(\hat{t}_i^*-t_i^*)+o_p(1),$$

$$\int_{t_i^*}^{t_i^*+\Delta t_i} \{K(v + t_i - \hat{t}_i^*) - K(v + t_i - t_i^*)\} \, d\,\hat{A}_{i+1}(v) = -$$

$$\int_{t_i^*}^{t_i^*+\Delta t_i} \alpha_{i+1}(v) \, d\,k(v + t_i - t_i^*) \sqrt{n}(\hat{t}_i^* - t_i^*) + o_p(1),$$

$$\int_{t_i^*}^{t_i^*+\Delta t_i} K(v+t_i-t_i^*) \, d\{\hat{A}_{i+1}(v)-A_{i+1}(v)\} = \int_{t_i^*}^{t_i^*+\Delta t_i} k(v+t_i-t_i^*) \, dU_{i+1}(v)+o_p(1),$$

where $o_p(1) \xrightarrow{\text{P}} 0$ as $n \to \infty$. So the statistic T_n can be written in the form:

$$T_n = \int_0^{t_m} k(v)dU(v) + \sum_{i=1}^{m-1} k(t_i)\alpha_{i+1}(t_i^*)\sqrt{n}(\hat{t}_i^* - t_i^*)-$$

$$\sum_{i=1}^{m-1} k(t_i + \Delta t_i) \, \alpha_{i+1}(t_i^* + \Delta t_i) \sqrt{n}(\hat{t}_i^* - t_i^*)+$$

$$\sum_{i=1}^{m-1} \int_{t_i^*}^{t_i^*+\Delta t_i} \alpha_{i+1}(v) \, d\,k(v + t_i - t_i^*)\sqrt{n}(\hat{t}_i^* - t_i^*)$$

$$-\sum_{i=0}^{m-1} \int_{t_i^*}^{t_i^*+\Delta t_i} k(v + t_i - t_i^*) \, dU_{i+1}(v) + o_p(1). \qquad (12.17)$$

The lemma implies that

$$\sum_{i=1}^{m-1} \{k(t_i)\alpha_{i+1}(t_i^*)-k(t_i+\Delta t_i)\alpha_{i+1}(t_i^*+\Delta t_i)+\int_{t_i^*}^{t_i^*+\Delta t_i} \alpha_{i+1}(v)d\,k(v+t_i-t_i^*)\}\times$$

$$\sqrt{n}(\hat{t}_i^* - t_i^*) = \sum_{i=1}^{m-1} e_i \sum_{l=1}^{i} d_{il}\{U_l(t_{l-1}^* + \Delta t_{l-1}) - U_{l+1}(t_l^*)\} + o_p(1)$$

$$= \sum_{l=1}^{m-1} \left(\sum_{i=l}^{m-1} e_i d_{il}\right) \{U_l(t_{l-1}^* + \Delta t_{l-1}) - U_{l+1}(t_l^*)\} + o_p(1)$$

$$= \sum_{l=1}^{m-1} f_l\{U_l(t_{l-1}^* + \Delta t_{l-1}) - U_{l+1}(t_l^*)\} + o_p(1)$$

$$= \sum_{i=0}^{m-2} f_{i+1} U_{i+1}(t_i^* + \Delta t_i) - \sum_{i=1}^{m-1} f_i U_{i+1}(t_i^*) + o_p(1). \qquad (12.18)$$

The formulas (12.17) and (12.18) imply the result of the theorem.

Corollary 12.1. Under the assumptions of the theorem

$$T_n \xrightarrow{D} N(0, \sigma_T^2),$$

with

$$\sigma_T^2 = \int_0^{t_m} k^2(v)\, d\sigma^2(v) + \sum_{i=0}^{m-1} \{f_{i+1}^2 \sigma_{i+1}^2(t_i^* + \Delta t_i) + f_i(f_i - 2f_{i+1})\sigma_{i+1}^2(t_i^*) -$$

$$2f_{i+1}\int_{t_i^*}^{t_i^*+\Delta t_i} k(v + t_i - t_i^*)\, d\sigma_{i+1}^2(v) + \int_{t_i^*}^{t_i^*+\Delta t_i} k^2(v + t_i - t_i^*)\, d\sigma_{i+1}^2(v)\}.$$

$$(12.19)$$

If $m = 2$ then

$$\sigma_T^2 = \int_0^{t_m} k^2(v)\, d\sigma^2(v) + f_1^2(\sigma_1^2(t_1) + \sigma_2^2(t_1^*)) - 2f_1\int_0^{t_1} k(v)\, d\sigma_1^2(v) +$$

$$\int_0^{t_1} k^2(v)\, d\sigma_1^2(v) + \int_{t_1^*}^{\infty} k^2(v + t_1 - t_1^*)\, d\sigma_2^2(v).$$

Remark 12.1. The variance σ_T^2 can be consistently estimated by the statistic

$$\hat\sigma_T^2 = \int_0^{t_m} \hat k^2(v) d\hat\sigma^2(v) + \sum_{i=0}^{m-1} \{\hat f_{i+1}^2 \hat\sigma_i^2(\hat t_i^* + \Delta t_i) + \hat f_i(\hat f_i - 2\hat f_{i+1})\hat\sigma_i^2(\hat t_i^*) -$$

$$2\hat f_{i+1}\int_{\hat t_i^*}^{\hat t_i^*+\Delta t_i} \hat k(v + t_i - \hat t_i^*) d\hat\sigma_i^2(v) + \int_{\hat t_i^*}^{\hat t_i^*+\Delta t_i} \hat k^2(v + t_i - \hat t_i^*)\, d\hat\sigma_{i+1}^2(v)\},$$

where

$$\hat k(v) = K(v)/\sqrt{n}, \quad \hat\sigma^2(v) = \frac{n}{n_0}\left(\frac{1}{\hat S(v)} - 1\right), \quad \hat\sigma_i^2(v) = \frac{n}{n_i}\left(\frac{1}{\hat S_i(v)} - 1\right),$$

$\hat S$ and $\hat S_i$ are the empirical survival functions,

$$\hat f_0 = \hat f_m = 0, \quad \hat f_i = \sum_{s=i}^{2m-1} \hat e_s \hat d_{si} \quad (i = 1, \cdots, m-1),$$

$$\hat e_s = \hat a_s\left(\hat k(t_s) - \hat k(t_s + \Delta t_s) - \hat k(t_s)\,\hat\alpha_{s+1}(\hat t_s^*) + \hat k(t_s + \Delta t_s)\,\hat\alpha_{s+1}(\hat t_s^* + \Delta t_s)\right.$$

$$\left. - \int_{\hat t_s^*}^{\hat t_s^*+\Delta t_s} \hat k(v + t_s - \hat t_s^*) d\hat A_{s+1}(v)\right),$$

$$\hat{d}_{si} = \prod_{l=i}^{s-1} \hat{c}_l, \quad i = 1, ..., s-1, \quad \hat{d}_{ss} = 1,$$

$$\hat{c}_l = \frac{\hat{\alpha}_{l+1}(\hat{t}_l^* + \Delta t_l)}{\hat{\alpha}_{l+1}(\hat{t}_l^*)}, \quad \hat{a}_s = \frac{1}{\hat{\alpha}_{s+1}(\hat{t}_s^*)},$$

and $\hat{\alpha}_{s+1}(\hat{t}_s^*)$, $\hat{\alpha}_{s+1}(\hat{t}_s^* + \Delta t_s)$ are the kernel estimators:

$$\hat{\alpha}_{s+1}(\hat{t}_s^*) = \frac{1}{b} \int_0^{t_m} Ker\left(\frac{\hat{t}_i^* - u}{b}\right) d\hat{A}_{i+1}(u),$$

$$\hat{\alpha}_{s+1}(\hat{t}_s^* + \Delta t_s) = \frac{1}{b} \int_0^{t_m} Ker\left(\frac{\hat{t}_i^* + \Delta t_s - u}{b}\right) d\hat{A}_{i+1}(u);$$

here Ker is a kernel function.

12.1.3 The test

The hypothesis

$$H_0 : GS \ model \ holds \ on \ E = \{x_1, \cdots, x_m, x(\cdot)\}$$

is rejected with the approximative significance level α, if

$$\left(\frac{T}{\hat{\sigma}_T}\right)^2 > \chi_{1-\alpha}^2(1),$$

where $\chi_{1-\alpha}^2(1)$ is the $(1-\alpha)$-quantile of the chi-square distribution with one degree of freedom.

12.1.4 Consistency and the power of the test against approaching alternatives

Let us find the power of the test against the following alternatives:

$$H_1 : PH \ model \ with \ specified \ non\text{-}exponential \ time\text{-}to\text{-}failure$$

$$distributions \ under \ constant \ stresses$$

Under H_1

$$\hat{A}^{(1)}(v) \xrightarrow{\mathbf{P}} A_*^{(1)}(v) = A_i(v), \ v \in [t_i, t_{i+1}) \ (i = 0, \cdots, m-1)$$

$$\hat{A}^{(2)}(v) \xrightarrow{\mathbf{P}} A^{(2)}(v) = A_i(v - t_i + t_i^*), \ v \in [t_i, t_{i+1}) \ (i = 0, \cdots, m-1),$$

and

$$\frac{1}{\sqrt{n}} K(v) \xrightarrow{\mathbf{P}} k^*(v),$$

where for $v \in [t_i, t_{i+1})$

$$k^*(v) = \frac{l_0 l_{i+1} S_*^{(1)}(v) S_i(v - t_i + t_i^*)}{l_0 S_*^{(1)}(v) + l_{i+1} S_i(v - t_i + t_i^*)} g\left(l_0 S_*^{(1)}(v) + l_{i+1} S_i(v - t_i + t_i^*)\right),$$

and $S_*^{(1)}(v) = exp\{-A_*^{(1)}(v)\}$. Convergence is uniform on $[0, t_m]$.

Proposition 12.1. *Assume that Assumptions A hold under H_1 and*

$$\Delta^* = \int_0^{t_m} k^*(v)\, d\{A^{(1)}(v) - A(v)\} \neq 0.$$

Then the test is consistent against H_1.

Proof. Write the test statistic in the form

$$T_n = \int_0^\infty K(v)\, d\{\hat{A}^{(1)}(v) - A_*^{(1)}(v)\} - \int_0^\infty K(v)\, d\{\hat{A}^{(2)}(v) - A^{(2)}(v)\} +$$

$$\int_0^\infty K(v)\, d\{A_*^{(1)}(v) - A^{(2)}(v)\} = T_{1n} + T_{2n} + T_{3n}. \qquad (12.20)$$

Analogously as in the case when seeking the limit distribution of the statistic T_n under the hypothesis H_0, we obtain that under H_1

$$T_{1n} + T_{2n} \xrightarrow{D} N(0, \sigma_T^{*2}),$$

where σ_T^{*2} has the same form (12.19) with only difference that k is replaced by k^* and $\sigma^2(t)$ is replaced by

$$(\sigma^{(1)})^2(t) = \frac{1}{l_0}\left(\frac{1}{S^{(1)}(t)} - 1\right).$$

Under H_1 we have

$$\hat{\sigma}_T^2 \xrightarrow{P} \sigma_T^{*2} \qquad (12.21)$$

and

$$\frac{T_{1n} + T_{2n}}{\hat{\sigma}_T} \xrightarrow{D} N(0,1). \qquad (12.22)$$

The third member in (12.20) can be written in the form

$$T_{3n} = \sum_{i=1}^{m-1} \int_{t_i}^{t_{i+1}} K(v)\{\alpha_{i+1}(v) - \alpha_{i+1}(v - t_i + t_i^*)\}\, dv. \qquad (12.23)$$

The assumptions of the proposition and the equalities (12.20)-(12.23) imply that under H_1

$$\frac{1}{\sqrt{n}}T_{3n} \xrightarrow{P} \Delta^*, \qquad \frac{T_n}{\hat{\sigma}_T} \xrightarrow{P} \infty.$$

Thus under H_1

$$\mathbf{P}\left\{\left(\frac{T}{\hat{\sigma}_T}\right)^2 > \chi_{1-\alpha}^2(1)\right\} \to 1.$$

The proposition is proved.

Remark 12.2. *If α_i are increasing (decreasing) then the test is consistent against H_1.*

Proof. We shall show by recurrence that $t_i > t_i^*$ for all i. Indeed, the inequalities $x_1 < \cdots < x_m$ imply that

$$S_1(t_1^*) > S_2(t_1^*) = S_1(t),$$

which give $t_1 > t_1^*$. If we assume that $t_{i-1} > t_{i-1}^*$ then

$$S_{i+1}(t_i^*) = S_i(t_i - t_{i-1} + t_{i-1}^*) > S_i(t_i - t_{i-1} + t_{i-1}) = S_i(t_i) > S_{i+1}(t_i),$$

which imply $t_i > t_i^*$. If α_i are increasing (decreasing) then $\Delta^* > 0$ ($\Delta^* < 0$) under H_1. The proposition implies the consistency of the test.

Let us consider the sequence of approaching alternatives

$$H_n : \; PH \; with \; \alpha_i(t) = \left(\frac{t}{\theta_i}\right)^{\frac{\varepsilon}{\sqrt{n}}}$$

with fixed $\varepsilon > 0$ ($i = 1, \cdots, m$). Then

$$T_{3n} \xrightarrow{\mathbf{P}} \mu = -\varepsilon \sum_{i=1}^{m-1} \int_{t_i}^{t_{i+1}} k^*(v) \ln(1 + \frac{t_i^* - t_i}{v}) dv > 0$$

and

$$\frac{T_n}{\hat{\sigma}_T} \xrightarrow{\mathcal{D}} N(a, 1), \quad \left(\frac{T}{\hat{\sigma}_T}\right)^2 \xrightarrow{\mathcal{D}} \chi^2(1, a),$$

where $a = \mu/\sigma_T^*$ and $\chi^2(1, a)$ denotes the chi-square distribution with one degree of freedom and the noncentrality parameter a (or a random variable having such distribution).

The power function of the test is approximated by the function

$$\beta = \lim_{n \to \infty} \mathbf{P} \left\{ \left(\frac{T}{\hat{\sigma}_T}\right)^2 > \chi_{1-\alpha}^2(1) \mid H_n \right\} = \mathbf{P}\left\{\chi^2(1, a) > \chi_{1-\alpha}^2(1)\right\}. \quad (12.24)$$

Let us find the power of the test against the following alternatives:

$$H_2 : the \; model \; (12.4) \; with \; specified \; time\text{-}to\text{-}failure$$

$$distributions \; under \; constant \; stresses$$

Under H_2

$$\hat{A}^{(1)}(v) \xrightarrow{\mathbf{P}} A_{**}^{(1)}(v) = A_i(v - t_i + t_i^{**}), \; v \in [t_i, t_{i+1}) \; (i = 0, \cdots, m-1),$$

$$\hat{A}^{(2)}(v) \xrightarrow{\mathbf{P}} A^{(2)}(v) = A_i(v - t_i + t_i^*), \; v \in [t_i, t_{i+1}) \; (i = 0, \cdots, m-1),$$

and

$$\frac{1}{\sqrt{n}}K(v) \xrightarrow{\mathbf{P}} k^{**}(v),$$

where for $v \in [t_i, t_{i+1})$

$$k^{**}(v) = \frac{l_0 l_{i+1} S_{**}^{(1)}(v) S_i(v - t_i + t_i^*)}{l_0 S_{**}^{(1)}(v) + l_{i+1} S_i(v - t_i + t_i^*)} g\left(l_0 S_{**}^{(1)}(v) + l_{i+1} S_i(v - t_i + t_i^*)\right),$$

and $S_{**}^{(1)}(v) = exp\{-A_{**}^{(1)}(v)\}$. Convergence is uniform on $[0, t_m]$.

Proposition 12.2. *Assume that Assumptions A hold under H_2 and*

$$\Delta^{**} = \int_0^\infty k^{**}(v)\, d\{A^{(1)}(t) - A(t)\} \neq 0.$$

Then the test is consistent against H_2.

Proof. Write the test statistic in the form (12.20). Analogously as in the case when seeking the limit distribution of the statistic T_n under the hypothesis H_0, we obtain that under H_2

$$T_{1n} + T_{2n} \xrightarrow{D} N(0, \sigma_T^{**2}),$$

where σ_T^{*2} has the same form (12.19) with only difference that k is replaced by k^{**} and $\sigma^2(t)$ is replaced by

$$(\sigma^{(1)})^2(t) = \frac{1}{l_1}\left(\frac{1}{S^{(1)}(t)} - 1\right).$$

The third member in (12.20) can be written in the form

$$T_{3n} = \sum_{i=1}^{m-1} \int_{t_i}^{t_{i+1}} K(v)\{\alpha_{i+1}(v - t_i + t_i^{**}) - \alpha_{i+1}(v - t_i + t_i^*)\}dv.$$

The assumptions of the proposition and the the last equality imply that

$$\frac{1}{\sqrt{n}}T_{3n} \xrightarrow{P} \Delta^{**}, \qquad \frac{T_n}{\hat{\sigma}_T} \xrightarrow{P} \infty.$$

Remark 12.3. *If α_i are increasing (decreasing) then the test is consistent against H_2.*

Proof. Let us show by recurrence that $t_i^{**} > t_i^*$. Really, the inequalities $x_1 < \cdots < x_m$ imply that

$$S_2(t_1^{**}) = S_1(t_1)\delta_1 < S_1(t_1) = S_2(t_1^*),$$

which give $t_1^{**} > t_1^*$. If we assume that $t_{i-1}^{**} > t_{i-1}^*$ then

$$S_{i+1}(t_i^{**}) = S_i(t_i - t_{i-1} + t_{i-1}^{**})\delta_i < S_i(t_i - t_{i-1} + t_{i-1}^{**}) < S_i(t_i - t_{i-1} + t_{i-1}^*) = S_{i+1}(t_i^*),$$

which imply $t_i^{**} > t_i^*$. If α_i are increasing (decreasing) then $\Delta > 0$ ($\Delta < 0$) under H_2. Proposition 12.2 implies the consistency of the test.

Let us consider the sequence of approaching alternatives

H_n : *the model (12.4) with specified time-to-failure distributions*

under constant stresses and $\delta_i = 1 - \dfrac{\varepsilon_i}{\sqrt{n}}$.

Let us find the limit of $\sqrt{n}(t_i^{**} - t_i^*)$ by recurrence. If $i = 1$, then

$$\sqrt{n}(t_1^{**} - t_1^*) = \sqrt{n}\left\{S_2^{-1}\left(S_1(t_1)(1 - \frac{\varepsilon_1}{\sqrt{n}})\right) - S_2^{-1}(S_1(t_1))\right\}$$

$$\longrightarrow -\frac{\varepsilon_1}{\alpha_2(t_1^*)} = -a_1\varepsilon_1$$

Suppose that

$$\sqrt{n}(t_i^{**} - t_i^*) \longrightarrow a_i \sum_{j=1}^{i} d_{ij}\varepsilon_j, \qquad (12.25)$$

where a_i, d_{ij} are defined in the formulation of Lemma 2.1. Then

$$\sqrt{n}(t_{i+1}^{**} - t_{i+1}^*) = \sqrt{n}(S_{i+2}^{-1}\left(S_{i+1}(t_{i+1} - t_i + t_i^{**})(1 - \frac{\varepsilon_{i+1}}{\sqrt{n}})\right) -$$

$$S_{i+2}^{-1}(S_{i+1}(t_{i+1} - t_i + t_i^*)) = \sqrt{n}\frac{1}{p_{i+2}(t_{i+1}^*)} \times$$

$$\left\{p_{i+1}(t_{i+1} - t_i + t_i^*)(t_i^{**} - t_i^*) - S_{i+1}(t_{i+1} - t_i + t_i^*)\frac{\varepsilon_{i+1}}{\sqrt{n}}\right\} + o(1) =$$

$$\frac{1}{p_{i+2}(t_{i+1}^*)}\left\{-p_{i+1}(t_{i+1} - t_i + t_i^*)a_i \sum_{j=1}^{i} d_{ij}\varepsilon_j - S_{i+1}(t_{i+1} - t_i + t_i^*)\varepsilon_{i+1}\right\}$$

$$+o(1) = a_{i+1}\sum_{j=1}^{i+1} d_{i+1,j}\varepsilon_j + o(1).$$

We note p_i the densities of T_{x_i}. Thus the convergence (12.25) holds for all i. It implies that

$$T_{3n} \xrightarrow{\mathbf{P}} \mu = -\sum_{i=1}^{m-1} a_i \sum_{j=1}^{i} d_{ij}\,\varepsilon_j \int_{t_i}^{t_{i+1}} k(v)\,d\,\alpha_{i+1}(v - t_i + t_i^*)$$

and

$$\frac{T_n}{\hat{\sigma}_T} \xrightarrow{\mathcal{D}} N(a, 1), \qquad \left(\frac{T_n}{\hat{\sigma}_T}\right)^2 \xrightarrow{\mathcal{D}} \chi^2(1, |\,a\,|),$$

where $a = \mu/\sigma_T^{**}$.

The parameter μ is positive (negative) if the functions α_i are convex (concave).

The power function of the test is approximated by the function (12.24) with $a = \mu/\sigma_T^{**}$.

One can see also Bagdonavičius and Nikoulina (1997), Bagdonavičius and Nikulin (1995, 1998, 2001, 2001a), Nikulin and Solev (1999, 2001).

12.2 Goodness-of-fit for the model with absence of memory

If the PH model holds on a set E of time-varying explanatory variables then for any $x(\cdot) \in E$

$$\alpha_{x(\cdot)}(t) = \alpha_{x_t}(t), \qquad (12.26)$$

where x_t is constant explanatory variable equal to the value of time-varying explanatory variable $x(\cdot)$ at the moment t. For *any* t the hazard rate under the time-varying stress $x(\cdot)$ at the moment t *does not depend on the values of the stress $x(\cdot)$ before the moment t* but only on the value of stress at this moment. It is not natural when the hazard rates are not constant under constant stresses.

The equality (12.26) defines a model which means that the hazard rate under any time-varying stress at any moment t does not depend on the values of stress before this moment.

Let us call this model the *absence of memory* (AM) model. This model is wider than the PH model because it does not specify relations between survival distributions under different constant stresses. The PH model is a submodel of it.

The AM model (and the PH model) is not natural for aging units and its application should be carefully studied. A formal goodness-of-fit test would be useful.

The most used time-varying stresses in accelerated life testing are the step-stresses: units are placed on test at an initial low stress and if they do not fail in a predetermined time t_1, the stress is increased. If they do not fail in a predetermined time $t_2 > t_1$, the stresses is increased once more, and so on.

Let us consider a set E_m of step-stresses of the form (12.1).

If the AM model holds on E_m and $x(\cdot) \in E_m$ then

$$\alpha_{x(\cdot)}(t) = \alpha_{x_i}(t), \quad \text{if} \quad t \in [t_{i-1}, t_i), \; (i = 1, 2, \dots, m). \qquad (12.27)$$

It can be written in terms of the cumulative hazards $A_{x(\cdot)}$ and A_{x_i}:

$$A_{x(\cdot)}(t) = A_{x_i}(t) - A_{x_i}(t_{i-1}) + \mathbf{1}_{\{i \geq 2\}} \sum_{j=1}^{i-1} (A_{x_j}(t_j) - A_{x_j}(t_{j-1})), \; t \in [t_{i-1}, t_i)$$

$$(i = 1, \dots, m). \qquad (12.28)$$

A very possible alternative to this model is the generalized Sedyakin (GS) model:

$$\alpha_{x(\cdot)}(t) = g\left(x(t), S_{x(\cdot)}(t)\right).$$

Set

$$\alpha_i = \alpha_{x_i}, \quad \alpha = \alpha_{x(\cdot)}, \quad A_i = A_{x_i}, \quad A = A_{x(\cdot)} \quad (i = 1, \dots, m).$$

If the GS model holds on E_m and $x(\cdot) \in E_m$ then the formulas (12.6) and (12.7) hold.

Let us consider goodness-of-fit tests given by Bagdonavičius and Levulienė (2001).

12.2.1 Logrank-type test statistic for the AM model

Suppose that a group of n_0 units is tested under the step-stress (12.1) and m groups of $n_1, \cdots, n_m$ units are tested under constant in time stresses $x_1 \cdots, x_m$, respectively.

Suppose that at first that $x_1 < \cdots < x_m$. The units are observed time t_m given for the experiment.

The idea of goodness-of-fit is based on comparing two estimators $\hat{A}^{(1)}_{x(\cdot)}$ and $\hat{A}^{(2)}_{x(\cdot)}$ of the cumulative hazard rate $A_{x(\cdot)}$. One estimator can be obtained from the experiment under step-stress (12.1) and another from the experiments under stresses $x_1, \cdots, x_m$ by using the equalities (12.28).

Denote by $N_i(t)$ and $Y_i(t)$ the number of observed failures in the interval $[0, t]$ and the number of units at risk just prior the moment t, respectively, for the group of units tested under the stress x_i and $N(t)$, $Y(t)$ the analogous numbers for the group of units tested under the stress $x(\cdot)$.

The first estimator $\hat{A}^{(1)}$ of the accumulated hazard A is the Nelson -Aalen estimator obtained from the experiment under the step-stress (12.1):

$$\hat{A}^{(1)}(t) = \int_0^t \frac{dN(v)}{Y(v)}.$$

The second is suggested by the AM model (formula (12.28)) and is obtained from the experiments under the constant stresses:

$$\hat{A}^{(2)}(t) = \hat{A}_i(t) - \hat{A}_i(t_{i-1}) + \mathbf{1}_{\{i \geq 2\}} \sum_{j=1}^{i-1} (\hat{A}_j(t_j) - \hat{A}_j(t_{j-1})), \ t \in [t_{i-1}, t_i),$$

(12.29)

where

$$\hat{A}_i(t) = \int_0^t \frac{dN_i(v)}{Y_i(v)} \quad (i = 1, ..., m).$$

The first test is based on the logrank-type statistic

$$T_n = T_n(t_m), \quad \text{where} \quad T_n(t) = \int_0^t K(v) \, d\{\hat{A}^{(1)}(t) - \hat{A}^{(2)}(t)\}; \quad (12.30)$$

here K is the weight function.

Similarly as in the case of classical logrank tests (see Fleming and Harrington (1991)), we shall consider the weight functions of the following type: for $v \in [t_{i-1}, t_i)$

$$K(v) = \frac{1}{\sqrt{n}} \frac{Y(v)Y_i(v)}{Y(v) + Y_i(v)} g\left(\frac{Y(v) + Y_i(v)}{n}\right),$$

where $n = \sum_{i=0}^{m} n_i$ and g is a nonnegative bounded continuous function with bounded variation on $[0, 1]$.

12.2.2 *Asymptotic distribution of the logrank-type test statistic*

Assumptions A.
   a) *The hazard rates α_i are positive and continuous on $(0, \infty)$;*
   b) *$A_i(t_m) < \infty$;*
   c) *$n \to \infty$, $n_i/n \to l_i$, $l_i \in (0, 1)$.*

Under Assumptions A for any $t \in (0, t_m]$ the estimators $\hat{A}_i$ and $\hat{A}^{(1)}$ are uniformly consistent on $[0, t]$, and

$$\sqrt{n}(\hat{A}_i - A_i) \xrightarrow{D} U_i, \quad \sqrt{n}(\hat{A}^{(1)} - A) \xrightarrow{D} U$$

on $D[0, t]$. Here U and $U_1, \cdots, U_m$ are independent Gaussian martingales with $U_i(0) = U(0) = 0$, and

$$\mathbf{Cov}\,(U_i(s_1), U_i(s_2)) = \frac{1}{l_i} \frac{1 - S_i(s_1 \wedge s_2)}{S_i(s_1 \wedge s_2)} := \sigma_i^2(s_1 \wedge s_2),$$

$$\mathbf{Cov}\,(U(s_1), U(s_2)) = \frac{1}{l_0} \frac{1 - S(s_1 \wedge s_2)}{S(s_1 \wedge s_2)} := \sigma^2(s_1 \wedge s_2),$$

with $S_i = \exp\{-A_i\}$, $S = \exp\{-A\}$.

Let us consider the limit distribution of the stochastic process $T_n(t), t \in [0, t_m]$. Note that

$$\frac{K(v)}{\sqrt{n}} \xrightarrow{P} k(v) = \frac{l_0 l_i S(v) S_i(v)}{l_0 S(v) + l_i S_i(v)} g\left(l_0 S(v) + l_i S_i(v)\right), \quad v \in [t_{i-1}, t_i).$$

The convergence is uniform on $[0, t_m]$.

Proposition 12.3. *Under Assumptions A*

$$T_n(t) \xrightarrow{D} V_k(t) = \int_0^t k(v) dU(v) - \mathbf{1}\{i \geq 2\} \sum_{j=1}^{i-1} \int_{t_{j-1}}^{t_j} k(v) dU_j(v) -$$

$$\int_{t_{i-1}}^t k(v) dU_i(v), \quad t \in [t_{i-1}, t_i), \ i = 1, \cdots, m, \ t_0 = 0 \quad \text{on} \quad D[0, t_m].$$

$$(12.31)$$

Proof. For $t \in [t_{i-1}, t_i), \ i = 1, \cdots, m$, write the statistic (12.30) in the form

$$T_n(t) = \int_0^t K(v) \, d\{\hat{A}^{(1)}(t) - A(t)\} - \int_0^t K(v) \, d\{\hat{A}^{(2)}(t) - A(t)\}$$

$$= \int_0^t J(v)K(v) \frac{dM(v)}{Y(v)} - \mathbf{1}\{i \geq 2\} \sum_{j=1}^{i-1} \int_0^t J_j(v)K(v) \frac{dM_j(t)}{Y_j(t)} -$$

$$\int_{t_{i-1}}^t J_i(v)K(v) \frac{dM_i(t)}{Y_i(t)},$$

where

$$M(t) = N(t) - \int_0^t Y(u)dA(u), \quad M_i(t) = N_i(t) - \int_0^t Y_i(u)dA_i(u),$$

$$J(t) = \mathbf{1}_{\{Y(t)>0\}}, \quad J_i(t) = \mathbf{1}_{\{Y_i(t)>0\}}.$$

Note that

$$< \int_0^t J(v)K(v)\frac{dM(v)}{Y(v)} > = \int_0^t J(v)K^2(v)\frac{dA(v)}{Y(v)} \xrightarrow{P} \int_0^t k^2(v)\frac{dA(v)}{l_0 S(v)}$$

and for any $\varepsilon > 0$:

$$< \int_0^t J(v)\frac{K(v)}{Y(v)}\mathbf{1}_{\{|\frac{K(v)}{Y(v)}|\geq\varepsilon\}}dM(v) > = \int_0^t J(v)\frac{K^2(v)}{Y(v)}\mathbf{1}_{\{|\frac{K(v)}{Y(v)}|\geq\varepsilon\}}dA(v) \xrightarrow{P} 0$$

on $D[0, t_m]$. The Theorem A7 implies that

$$\int_0^t J(v)K(v)\frac{dM(v)}{Y(v)} \xrightarrow{D} \int_0^t k(v)\left(\frac{\alpha(v)}{l_0 S(v)}\right)^{1/2} dW(v),$$

on $D[0, t_m]$; here W is the standard Wiener process. The limit process has the same variance-covariance structure as the Gaussian process

$$\int_0^t k(v)dU(v).$$

So

$$\int_0^t J(v)K(v)\frac{dM(v)}{Y(v)} \xrightarrow{D} \int_0^t k(v)dU(v).$$

Analogously it is obtained that

$$\int_0^t J_i(v)K(v)\frac{dM_i(v)}{Y_i(v)} \xrightarrow{D} \int_0^t k(v)dU_i(v)$$

on $D[0, t_m]$.

Corollary 12.2. *Under the assumptions of the theorem*

$$\mathbf{Cov}(V_k(s), V_k(t)) = \sigma_{V_k}^2(s \wedge t)$$

where

$$\sigma_{V_k}^2(t) = \int_0^t k^2(v)d\sigma^2(v) + \mathbf{1}\{i \geq 2\}\sum_{j=1}^{i-1}\int_{t_{j-1}}^{t_j} k^2(t)d\sigma_j^2(v)$$

$$+ \int_{t_{i-1}}^t k^2(t)d\sigma_i^2(v)$$

$$= \int_0^t \frac{k^2(v)}{l_0 S(v)}dA(v) + \mathbf{1}\{i \geq 2\}\sum_{j=1}^{i-1}\int_{t_{j-1}}^{t_j}\frac{k^2(t)}{l_j S_j(v)}dA_j(v)\mathbf{1}\{i \geq 2\}$$

$$+ \int_{t_{i-1}}^t \frac{k^2(t)}{l_i S_i(v)}dA_i(v), \quad t \in [t_{i-1}, t_i), \ i = 1, \cdots, m, \ t_0 = 0,$$

and

$$T_n \xrightarrow{D} N(0, \sigma_{V_k}^2(t_m)),$$

Proposition 12.4. *The variance $\sigma_{V_k}^2(t_m)$ can be consistently estimated by* *the statistic*

$$\hat{\sigma}_{V_k}^2(t_m) = \int_0^{t_m} K^2(v)\frac{dN(v)}{Y^2(v)} + \sum_{i=1}^{m} \int_{t_{i-1}}^{t_i} K^2(v)\frac{dN_i(v)}{Y_i^2(v)}.$$

Proof. Let us consider the difference

$$\int_0^{t_m} K^2(v)\frac{dN(v)}{Y^2(v)} - \int_0^{t_m} k^2(v)\frac{dA(v)}{l_0 S(v)} =$$

$$\int_0^{t_m} J(v)K^2(v)\frac{Y(v)dA(v) + dM(v)}{Y^2(v)} - \int_0^{t_m} k^2(v)\frac{dA(v)}{l_0 S(v)} =$$

$$\int_0^{t_m} J(v)\left(\frac{K^2(v)/n}{(n_0/n)Y(v)} - \frac{k^2(v)}{l_0 S(v)}\right) dA(v) +$$

$$\int_0^{t_m} J(v)K^2(v)\frac{dM(v)}{Y^2(v)} + \int_0^{t_m} (1 - J(v))\frac{dA(v)}{l_0 S(v)} = B_1 + B_2 + B_3.$$

We have

$$|B_1| \le \sup_{[0,t_m]} \left| \frac{K^2(v)/n}{(n_0/n)Y(v)} - \frac{k^2(v)}{l_0 S(v)} \right| A(t_m) \overset{\mathbf{P}}{\to} 0,$$

and

$$<B_2> = <\int_0^{t_m} J(v)K^2(v)\frac{dM(v)}{Y^2(v)}> =$$

$$= \int_0^{t_m} J(v)K^4(v)\frac{dA(v)}{Y^3(v)} \le \frac{1}{n} \sup_{[0,t_m]} \frac{(K(v)//\sqrt{n})^4}{(Y(v)/n)^3} A(t_m) \overset{\mathbf{P}}{\to} 0,$$

which imply that $B_i \overset{\mathbf{P}}{\to} 0$ $(i = 1, 2)$. Convergence $B_3 \overset{\mathbf{P}}{\to} 0$ is evident.

12.2.3 Logrank-type test

The hypothesis

$$H_0 : \alpha_{x(\cdot)}(t) = \alpha_{x_i}(t), t \in [t_{i-1}, t_i) \ (i = 1, \cdots, m)$$

(or the AM model) is rejected with the approximative significance level α, if

$$\left(\frac{T_n}{\hat{\sigma}_{V_k}(t_m)}\right)^2 > \chi_{1-\alpha}^2(1),$$

where $\chi_{1-\alpha}^2(1)$ is the $(1 - \alpha)$-quantile of the chi-square distribution with one degree of freedom.

12.2.4 Consistency and the power of the test against the approaching alternatives

Let us find the power of the test against the following alternatives:

$$H_1 : GS \ model \ with \ specified \ non-exponential \ time-to-failure$$

distributions under constant stresses

Under H_1

$$\hat{A}^{(1)}(v) \xrightarrow{\mathbf{P}} A_*^{(1)}(v) = A_i(v - t_{i-1} + t_{i-1}^*), \; v \in [t_{i-1}, t_i) \; (i = 1, \cdots, m),$$

where t_i^* can be found by solving the equations

$$A_1(t_1) = A_2(t_1^*), \cdots, A_i(t_i - t_{i-1} + t_{i-1}^*) = A_{i+1}(t_i^*)$$

$$(i = 1, \cdots, m-1),$$

$$\hat{A}^{(2)}(v) \xrightarrow{\mathbf{P}} A^{(2)}(v) = A_i(t) - A_i(t_{i-1}) + \mathbf{1}_{\{i \geq 2\}} \sum_{j=1}^{i-1} (A_j(t_j) - A_j(t_{j-1})),$$

$$v \in [t_{i-1}, t_i) \; (i = 1, ..., m),$$

and

$$\frac{1}{\sqrt{n}} K(v) \xrightarrow{\mathbf{P}} k_*(v), \quad Y(v)/n_0 \xrightarrow{\mathbf{P}} S_*^{(1)}(v)$$

where $S_*^{(1)}(v) = exp\{-A_*^{(1)}(v)\}$, and for $v \in [t_{i-1}, t_i)$

$$k_*(v) = \frac{l_0 l_i S_*^{(1)}(v) S_i(v)}{l_0 S_*^{(1)}(v) + l_i S_i(v)} g\left(l_0 S_*^{(1)}(v) + l_i S_i(v)\right)$$

$$= \frac{l_0 l_i S_i(v) S_i(v - t_{i-1} + t_{i-1}^*)}{l_0 S_i(v - t_{i-1} + t_{i-1}^*) + l_i S_i(v)} g\left(l_0 S_i(v - t_{i-1} + t_{i-1}^*) + l_i S_i(v)\right).$$

Convergence is uniform on $[0, t_m]$.

Proposition 12.5. *Suppose that Assumptions A hold and*

$$\Delta^* = \Delta^*(t_m) \neq 0,$$

where

$$\Delta^*(t) = \mathbf{1}_{\{i \geq 2\}} \sum_{j=1}^{i-1} \int_{t_{j-1}}^{t_j} k_*(v) \{\alpha_j(v - t_{j-1} + t_{j-1}^*) - \alpha_j(v)\} dv +$$

$$\int_{t_{i-1}}^{t} k_*(v) \{\alpha_i(v - t_{i-1} + t_{i-1}^*) - \alpha_i(v)\} dv.$$

Then the test is consistent against H_1.

Proof. Write the test statistic in the form

$$T_n = \int_0^{t_m} K(v) \, d\{\hat{A}^{(1)}(v) - A_*^{(1)}(v)\} - \int_0^{t_m} K(v) \, d\{\hat{A}^{(2)}(v) - A^{(2)}(v)\} +$$

$$\int_0^{t_m} K(v) \, d\{A_*^{(1)}(v) - A^{(2)}(v)\} = T_{1n} + T_{2n} + T_{3n}. \quad (12.32)$$

As in the case when seeking the limit distribution of the statistic T_n under the hypothesis H_0, we obtain that under H_1

$$T_{1n} + T_{2n} \xrightarrow{D} N(0, \sigma_{V_k}^{*\,2}(t_m)),$$

where $\sigma_{V_k}^{*\,2}(t)$ has the same form as in Corollary 12.2 with only difference that $k(v)$ is replaced by $k_*(v)$ and $\sigma^2(v)$ is replaced by

$$(\sigma^{(1)})^2(v) = \frac{1}{l_0}\left(\frac{1}{S_*^{(1)}(v)} - 1\right),$$

i.e.

$$\sigma_{V_k}^{*\,2}(t) = \int_0^t \frac{k_*^2(v)}{l_0 S_*(v)}\, d\,A_*(v) + \mathbf{1}\{i \geq 2\}\sum_{j=1}^{i-1}\int_{t_{j-1}}^{t_j}\frac{k_*^2(t)}{l_j S_j(v)}dA_j(v)$$

$$+ \int_{t_{i-1}}^t \frac{k^2(t)}{l_i S_i(v)}dA_i(v), \quad t \in [t_{i-1}, t_i),\ i = 1,\cdots, m,\ t_0 = 0. \tag{12.33}$$

Under H_1 we have

$$\hat{\sigma}_{V_k}^2(t) \xrightarrow{\mathbf{P}} \sigma_{V_k}^{*\,2}(t) \tag{12.34}$$

uniformly on $D[0, t_m]$, and

$$\frac{T_{1n} + T_{2n}}{\sigma_{V_k}^*(t_m)} \xrightarrow{\mathcal{D}} N(0, 1). \tag{12.35}$$

The third member in (12.32) can be written in the form

$$T_{3n} = \sum_{i=1}^m \int_{t_{i-1}}^{t_i} K(v)\left\{\alpha_i(v - t_{i-1} + t_{i-1}^*) - \alpha_i(v)\right\} dv. \tag{12.36}$$

The assumptions of the proposition and the equalities (12.32)-(12.36) imply that under H_1

$$\frac{1}{\sqrt{n}}T_{3n} \xrightarrow{\mathbf{P}} \Delta^*, \qquad \frac{T_n}{\hat{\sigma}_{V_k}(t_m)} \xrightarrow{\mathbf{P}} \infty.$$

Thus under H_1

$$\mathbf{P}\left\{\left(\frac{T}{\hat{\sigma}_T}\right)^2 > \chi_{1-\alpha}^2(1)\right\} \to 1.$$

Proposition 12.6. *If α_i are increasing (decreasing) then the test is consistent against H_1.*

Proof. We shall show by recurrence that $t_i > t_i^*$ for all i. The inequalities $x_1 < \cdots < x_m$ imply that

$$S_1(t_1{}^*) > S_2(t_1{}^*) = S_1(t),$$

which give $t_1 > t_1{}^*$. If we assume that $t_{i-1} > t_{i-1}^*$ then

$$S_{i+1}(t_i^*) = S_i(t_i - t_{i-1} + t_{i-1}^*) > S_i(t_i - t_{i-1} + t_{i-1}) = S_i(t_i) > S_{i+1}(t_i),$$

which imply $t_i > t_i^*$. If α_i are increasing (decreasing) then $\Delta^* > 0$ ($\Delta^* < 0$) under H_1. The proposition implies the consistency of the test.

Let us consider the sequence of the approaching alternatives

$$H_n:\ GS\ with\ \alpha_i(t) = \left(\frac{t}{\theta_i}\right)^{\frac{\varepsilon_i}{\sqrt{n}}} \tag{12.37}$$

with fixed $\varepsilon_i > 0$ $(i = 1, \cdots, m)$. Then

$$T_{3n} \xrightarrow{\mathbf{P}} \mu = \sum_{i=1}^{m} \varepsilon_i \int_{t_{i-1}}^{t_i} k_*(v) \ln(1 + \frac{t_{i-1}^* - t_{i-1}}{v})dv < 0,$$

and

$$\frac{T_n}{\hat{\sigma}_{V_k}(t_m)} \xrightarrow{D} N(a, 1), \quad \left(\frac{T}{\hat{\sigma}_{V_k}(t_m)}\right)^2 \xrightarrow{D} \chi^2(1, a),$$

where $a = -\mu/\sigma_T^*$, and $\chi^2(1, a)$ denotes the chi-square distribution with one degree of freedom and the noncentrality parameter a (or the random variable having such distribution).

The power function of the test is approximated by the function

$$\beta = \lim_{n \to \infty} \mathbf{P}\left\{\left(\frac{T}{\hat{\sigma}_{V_k}(t_m)}\right)^2 > \chi_{1-\alpha}^2(1) \mid H_n\right\} = \mathbf{P}\left\{\chi^2(1, a) > \chi_{1-\alpha}^2(1)\right\}.$$

(12.38)

12.2.5 Kolmogorov-type test

The logrank-type test may be bad if the step-explanatory variable is not monotone, i.e. the condition $x_1 < \cdots < x_m$ is not satisfied. In such a case the following Kolmogorov-type test may be used.

The limit process $V_k(t)$ obtained in the Proposition 12.3 is a zero mean Gaussian martingale with the covariance function

$$\mathbf{Cov}(V_k(s), V_k(t)) = \sigma_{V_k}^2(s \wedge t)$$

It implies that $V_k(t) = W(\sigma_{V_k}^2(t))$, where W is the standard Wiener process. We have

$$\frac{1}{\sigma_{V_k}(t_m)} \sup_{0 \le t \le t_m} \mid V_k(t) \mid = \sup_{0 \le t \le t_m} \mid W\left(\frac{\sigma_{V_k}^2(t)}{\sigma_{V_k}^2(t_m)}\right) \mid = \sup_{0 \le u \le 1} \mid W(u) \mid.$$

(12.39)

The variance $\sigma_{V_k}^2(t)$ is consistently estimated by the statistic

$$\hat{\sigma}_{V_k}^2(t) = \int_0^t \frac{K^2(v)}{Y^2(v)} dN(v) + \sum_{j=1}^{i-1} \int_{t_{j-1}}^{t_j} \frac{K^2(t)}{Y_j^2(v)} dN_j(v) \mathbf{1}\{i \ge 2\}$$

$$+ \int_{t_{i-1}}^t \frac{K^2(t)}{Y_i^2(v)} dA_i(v), \quad t \in [t_{i-1}, t_i), \; i = 1, \cdots, m.$$

So the test statistic is

$$Z_K = \frac{1}{\hat{\sigma}_{V_k}(t_m)} \sup_{t \in [0, t_m]} \mid \int_0^t K(v) d\{\hat{A}^{(1)}(v) - \hat{A}^{(2)}(v)\} \mid.$$

(12.40)

If $n \to \infty$ then

$$Z_K \xrightarrow{D} \sup_{0 \le u \le 1} \mid W(u) \mid.$$

Denote by $W_{1-\alpha}$ the $(1-\alpha)$-quantile of the supremum of the Wiener process on the interval $[0,1]$. The hypothesis H_0 is rejected with approximative significance level α if $Z_K > W_{1-\alpha}$.

Consistence of the first test against H_1 implies consistence of this test against H_1 because the convergence $T_n \overset{\mathbf{P}}{\to} \infty$ implies the convergence $Z_K \overset{\mathbf{P}}{\to} \infty$.

Let us consider the sequence of the approaching alternatives (2.39). Similarly as in the case of the hypothesis H_0 we have

$$T_n(t) = \int_0^t K(v)\,d\{\hat{A}^{(1)}(v) - A_*^{(1)}(v)\} - \int_0^t K(v)\,d\{\hat{A}^{(2)}(v) - A^{(2)}(v)\}+$$

$$\int_0^t K(v)\,d\{A_*^{(1)}(v) - A^{(2)}(v)\} = T_{1n}(t) + T_{2n}(t) + T_{3n}(t) \overset{D}{\to} V_k^*(t) + \Delta^*(t)$$

on $D[0, t_m]$, where

$$V_k^*(t) = \int_0^t k_*(v)\,dU^*(v) - \sum_{j=1}^{i-1} \int_{t_{j-1}}^{t_j} k_*(v)\,dU_j(v)\mathbf{1}\{i \geq 2\} - \int_{t_{i-1}}^t k(v)\,dU_i(v)$$

$$t \in [t_{i-1}, t_i), \ i = 1, \cdots, m, \ t_0 = 0,$$

U^* is a Gaussian martingale with $U^*(0) = 0$, and

$$\mathbf{Cov}\,(U^*(s_1), U^*(s_2)) = \frac{1}{l_0} \frac{1 - S_*(s_1 \wedge s_2)}{S_*(s_1 \wedge s_2)},$$

with $S_* = \exp\{-A_*\}$,

$$\Delta^*(t) = \mathbf{1}_{\{i \geq 2\}} \sum_{j=1}^{i-1} \varepsilon_j \int_{t_{j-1}}^{t_j} k_*(v)\ln(1 + \frac{t_{j-1}^* - t_{j-1}}{v})dv+$$

$$\varepsilon_i \int_{t_{i-1}}^t k_*(v)\ln(1 + \frac{t_{i-1}^* - t_{i-1}}{v})dv < 0, \ t \in [t_{i-1}, t_i).$$

Analogously as in the case of the hypothesis H_0,

$$V_k^*(t) = W(\sigma_{V_k}^2(t)),$$

and

$$\frac{1}{\hat{\sigma}_{V_k}(t_m)} \sup_{t \in [0, t_m]} |T_n(t)| \overset{D}{\to} \sup_{t \in [0, t_m]} |W\left(\frac{\sigma_{V_k}^2(t)}{\sigma_{V_k}^2(t_m)}\right) + \frac{\Delta^*(t)}{\sigma_{V_k}(t_m)}|$$

$$= \sup_{0 \leq u \leq 1} |W(u) + \frac{1}{c}\Delta^*(h(cu))|,$$

where $h(s)$ is the function inverse to $\sigma_{V_k}(t)$, and $c = \sigma_{V_k}(t_m)$.

Bagdonavičius and Levulienė (2001) investigated by simulation the properties of both tests under various alternatives. They showed that in the case of monotone stresses the logrank-type test is slightly more powerful then the Kolmogorov-type test.

Their results show that it is possible to find such plan of experiment with nonmonotone explanatory variables that a logrank-type test does not distinguish the hypothesis H_0 from the alternatives and is even biased. The Kolmogorov-type test can be used for such explanatory variables. The power of the test increases when the size of the data increases or the alternatives go away from the null hypothesis.

So the practical recommendation: use the logrank-type test for monotone stresses and the Kolmogorov-type test for nonmonotone stresses.

12.3 Goodness-of-fit for the AFT model

Suppose that the first two moments of the failure times under the explanatory variables $x(\cdot) \in E$ exist. Let $x^{(i)}(\cdot) \in E_1$ be explanatory variables of the form

$$
x^{(i)}(\tau) = \begin{cases} x_{n(1,i)}, & 0 \leq \tau < t_{i1}, \\ x_{n(2,i)}, & t_{i1} \leq \tau < t_{i2}, \\ \cdots & \cdots \quad \cdots \quad \cdots \\ x_{n(m,i)}, & \tau \geq t_{i,m-1}, \end{cases}
$$

where $n(1,i), \cdots, n(m,i)$ are the permutations of numbers $1, 2, \cdots, m$. $x^i \in E$ are constant explanatory variables; $t_{ij} \in [0, +\infty)$ are the moments of switching over from one constant explanatory variable to another one ($i = 1, \cdots, N; j = 1, \cdots, m$). In the particular case when $t_{i1} = \infty$, explanatory variable $x^{(i)}(\cdot)$ can be constant.

Suppose that $s \, (s > m)$ experiments are carried out and n_i units are tested under explanatory variable $x^{(i)}(\cdot)$ in the ith experiment ($i = 1, 2, \cdots, s$).

Let $T_{j1}^{(i)}, \cdots, T_{jn_i}^{(i)}$ be the lives of units under the constant explanatory variable x_j in the ith experiment.

If the AFT model holds, the following equalities are true:

$$
\sum_{j=1}^{m} r_j T_{jk}^{(i)} - 1 = \sigma R_{ik} \quad (i = 1, 2, \cdots, s)
$$

where $r_j = r(x_j)/a$; $\sigma > 0$; R_{ik} are independent identically distributed random variables with means $\mathbf{E}(R_{ik}) = 0$ and variances $\mathbf{Var}(R_{ik}) = 1$ ($k = 1, \cdots, n_i$). The means $\tau_j^{(i)} = \mathbf{E}(T_{jk}^{(i)})$ satisfy the equations

$$
\sum_{j=1}^{m} r_j \tau_j^{(i)} - 1 = 0 \quad (i = 1, 2, \cdots, s).
$$

Set

$$
T_k^{(i)} = \left(T_{1k}^{(i)}, \cdots, T_{mk}^{(i)} \right)^{\mathrm{T}}, \quad T_{j\cdot}^{(i)} = (1/n_i) \sum_{k=1}^{n_i} T_{jk}^{(i)}, \quad T_{\cdot}^{(i)} = \left(T_{1\cdot}^{(i)}, \cdots, T_{m\cdot}^{(i)} \right)^{\mathrm{T}},
$$

$$S_0^{(i)} = (1/n_i) \sum_{k=1}^{n_i} \left(T_k^{(i)} - T_\cdot^{(i)} \right) \left(T_k^{(i)} - T_\cdot^{(i)} \right)^{\mathrm{T}}, \quad \tau^{(i)} = \mathbf{E}(T_k^{(i)})$$

and define the estimates of parameters $r = (r_1, \cdots, r_m)^{\mathrm{T}}$ by minimizing the sum

$$\sum_{i=1}^{s} n_i \left(r^{\mathrm{T}} T_\cdot^{(i)} - 1 \right)^2.$$

The normal equations have the form

$$\sum_{i=1}^{s} n_i T_\cdot^{(i)} \left(T_\cdot^{(i)} \right)^{\mathrm{T}} r = \sum_{i=1}^{s} n_i T_\cdot^{(i)}. \tag{12.41}$$

Suppose that a system of vectors $\tau^{(i)}$, $i = 1, 2, \cdots, s$ has a rank m. This condition is satisfied practically in all cases when $x^{(i)}(\cdot)$ are different explanatory variables. Under the AFT model the solution of normal equations $\hat{r}$ converges with probability one (as $min\, n_i \to \infty$, $n_i/ max\, n_i \to l_i > 0$) to the parameters r satisfying a system of equations $r^{\mathrm{T}} \tau^{(i)} - 1 = 0$ $(i = 1, \cdots, s)$ and the estimator

$$\hat{\sigma}^2 = \left(1 / \left(\sum_{i=1}^{s} n_i - s \right) \right) \sum_{i=1}^{s} \sum_{k=1}^{n_i} \left\{ \sum_{j=1}^{m} \hat{r}_j (T_{jk}^{(i)} - T_{j\cdot}^{(i)}) \right\}^2$$

converges with probability one to a parameter σ^2 satisfying the equation

$$r^{\mathrm{T}} B^{(i)} r = \sigma^2,$$

where $(B^{(i)} = \mathbf{E}(S_0^{(i)}))$.

Theorem 12.2. *Assume that*

1) *the AFT model holds on E;*
2) *there exist two moments of the random variables $T_{(\cdot)}$, $x(\cdot) \in E$;*
3) *the system of vectors $\tau^{(i)}$, $i = 1, \cdots, s$ has a rank m.*

Then the asymptotical distribution of the statistic

$$Y^2 = (1/\hat{\sigma}^2) \sum_{i=1}^{s} n_i \left(\hat{r}^{\mathrm{T}} T^{(i)} - 1 \right)^2$$

is chi-square with $s - m$ degrees of freedom as

$$min\, n_i \to \infty, \quad \frac{n_i}{max\, n_i} \to l_i > 0.$$

Proof. Set

$$n = max\, n_i; \quad L = \left(\sqrt{l_1}, \cdots, \sqrt{l_s} \right)^{\mathrm{T}}; \quad T = (\sqrt{l_i} T_{j\cdot}^{(i)}),$$

the $s \times m$ matrix; $Z(r) = \sqrt{K}(Tr - L)$. In such a case

$$\sum_{i=1}^{s} n_i \left(r^{\mathrm{T}} T^{(i)} - 1 \right)^2 = (Z(r))^{\mathrm{T}} Z(r).$$

Let C be a $s \times m$ matrix with elements $\sqrt{l_i} \tau_j^{(i)}$ $(i = 1, \cdots, s; j = 1, \cdots, m)$.
From the assumption 3) of the theorem it follows that the rank of the matrix C is equal to m. The equalities

$$\mathrm{rank}\ C^{\mathrm{T}} C = \mathrm{rank}\ C = M$$

implies that a random matrix $T^{\mathrm{T}} T \to C^{\mathrm{T}} C$ with probability one. Therefore, the solution of the equation (12.41) is the statistic

$$\hat{r} = \left(T^{\mathrm{T}} T\right)^{-1} T^{\mathrm{T}} L$$

provided $min\ n_i$ is sufficiently large. The distribution of a random variable $(1/\sigma^2) \left(Z(r)\right)^{\mathrm{T}} Z(r)$ converges to a chi-square distribution with s degrees of freedom.

Let us consider the limit distribution of a random variable

$$\min_{r} \left(Z(r)\right)^{\mathrm{T}} Z(r) = (Z(\hat{r}))^{\mathrm{T}} Z(\hat{r}).$$

It is easy to show that the random vector $Z(\hat{R})$ can be expressed in the form:

$$Z(\hat{r}) = (E - T(T^{\mathrm{T}} T)^{-1} T^{\mathrm{T}}) \sqrt{n}(Tr - L),$$

where E is a $s \times s$ identity matrix. The former implies that the asymptotic distributions of random variables

$$Z(\hat{r}) \quad \text{and} \quad (E - C(C^{\mathrm{T}} C)^{-1} C^{\mathrm{T}}) Z(r)$$

are the same. The matrix

$$A = E - C(C^{\mathrm{T}} C)^{-1} C^{\mathrm{T}}$$

is idempotent, i.e. $AA = A$. This implies that asymptotic distributions of random variables $(Z(\hat{r}))^{\mathrm{T}} Z(\hat{r})$ and $(Z(r))^{\mathrm{T}} A Z(r)$ are the same. The random variable

$$\frac{1}{\hat{\sigma}^2} \left(Z(r)\right)^{\mathrm{T}} A Z(r)$$

has asymptotically a chi-square distribution with $s - m$ degrees of freedom. This follows from the equality

$$\mathrm{rank}\ A = \mathrm{tr}(A) = s - m$$

and from the fact that $\hat{\sigma} \to \sigma$ with probability one. Thus, the proof is completed.

Corollary 12.3. *Under assumptions of the theorem the asymptotic distribution of the statistic*

$$\chi^2 = \sum_{i=1}^{s} n_i (\hat{r}^{\mathrm{T}} T^{(i)} - 1)^2 / (\hat{r}^{\mathrm{T}} S_0^{(i)} \hat{r})$$

is chi-square with $s - m$ degrees of freedom.

If the AFT model is not true, the estimators $\hat{r}$ converges, on the whole, with probability one to some value r_0 but some of the equalities

$$r_0 \tau^{(i)} - 1 = 0 \quad \text{or} \quad r_0^{\mathrm{T}} B^{(i)} r_0 = \sigma^2 \ (i = 1, .., s)$$

take no place. If the sequence of alternatives is such that $\mid \nu_i \mid < const$, where

$$\nu_i = \sqrt{n_i}(r_0^{\mathrm{T}} \tau^{(i)} - 1)/(r_0^{\mathrm{T}} B^{(i)} r_0)^{1/2},$$

the asymptotic distribution of the statistic χ^2 is noncentral chi-square with $s - m$ degrees of freedom and the noncentrality parameter $\sum\limits_{i=1}^{s} \nu_i^2$.

The statistic χ^2 can be used as a test statistic when samples are sufficiently large.

12.4 Goodness-of-fit for the PH model

Let us consider tests for the PH hypothesis

$$H_0 : \alpha_{x(\cdot)}(t) = e^{\beta^T x(t)} \alpha(t),$$

where β and α are unknown regression parameter and baseline function, respectively.

Suppose that n units are tested. The ith unit is tested under the explanatory variable $x^{(i)}(\cdot)$.

Suppose that the data are right censored. Denote by T_i and C_i the failure and censoring times,

$$X_i = T_i \wedge C_i, \quad \delta_i = \mathbf{1}_{\{T_i \leq C_i\}}, \quad N_i(t) = \mathbf{1}\{T_i \leq t, \delta_i = 1\},$$

$$Y_i(t) = \mathbf{1}_{\{X_i \geq t\}}, \quad N(t) = \sum_{i=1}^{n} N_i(t) \quad \text{and} \quad Y(t) = \sum_{i=1}^{n} Y_i(t).$$

We assume that at the nonrandom moment τ all censoring and failure processes are censored.

12.4.1 Score tests

Let us consider general score tests for checking the adequacy of the PH model based on likelihood functions under more general models including this model. Examples of such models are:

GPH_{GW} model (any ratio of the hazard rates under constant explanatory variables increases (or decreases) from a finite value to a finite value but the hazard rates do not intersect or meet):

$$\alpha_{x(\cdot)}(t) = e^{\beta^T x(t)}(1 + A_{x(\cdot)}(t))^\gamma \alpha(t);$$

GPH_{GLL} model (any ratio of hazard rates increases or decreases from a finite value to 1, i.e. the hazard rates meet at infinity):

$$\alpha_{x(\cdot)}(t) = e^{\beta^T x(t) + \gamma A_x(t)} \, \alpha(t);$$

IGF model (the ratio of hazard increases (or decreases) from 1 to a finite value, i.e. the hazard rates meet at zero):

$$\alpha_{x(\cdot)}(t) = e^{\beta^T x(t)} \, \frac{\alpha(t)}{1 + \gamma A_{x(\cdot)}(t)};$$

two GPH2 models with cross-effects of the hazard rates:

$$\alpha_{x(\cdot)}(t) = e^{\beta^T x(t)} (1 + A_{x(\cdot)}(t))^{\gamma^T x(t)} \, \alpha(t),$$

$$\alpha_{x(\cdot)}(t) = e^{(\beta+\gamma)^T x(t)} (A(t))^{e^{\gamma^T x(t)} - 1} \, \alpha(t);$$

PH model with time-varying regression coefficients:

$$\alpha_{x(\cdot)}(t) = e^{\beta^T(t) x(t)} \alpha(t) = \exp\{\sum_{j=1}^{m} \beta_j(t) \, x_j(t)\} \, \alpha(t),$$

where $\beta_j(t) = \beta_j + \gamma_j \, g_j(t)$, β_j, γ_j are unknown parameters and $g_j(\cdot)$ are specified deterministic functions or paths of predictable processes. For example, $g_j(t) = t$; $\ln t$; $\mathbf{1}_{\{t \le t_0\}}$; $N(t-)$; $\hat{F}(t-)$, etc.

Set

$$A(t) = \int_0^t \alpha(u) du, \quad \theta = (\theta_1, \cdots, \theta_{m+r})^T = (\beta_1, \cdots, \beta_m, \gamma_1, \cdots, \gamma_r)^T,$$

where $r = 1$, when γ is one-dimensional (GPH1 models), and $r = m$, when γ is m-dimensional (in the case of GPH2 models and models with time-varying coefficients). Each of above mentioned models is an alternative to the PH model if $\gamma \ne 0$.

Test statistics

If A is completely known then under any of the above mentioned alternatives the parametric maximum likelihood estimator of the parameter θ is obtained by solving the system of equations

$$U_j(\tau; \theta, A) = 0, \quad (j = 1, \cdots, m + r),$$

where

$$U_j(t; \theta, A) = \sum_{i=1}^{n} \int_0^t \frac{\partial}{\partial \theta_j} \log\{\alpha_i(t, \theta)\}\{dN_i(u) - Y_i(u)\alpha_i(u, \theta) du\}.$$

Note that if A is unknown, then the partial derivative

$$\frac{\partial}{\partial \theta_j} \log\{\alpha_i(t, \theta)\} = w_j^{(i)}(t, \theta, A)$$

depends on a finite dimensional parameter $\theta = (\beta^T, \gamma^T)^T$ and a infinite dimensional parameter A. Under the PH model $\gamma = 0$ and A is the baseline hazard corresponding to this model.

Set
$$\hat{U}_j = \hat{U}_j(\tau), \quad \text{where} \quad \hat{U}_j(t) = U_j(t, (\hat{\beta}^T, 0^T)^T, \hat{A}),$$

where $\hat{\beta}$ is the partial likelihood estimator of the regression parameter β and $\hat{A}$ is the Breslow (1975b) estimator of the baseline cumulative hazard under the PH model.

Note that under PH model
$$\hat{U}_j = 0, \quad (j = 1, \cdots, m)$$

and

$$\hat{U}_j(t) = \sum_{i=1}^{n} \int_0^t \hat{w}_j^{(i)}(u) d\hat{M}_i(u), \quad (j = m+1, \cdots, m+r), \qquad (12.42)$$

where $\hat{M}_i(t)$ are the martingale residuals corresponding to the PH model:

$$\hat{M}_i(t) = N_i(t) - \int_0^t Y_i(u)\, e^{\hat{\beta}^T x^{(i)}(u)} d\hat{A}(u) = N_i(t) - \int_0^t Y_i(u)\, e^{\hat{\beta}^T x^{(i)}(u)} \frac{dN(u)}{S^{(0)}(u, \hat{\beta})},$$

$S^{(0)}(t, \beta) = \sum_{i=1}^{n} Y_i(t)\, e^{\beta^T x^{(i)}(u)}$, $\hat{w}_j^{(i)}(t) = w_j^{(i)}(t, (\hat{\beta}^T, 0^T)^T, \hat{A})$.

In dependence on the alternative the weights $\hat{w}_j^{(i)}(t)$ have the following forms:

for the GPH_{GW} model:
$$\hat{w}_j^{(i)}(t) = \ln\{1 + e^{\hat{\beta}^T x^{(i)}(t)} \hat{A}(t)\}; \qquad (12.43)$$

for the GPH_{GLL} model:
$$\hat{w}_j^{(i)}(t) = e^{\hat{\beta}^T x^{(i)}(t)} \hat{A}(t); \qquad (12.44)$$

for the IGF model:
$$\hat{w}_j^{(i)}(t) = e^{\hat{\beta}^T x^{(i)}(t)} \hat{A}(t); \qquad (12.45)$$

for the first GPH2 model:
$$\hat{w}_j^{(i)}(t) = x_j^{(i)}(t) \ln\{1 + e^{\hat{\beta}^T x^{(i)}(t)} \hat{A}(t)\}; \qquad (12.46)$$

for the second GPH2 model:
$$\hat{w}_j^{(i)}(t) = x_j^{(i)}(t) \ln\{1 + \hat{A}(t)\}; \qquad (12.47)$$

for the models with time-varying coefficients
$$\hat{w}_j^{(i)}(t) = x_j^{(i)}(t) g_j(t). \qquad (12.48)$$

The test is based on the statistics
$$\hat{U} = (\hat{U}_{m+1}, \cdots, \hat{U}_{m+r})^T. \qquad (12.49)$$

To construct a test we need the asymptotic distribution of $\hat{U}$ under the PH model.

Asymptotic properties of the test statistics

Set

$$\hat{w}^{(i)}(t) = (\hat{w}^{(i)}_{m+1}(t), \cdots, \hat{w}^{(i)}_{m+r}(t))^{T},$$

$$S^{(1)}(t, \beta) = \sum_{i=1}^{n} x^{(i)}(t) Y_i(t)\, e^{\beta^T x^{(i)}(t)},$$

$$S^{(2)}(t, \beta) = \sum_{i=1}^{n} (x^{(i)}(t))^{\otimes 2} Y_i(t)\, e^{\beta^T x^{(i)}(t)},$$

$$E(t, \beta) = \frac{S^{(1)}(t, \beta)}{S^{(0)}(t, \beta)}, \quad \tilde{S}^{(1)}(t, \beta) = \sum_{i=1}^{n} \hat{w}^{(i)}(t) Y_i(t)\, e^{\beta^T x^{(i)}(t)},$$

$$\tilde{E}(t, \beta) = \frac{\tilde{S}^{(1)}(t, \beta)}{S^{(0)}(t, \beta)}, \quad \tilde{S}^{(2)}(t, \beta) = \sum_{i=1}^{n} \hat{w}^{(i)}(t)(x^{(i)}(t))^{T} Y_i(t)\, e^{\beta^T x^{(i)}(t)},$$

$$\tilde{\tilde{S}}^{(2)}(t, \beta) = \sum_{i=1}^{n} (\hat{w}^{(i)}(t))^{\otimes 2} Y_i(t)\, e^{\beta^T x^{(i)}(t)}.$$

Denote by β_0 the true value of β.

Suppose that Conditions A of Chapter 5 are verified. Suppose also that conditions analogous to the conditions a)-d) not only for $S^{(i)}$ but also for $\tilde{S}^{(i)}$, $(i = 1, 2)$ and $\tilde{\tilde{S}}^{(2)}$ are verified.

The Doob-Meier decomposition, the formulas (7.60), Theorem 7.1 and the delta method imply that

$$n^{-1/2}\hat{U}(t) = n^{-1/2} \sum_{i=1}^{n} \int_0^t \{\hat{w}^{(i)}(u) - \tilde{E}(u, \hat{\beta})\} dN_i(u) =$$

$$n^{-1/2} \sum_{i=1}^{n} \int_0^t \{\hat{w}^{(i)}(u) - \tilde{E}(u, \hat{\beta})\} dM_i(u) +$$

$$n^{-1/2} \int_0^t \{\tilde{E}(u, \beta_0) - \tilde{E}(u, \hat{\beta})\} S^{(0)}(t, \beta_0) dA(u) =$$

$$n^{-1/2} \sum_{i=1}^{n} \int_0^t \{\hat{w}^{(i)}(u) - \tilde{E}(u, \beta_0)\} dM_i(u) -$$

$$n^{-1} \int_0^t \frac{\partial \tilde{E}(u, \beta_0)}{\partial \beta} S^{(0)}(u, \beta_0) dA(u)\, n^{1/2}(\hat{\beta} - \beta_0) + o_p(1) =$$

$$n^{-1/2} \sum_{i=1}^{n} \int_0^t \{\hat{w}^{(i)}(u) - \tilde{E}(u, \beta_0)\} dM_i(u) -$$

$$\Sigma^*(t)\Sigma^{-1}(\tau)n^{-1/2}\sum_{i=1}^{n}\int_0^\tau \{x^{(i)}(u) - E(u, \beta_0)\}dM_i(u) + o_p(1), \qquad (12.50)$$

where $\Sigma(t)$ and $\Sigma^*(t)$ are the limits in probability of the random matrices (cf. (7.58)),

$$\hat{\Sigma}(t) = n^{-1}\int_0^t V(u, \hat{\beta})dN(u), \quad \hat{\Sigma}^*(t) = n^{-1}\int_0^t \tilde{V}(u, \hat{\beta})dN(u),$$

$$V(t, \beta) = \frac{S^{(2)}(t, \beta)}{S^{(0)}(t, \beta)} - (E(t, \beta))^{\otimes 2}, \quad \tilde{V}(t, \beta) = \frac{\tilde{S}^{(2)}(t, \beta)}{S^{(0)}(t, \beta)} - \tilde{E}(t, \beta)E^T(t, \beta).$$

$$(12.51)$$

So

$$< n^{-1/2}\hat{U} > (t) = n^{-1}\sum_{i=1}^{n}\int_0^t \{\hat{w}^{(i)}(u) - \tilde{E}(u, \beta_0)\}^{\otimes 2}e^{\beta_0^T x^{(i)}(u)}Y_i(u)dA(u) -$$

$$2\Sigma^*(t)\Sigma^{-1}(\tau)n^{-1}\sum_{i=1}^{n}\int_0^t \{\hat{w}^{(i)}(u) - \tilde{E}(u, \beta_0)\}\{x^{(i)}(u) - E(u, \beta_0)\}^T e^{\beta_0^T x^{(i)(u)}} \times$$

$$Y_i(u)dA(u) + \Sigma^*(t)\Sigma^{-1}(\tau)n^{-1}\sum_{i=1}^{n}\int_0^\tau \{x^{(i)}(u) - E(u, \beta_0)\}^{\otimes 2}e^{\beta_0^T x^{(i)}(u)} \times$$

$$Y_i(u)dA(u)\Sigma^{-1}(\tau)(\Sigma^*(t))^T + o_p(1) = \Sigma^{**}(t) - \Sigma^*(t)\Sigma^{-1}(\tau)(\Sigma^*(t))^T + o_p(1),$$

where $\Sigma^{**}(t)$ is the limit in probability of the random matrice

$$\hat{\Sigma}^{**}(t) = n^{-1}\int_0^t \tilde{\tilde{V}}(u, \hat{\beta})dN(u),$$

$$\tilde{\tilde{V}}(u, \beta) = \frac{\tilde{\tilde{S}}^{(2)}(t, \beta)}{S^{(0)}(t, \beta)} - \left(\tilde{E}(t, \beta)\right)^{\otimes 2}. \qquad (12.52)$$

Similarly as proving the Lindeberg condition (7.62) (see Andersen, Borgan, Gill and Keiding (1993)) it can be shown that

$$n^{-1}\sum_{i=1}^{n}\int_0^\tau \{\hat{w}_j^{(i)}(u) - \tilde{E}_j(u, \beta_0)\}^2 1_{\{|\hat{w}_j^{(i)}(u) - \tilde{E}_j(u, \beta_0)| \geq \sqrt{n}\varepsilon\}}e^{\beta_0^T x^{(i)}}Y_i(u)dA(u) \xrightarrow{P} 0.$$

Theorem A.7 implies that the stochastic process $n^{-1/2}\hat{U}$ converges in distribution to a zero mean Gaussian process, in particular

$$n^{-1/2}\hat{U} \xrightarrow{D} N_r(0, D),$$

where $D = \Sigma^{**}(\tau) - \Sigma^*(\tau)\Sigma^{-1}(\tau)(\Sigma^*(\tau))^T$.

Test

The critical region of the chi-square type test with approximate significance level α is $T > \chi^2_{1-\alpha}(r)$, where

$$T = n^{-1}\hat{U}^T\hat{D}\hat{U}, \qquad (12.53)$$

$$\hat{D} = \hat{\Sigma}^{**}(\tau) - \hat{\Sigma}^{*}(\tau)\hat{\Sigma}^{-1}(\tau)(\hat{\Sigma}^{*}(\tau))^{T}. \qquad (12.54)$$

The calculation of the test statistic is simple:

$$U = \sum_{j=1}^{n} \delta_j \{\hat{w}^{(i)}(X_j) - \tilde{E}(X_j, \hat{\beta})\}, \quad \hat{\Sigma}(\tau) = n^{-1} \sum_{j=1}^{n} \delta_j V(X_j, \hat{\beta}),$$

$$\hat{\Sigma}^{*}(\tau) = n^{-1} \sum_{j=1}^{n} \delta_j \tilde{V}(X_j, \hat{\beta}), \quad \hat{\Sigma}^{**}(\tau) = n^{-1} \sum_{j=1}^{n} \delta_j \tilde{\tilde{V}}(X_j, \hat{\beta}), \qquad (12.55)$$

where $V, \tilde{V}, \tilde{\tilde{V}}$ are defined by the formulas (12.50),(12.51), and $\hat{w}^{(i)}$ by (12.43)-(12.48).

Note that if the alternative is the PH model with time-varying coefficients then

$$\tilde{V}(u,\beta) = G(u)V(u,\beta), \quad \tilde{\tilde{V}}(u,\beta) = G(u)V(u,\beta)G(u)^{T},$$

where $G(u)$ is a diagonal matrix with the elements $g_1(u), \cdots, g_m(u)$ on the diagonal. In this case the test statistic (12.52) coincides with the statistic of Grambsch and Therneau (1994) obtained using generalized least squares procedure. In dependence of the choice of g_j particular cases of such statistic are (or are equivalent) the test statistics given by Cox (1972), Moreau, O'Quigley, Mesbah (1985), Lin (1991), Nagelkerke, Oosting, Hart (1984), Gill and Schumacher (1987). The test with the weight (12.47) is equivalent to the test of Quantin et al (1996). In the case of univariate explanatory variable Grambsch and Therneau (1994) show how the function $\beta(t)$ can be visualized by smoothing plots of $V^{-1}(X_i, \hat{\beta})\hat{r}_i + \hat{\beta}$ with $\delta_i = 1$ versus X_i; here $\hat{r}_i = x^{(i)}(X_i) - E(X_i, \hat{\beta})$ are Schoenfeld residuals (1980), $\hat{\beta}$ is the partial likelihood estimator.

12.4.2 Tests based on estimated score process

The estimated score process under the PH model is defined as

$$\hat{U}(t) = U(t, \hat{\beta}) = \sum_{i=1}^{n} \int_0^t \{x^{(i)}(u) - E(u, \hat{\beta})\} \, dN_i(u) = \sum_{i=1}^{n} \int_0^t x^{(i)}(u) d\hat{M}_i(u).$$

So it is a special case of the process with the components (12.42) and $\hat{w}^{(i)}(u) = x^{(i)}(u)$. In this case $\tilde{V} = \tilde{\tilde{V}} = V$, $\Sigma^{**} = \Sigma^{*} = \Sigma$ and by (12.49)

$$n^{-1/2}\hat{U}(t) = n^{-1/2} \sum_{i=1}^{n} \int_0^t \{x^{(i)}(u) - E(u, \beta_0)\} dM_i(u) -$$

$$\Sigma(t)\Sigma^{-1}(\tau)n^{-1/2} \sum_{i=1}^{n} \int_0^\tau \{x^{(i)}(u) - E(u, \beta_0)\} dM_i(u) + o_p(1) =$$

$$n^{-1/2}U(t, \beta_0) - \Sigma(t)\Sigma^{-1}(\tau)n^{-1/2}U(\tau, \beta_0) + o_p(1) := B(t) + o_p(1). \qquad (12.56)$$

Denote by $\sigma_{jj'}(s)$ the elements of the matrix $\Sigma(t)$. By Theorem 7.1

$$n^{-1/2}\hat{U}(\cdot) \xrightarrow{D} Z(\cdot,\beta_0) - \Sigma(t)\Sigma^{-1}(\tau)Z(\tau,\beta_0) \quad \text{on } (D[0,\tau])^m, \qquad (12.57)$$

where Z is a m-variate Gaussian process having components with independent increments, $Z_j(0) = 0$ a.s. and for all $0 \leq s \leq t \leq \tau$:

$$\mathbf{cov}(Z_j(s), Z_{j'}(t)) = \sigma_{jj'}(s).$$

Note that $Z_j(t) = W(\sigma_{jj}(t))$, where W is the standard Wiener process.

Using the fact that that under *one-dimensional* explanatory variables the limit of the estimated score process can be transformed to the Brownian bridge, Wei (1984) proposed the following test for this case.

Set $a(t) = \sigma_{11}(t)$. The formula (12.55) implies that

$$n^{-1/2}\hat{U}(t) \xrightarrow{D} W(a(t)) - \frac{a(t)}{a(\tau)}W(a(\tau)) \quad \text{on} \quad D[0,\tau]. \qquad (12.58)$$

We have

$$\frac{\hat{U}(\cdot)}{\sqrt{na(\tau)}} \xrightarrow{D} W\left(\frac{a(\cdot)}{a(\tau)}\right) - \frac{a(\cdot)}{a(\tau)}W(1)$$

on $D[0,\tau]$. Then

$$\sup_{t\in[0,\tau]} \left| \frac{\hat{U}(t)}{\sqrt{na(\tau)}} \right| \xrightarrow{D} \sup_{u\in[0,1]} | W(u) - uW(1) | = K,$$

and

$$T = \sup_{t\in[0,\tau]} \left| \frac{\hat{U}(t)}{\sqrt{n\hat{a}(\tau)}} \right| \xrightarrow{D} K;$$

here

$$\hat{a}(\tau) = \frac{1}{n}\int_0^t V(u,\hat{\beta})dN(u).$$

So the statistic T converges in distribution to the supremum of the Brownian bridge, i.e. the limit distribution is Kolmogorov. Denote by K_α the α-quantile of the random variable K.

Test (univariate case)

The approximate critical region with the significance level α is

$$T > K_{1-\alpha}.$$

If $x^{(i)}(\cdot)$ are m-dimensional, then the components of the limit process (12.56) can not be transformed to the Brownian bridges because they are dependent. Lin et al. (1993) propose to use a statistic

$$T = \sup_t \sum_{j=1}^m \{\hat{\sigma}^{jj}(\tau)\}^{1/2} | \hat{U}_j(t) |, \qquad (12.59)$$

where $U_j(t)$ is the jth component of the estimated score statistic $\hat{U}(t)$ and $\hat{\sigma}^{jj}(\tau, \hat{\beta})$ are the diagonal elements of $\hat{\Sigma}^{-1}(\tau)$.

The limit distribution of the statistic T can be generated through simulation.

Indeed, the equalities

$$\mathbf{E}\{M_i(t)\} = 0 \quad \text{and} \quad \mathbf{Cov}\{M_i(s), M_i(t)\} = \mathbf{E}\{N_i(s)N_i(t)\}$$

imply that if V_i is a random variable which does not depend on $N_i(\cdot)$, then

$$\mathbf{E}\{V_i N_i(t)\} = \mathbf{E}\{M_i(t)\} = 0, \quad \mathbf{Cov}\{V_i N_i(s), V_i N_i(t)\} = \mathbf{Cov}\{M_i(s), M_i(t)\}.$$

So the covariance structure of the stochastic processes $V_i N_i(\cdot)$ and $N_i(\cdot)$ is the same.

By replacing $M_i(\cdot)$ with $N_i(\cdot)V_i$, where $V_1, \cdots, V_n$ are independent standard normal variables which are independent of $(Y_i(\cdot), N_i(\cdot), x^{(i)}(\cdot))$, and replacing β by $\hat{\beta}$, Σ by $\hat{\Sigma}$ in the expression of $B(t)$ given in (12.55), we obtain the stochastic process

$$\hat{B}(t) = n^{-1/2} \sum_{i=1}^{n} \int_0^t \{x^{(i)}(u) - E(u, \hat{\beta})\} V_i dN_i(u) -$$

$$\hat{\Sigma}(t)\hat{\Sigma}^{-1}(\tau) n^{-1/2} \sum_{i=1}^{n} \int_0^\tau \{x^{(i)}(u) - E(u, \hat{\beta})\} V_i dN_i(u).$$

The conditional (given $\{N_i(\cdot), Y_i(\cdot), x^{(i)}(\cdot)\}$) distribution of $\hat{B}(\cdot)$ has the same limit as the unconditional distribution of $B(\cdot)$, consequently, the same as the unconditional distribution of $n^{-1/2}\hat{U}(t)$ (see Appendix of Lin et al (1993)). So the distribution of the statistic T can be approximated by simulating samples $\{V_i; i = 1, \cdots, n\}$ while fixing the data $\{N_i(\cdot), Y_i(\cdot), x^{(i)}\}$.

Set

$$\hat{T} = \sup_t \sum_{j=1}^{m} \{\hat{\sigma}^{jj}(\tau)\}^{1/2} \mid n^{1/2}\hat{B}(t) \mid$$

Test (multivariate case)

Let t_0 be the observed value of T. Then the P-value $\mathbf{P}(T > t_0)$ can be approximated by $\mathbf{P}(\hat{T} > t_0)$, the latter probability being approximated through simulation of the values of $\hat{T}$.

Remark 12.4. Goodness-of-fit for the AH model based on the estimated score process are obtained similarly and are given in Kim *et al* (1998).

12.4.3 Tests based on linear combinations of martingale residuals with two variable weights

Lin et al. (1993) considered one-dimensional statistics of the form (we write the evident generalization of them when explanatory variables are time-varying)

$$U^{(1)}(t,x) = \sum_{i=1}^{n} \int_0^t \mathbf{1}_{\{x^{(i)}(u) \leq x\}} d\hat{M}_i(u),$$

$$U^{(2)}(t,x) = \sum_{i=1}^{n} \int_0^t \mathbf{1}_{\{\hat{\beta}^T x^{(i)}(u) \leq x\}} d\hat{M}_i(u),$$

where $\hat{M}_i$ are the martingale residuals, and the event $\{x^{(i)}(u) \leq x\}$ means that all the m components of $x^{(i)}(u)$ are not larger than the respective components of x. For the second statistic x is a one-dimensional variable.

Under each fixed x the statistics $U^{(1)}(t,x)$ and $U^{(2)}(t,x)$ are special cases of the process with the components (12.42) and $\hat{w}^{(i)}(u,x) = \hat{w}^{(1i)}(u,x) = \mathbf{1}_{\{x^{(i)}(u) \leq x\}}$ and $\hat{w}^{(i)}(u,x) = \hat{w}^{(2i)}(u,x) = \mathbf{1}_{\{\hat{\beta}^T x^{(i)}(u) \leq x\}}$, respectively.

By (12.49)

$$n^{-1/2} U^{(j)}(t,x) = n^{-1/2} \sum_{i=1}^{n} \int_0^t \{\hat{w}^{(ji)}(u,x) - \tilde{E}^{(j)}(u,x,\beta_0)\} dM_i(u)$$

$$-\Sigma_j^*(t,x)\Sigma^{-1}(\tau) n^{-1/2} \sum_{i=1}^{n} \int_0^\tau \{x^{(i)}(u) - E(u,\beta_0)\} dM_i(u) + o_p(1),$$

where $\Sigma_j^*(t,x)$ is the limit in probability of the random matrix $\hat{\Sigma}_j^*(t,x)$. Note that $\hat{\Sigma}_j^*$ and $E^{(j)}$ are the special cases of $\hat{\Sigma}^*$ and $\tilde{E}$ taking $\hat{w}^{(i)} = \hat{w}^{(ji)}$ in all definitions. So, as in previous section, the distribution of $n^{-1/2} U^{(j)}(t,x)$ is approximated by the distribution of the statistic

$$n^{-1/2} \hat{U}^{(j)}(t,x) = n^{-1/2} \sum_{i=1}^{n} \int_0^t \{\hat{w}^{(ji)}(u) - \tilde{E}^{(j)}(u,\hat{\beta})\} V_i dN_i(u) -$$

$$\hat{\Sigma}_j^*(t)\hat{\Sigma}^{-1}(\tau) n^{-1/2} \sum_{i=1}^{n} \int_0^\tau \{x^{(i)}(u) - E(u,\hat{\beta})\} V_i dN_i(u).$$

The global test statistic for assessment of the PH model is

$$T = \sup_{t,x} | n^{-1/2} U^{(1)}(t,x) |.$$

Set

$$\hat{T} = \sup_{t,x} | n^{-1/2} \hat{U}^{(1)}(t,x) |.$$

Let t_0 be the observed value of T. Then the P-value $\mathbf{P}(T > t_0)$ can be approximated by $\mathbf{P}(\hat{T} > t_0)$, the latter probability being approximated through simulation of the values of $\hat{T}$.

Computing of the P-value may be time consuming if the explanatory variables are not constant or step functions with a small number of different values.

12.5 Goodness-of-fit for the GPH models

12.5.1 Estimators used for goodness-of-fit test construction

We shall consider construction of goodness-of-fit tests for the GPH1 models in two cases when

1. q is specified, i.e. tests for some relatively narrow models (logistic regression model being an example);

2. q is parametrized via the parameter γ, i.e. tests for wider models (such as GPH_{GW}, GPH_{GLL}, etc.).

Suppose that s groups of units are tested. The ith group of n_i units is tested under the explanatory variable $x_i(\cdot)$. The data are right censored.

Denote by T_{ij} and C_{ij} the failure and censoring times,

$$X_{ij} = T_{ij} \wedge C_{ij}, \quad \delta_{ij} = I\{T_{ij} \leq C_{ij}\}, \quad N_{ij}(t) = \mathbf{1}_{\{T_{ij} \leq t, \delta_{ij}=1\}},$$

$$Y_{ij}(t) = \mathbf{1}_{\{X_{ij} \geq t\}}, \quad N_i(t) = \sum_{j=1}^{n_i} N_{ij}, \quad Y_i(t) = \sum_{j=1}^{n_i} Y_{ij},$$

$$N(t) = \sum_{i=1}^{s} N_i(t) \quad Y(t) = \sum_{i=1}^{s} Y_i(t).$$

We assume that at the nonrandom moment τ all experiments are censored.

We'll construct tests for the hypothesis

$$H_0 : \alpha_i(t) = e^{\beta^T x_i(t)} q(A_i(t); \gamma)\alpha_0(t),$$

where the function q is completely specified (models of the first level) or parametrized via some parameter γ (models of the second level), α_0 is an unknown baseline function.

The function q can depend on $x^{(i)}$ (as in the GPH2 models). We do not write this variable in the expressions of q but this possibility can be taken in mind. The formulas do not change.

We'll write all formulae in this chapter for the more general model of the second level which includes γ. If q is specified, all needed slight modifications will be indicated.

Let

$$\hat{A}_i(t) = \int_0^t \frac{dN_i(y)}{Y_i(y)}$$

be the Nelson-Aalen estimator of $A_i(t) = A_{x_i(\cdot)}(t)$. Denote $\theta = (\beta^T, \gamma^T)^T$.

The modified partial score function has the form $U(\theta) = U(\theta, \tau)$, where

$$U(\theta, t) = \sum_{i=1}^{s} \int_0^t \{\tilde{z}_i(u, \gamma) - \tilde{E}(u, \theta)\}dN_i(u),$$

where

$$\tilde{z}_i(u,\gamma) = (x_i^T(u), (\ln q(\hat{A}_i(u);\gamma))_\gamma')^T$$

$$\tilde{E}(u,\theta) = \frac{\tilde{S}^{(1)}(u,\theta)}{\tilde{S}^{(0)}(u,\theta)}, \quad \tilde{S}^{(0)}(u,\theta) = \sum_{i=1}^{s} \tilde{S}_i^{(0)}(u,\theta),$$

$$\tilde{S}_i^{(0)}(u,\theta) = e^{\beta^T x_i(u)} Y_i(u) q(\hat{A}_i(u),\gamma), \quad \tilde{S}^{(1)}(u,\theta) = \sum_{i=1}^{s} \tilde{S}_i^{(1)}(u,\theta),$$

$$\tilde{S}_i^{(1)}(u,\theta) = \tilde{z}_i(u;\gamma)\tilde{S}_i^{(0)}(u,\theta).$$

For the models of the first level $\tilde{z}_i(u,\gamma)$ is changed by $x_i(u)$ and the argument γ is absent in q.

Denote by $\hat{\theta}$ the estimator, verifying the equation $U(\hat{\theta}) = 0$ and

$$S_i^{(0)}(u,\theta) = e^{\beta^T x_i(u)} Y_i(u) q(A_i(u),\gamma), \quad S^{(0)}(u,\theta) = \sum_{i=1}^{s} S_i^{(0)}(u,\theta),$$

$$z_i(u,\gamma) = (x_i(u), (\ln q(A_i(u),\gamma))_\gamma')^T, \quad E(u,\theta) = \frac{S^{(1)}(u,\theta)}{S^{(0)}(u,\theta)},$$

$$S_i^{(1)}(u,\theta) = z_i(u,\gamma)S_i^{(0)}(u,\theta), \quad S^{(1)}(u,\theta) = \sum_{i=1}^{s} S_i^{(1)}(u,\theta),$$

$$\tilde{S}_{*i}^{(0)}(u,\theta) = e^{\beta^T x_i(u)} Y_i(u) q_1'(\hat{A}_i(u),\gamma), \quad \tilde{S}_{*i}^{(1)}(u,\theta) = \tilde{z}_i(u,\gamma)\tilde{S}_{*i}^{(0)}(u,\theta),$$

where q_1' is the (partial) derivative of the function q with respect to the first argument.

Denote by θ_0 the true value of θ.

Assumptions A:

a) there exist nonnegative functions y_i, continuous and positive on $[0,\tau]$ such that

$$\sup_{0 \le t \le \tau} |\frac{Y_i(t)}{n} - y_i(t)| \xrightarrow{P} 0, \quad \text{as} \quad n \to \infty, \quad \frac{n_i}{n} \to l_i \in (0,1);$$

b) $A_0(\tau) < \infty$;

c) there exist a neighborhood $U(\gamma_0)$ of γ_0 such that $q(u,\gamma)$ is positive and continuously differentiable on $[0,\tau] \times U(\gamma_0)$;

d) $S_{x_i(\cdot)}(\tau) > 0 \quad (i = 1,2)$.

For the models of the first level c) is changed by the condition of positiveness and continuous differentiability of q on $[0,\tau]$.

Set

$$s_i^{(0)}(u,\theta) = e^{\beta^T x_i(u)} y_i(u) q(A_i(u),\gamma); \quad s_i^{(1)}(u,\theta) = z_i(u,\gamma)s_i^{(0)}(u,\theta),$$

$$s_{*i}^{(0)}(u,\theta) = e^{\beta^T x_i(u)} y_i(u) q_1'(A_i(u),\gamma), \quad s_{*i}^{(1)}(u,\theta) = z_i(u,\gamma)s_{*i}^{(0)}(u,\theta),$$

$$s^{(j)}(u,\theta) = \sum_{i=1}^{s} s_i^{(j)}(u,\theta), \quad e(u,\theta) = \frac{s^{(1)}(u,\theta)}{s^{(0)}(u,\theta)}.$$

In the case of the models of the first level $z_i(u, \gamma) = x_i(u)$, $q_1' = q'$ and the argument γ is absent in all formulas.

Denote $B = B(\tau)$, where

$$B(t) = \sum_{i=1}^{s} \int_0^t \frac{\partial}{\partial \theta}(z_i(u, \gamma_0) - e(u, \theta_0))s_i^{(0)}(u, \theta_0)dA_0(u).$$

Proposition 12.7. *Suppose that the matrix B is non singular. Under Assumptions **A** there exists a neighbourhood of θ_0 within which, with probability tending to 1 as $n \to \infty$, the root $\hat{\theta}$ of $U(\theta, \tau) = 0$ is uniquely defined and*

$$n^{1/2}(\hat{\theta} - \theta_0) \xrightarrow{D} N(0, \Sigma^*),$$

where

$$\Sigma^* = B^{-1}\Sigma(B^{-1})^T, \quad \Sigma = \sum_{i=1}^{s} \int_0^\tau h_i(v; \theta_0)h_i(v; \theta_0)^T s_i^{(0)}(v; \theta_0)dA_0(v),$$

$$h_i(v; \theta_0) = z_i(v; \gamma_0) - e(v; \theta_0) +$$

$$\frac{1}{y_i(v)} \int_v^t \frac{s^{(1)}(u, \theta_0)s_{*i}^{(0)}(u, \theta_0) - s_{*i}^{(1)}(u, \theta_0)s^{(0)}(u, \theta_0)}{s^{(0)}(u, \theta_0)}dA_0(u).$$

The proposition is proved similarly as in Bagdonavičius & Nikulin (1997) for the generalized multiplicative models. Presence of the parameter γ (which does not take place in that paper) does not change the proof significantly.

12.5.2 Tests for the models with specified q

Let us consider models with specified q and unidimensional explanatory variables.

The estimated score process is

$$\hat{U}(t) = \sum_{j=1}^{s} \int_0^t (x_j(u) - \tilde{E}(u; \hat{\theta}))dN_j(u) = \sum_{j=1}^{s} \int_0^t (x_j(u) - \tilde{E}(u; \hat{\theta}))dM_j(u) -$$

$$\int_0^t (\tilde{E}(u; \hat{\theta}) - E(u; \theta_0))S^{(0)}(u, \theta_0)dA_0(u) - \int_0^t (\tilde{E}(u; \hat{\theta}) - E(u; \theta_0))dM(u),$$

$$(12.60)$$

where the martingale $M_i(t) = N_i(t) - \int_0^t \alpha_{x_i}(u)Y_i(u)du$, $M(t) = \sum_{i=1}^{s} M_i(t)$.

Using the Lenglart's inequality and Assumptions A we obtain:

for all $\delta > 0$, $\varepsilon > 0$, $t \in [0, \tau]$

$$\mathbf{P}\{n^{-1/2} \int_0^t (\tilde{E}(u; \hat{\theta}) - E(u; \theta_0))dM(u) \geq \varepsilon\} \leq$$

$$\frac{\delta}{\varepsilon}\mathbf{P}\{n^{-1} \int_0^\tau (\tilde{E}(u; \hat{\theta}) - E(u; \theta_0))^2 S^{(0)}(u, \theta_0)dA_0(u) \geq \delta\}.$$

So the normed third term in (9.58) converges in probability to zero uniformly on $[0, \tau]$.

Assumptions A imply

$$\sqrt{n}q(\hat{A}_i(u)) = \sqrt{n}q(A_i(u)) + q_1'(A_i(u))\sqrt{n}\int_0^u \frac{dM_i(v)}{Y_i(v)} + o_p(1)$$

uniformly on $[0, \tau]$. Consider the second term: we have

$$\sqrt{n}(\tilde{E}(u; \hat{\theta}) - E(u; \theta_0)) = \frac{\partial e_i(u, \theta_0)}{\partial \theta}\sqrt{n}(\hat{\theta} - \theta_0) +$$

$$\sum_{j=1}^s \frac{S_{*j}^{(1)}(u, \theta_0)S^{(0)}(u, \theta_0) - S_{*j}^{(0)}(u, \theta_0))S^{(1)}(u, \theta_0)}{(S^{(0)}(u, \theta_0))^2}\sqrt{n}\int_0^u \frac{dM_j(v)}{Y_j(v)} + o_p(1),$$

uniformly on $[0, \tau]$; here θ^* is on the line segment $[\theta_0, \hat{\theta}]$ and

$$n^{-1/2}\hat{U}(t) = n^{-1/2}\sum_{j=1}^s \int_0^t h_j(v, \theta_0)dM_j(v) - \frac{a(t)}{a(\tau)}\int_0^\tau h_j(v, \theta_0)dM_j(v) + o_p(1)$$

$$\xrightarrow{D} W(g(t)) - \frac{a(t)}{a(\tau)}W(g(\tau)) \quad \text{on} \quad D[0, \tau];$$

here

$$g(t) = \sum_{j=1}^s \int_0^t h_j^2(v, \theta_0)s_j^{(0)}(v, \theta_0)dA_0(v)),$$

$$a(t) = \int_0^t \frac{\partial}{\partial \theta}e(u, \theta_0)s^{(0)}(u, \theta_0)dA_0(u)$$

and W is the standard Wiener process. We have

$$\frac{\hat{U}(\cdot)}{\sqrt{ng(\tau)}} \xrightarrow{D} W\left(\frac{g(\cdot)}{g(\tau)}\right) - \frac{a(\cdot)}{a(\tau)}W(1) \quad \text{on} \quad D[0, \tau].$$

Set

$$\psi(u) = \frac{a(g^{-1}(g(\tau)u))}{a(\tau)}, \quad u \in [0, 1].$$

Then

$$\sup_{t \in [0, \tau]} |\frac{\hat{U}(t)}{\sqrt{ng(\tau)}}| \xrightarrow{D} \sup_{u \in [0, 1]} |W(u) - \psi(u)W(1)| = V_\psi.$$

In the case of PH model $a = g$, $\psi(u) = u$ and V_ψ is the Brownian bridge. The function $\psi : [0, 1] \to [0, 1]$ is increasing, $\psi(0) = 0$, $\psi(1) = 1$. We have

$$T = \sup_{t \in [0, \tau]} |\frac{\hat{U}(t)}{\sqrt{n\hat{g}(t)}}| \xrightarrow{D} V_\psi,$$

where

$$\hat{g}(t) = \frac{1}{n}\sum_{i=1}^s \int_0^t H_i^2(u; \hat{\theta})\tilde{S}_i^{(0)}(u, \hat{\theta})\frac{dN(u)}{\tilde{S}^{(0)}(u, \hat{\theta})}, \quad H_i(u; \hat{\theta}) = x_i(u) - \tilde{E}(u, \hat{\theta}) +$$

$$\frac{1}{Y_i(u)} \int_u^t \frac{(\tilde{S}^{(1)}(v,\hat{\theta})\tilde{S}^{(0)}_{*i}(v,\hat{\theta}) - \tilde{S}^{(1)}_{*i}(v,\hat{\theta})\tilde{S}^{(0)}(v,\hat{\theta}))}{(\tilde{S}^{(0)}(v,\hat{\theta}))^2} dN(v).$$

Denote by $V_{\psi,\alpha}$ the α-quantile of the random variable V_ψ,

$$\hat{\psi}(u) = \hat{a}(\hat{g}^{-1}(\hat{g}(\tau)u))/\hat{a}(\tau)),$$

where

$$\hat{g}^{-1}(s) = \sup\{u : \hat{g}(u) < s\}, \quad \hat{a}(t) = \int_0^t \frac{\partial}{\partial\hat{\theta}}\tilde{E}(a,\hat{\theta})dN(u).$$

The quantiles $V_{\psi,\alpha}$ can be approximated by $V_{\hat{\psi},\alpha}$ which can be obtained by simulating the standard Wiener process in the jump points of $\hat{\psi}$. The approximate critical region with the significance level α is $T > V_{\hat{\psi},1-\alpha}$.

In the case of the PH model the statistic T coincides with the statistic of Wei (1984):

$$\hat{g}(t) = \hat{a}(t) = n^{-1} \int_0^t \frac{\tilde{S}^{(2)}(u,\hat{\beta})S^{(0)}(u,\hat{\beta}) - (S^{(1)}(u,\hat{\beta}))^2}{(S^{(0)}(u,\hat{\beta}))^2} dN(u),$$

where $S^{(i)}(u,\hat{\beta}) = \sum x_j^i(u)e^{\hat{\beta}^T x_j(u)}Y_j(u)$ and V_ψ is the Brownian bridge.

12.5.3 Tests for the models with specified or parametrized q

Let us consider the models of the first or second level with possibly multidimensional explanatory variables.

The idea of goodness-of-fit statistic construction is similar to the idea of goodness-of-fit statistic construction for the PH model given by Lin (1991): consider a weighted score function

$$U_K(\theta,t) = \sum_{i=1}^k \int_0^t K(u)\{\tilde{z}_i(u,\gamma) - \tilde{E}(u,\theta)\}dN_i(u)$$

and an estimator $\hat{\theta}_K$, verifying the condition $U_K(\hat{\theta}_K,\tau) = 0$. The weight function $K(u)$ is a $\mathbb{F}$-predictable stochastic process that converges in probability to a nonnegative bounded function $k(u)$ uniformly in $u \in [0,\tau]$. For example, $K(u) = e^{-\hat{A}_0(u)}$. Under the hypothesis H_0 both estimators $\hat{\theta}_I$ (obtained when $K(u) = I(u) \equiv 1$) and $\hat{\theta}_K$ ($K \neq I$) are asymptotically normal with the same mean θ_0. Under alternatives both estimators $\hat{\theta}_I$ and $\hat{\theta}_K$ should be also asymptotically normal but with different means, so the test statistic may be constructed in terms of the difference $\hat{\theta}_I - \hat{\theta}_K$. In the case of the models of the first level (which are particular cases of the models of the second level) $\theta = \beta$ and γ is not present in the all following formulae.

Proposition 12.8. *Under Assumptions* A *and nonsingularity of* B_K *and* B_I

$$n^{1/2}(\hat{\theta}_K - \hat{\theta}_I) \xrightarrow{\mathcal{D}} N(0, \Sigma_{kI}^{**}),$$

where

$$\Sigma_{kI}^{**} = \Sigma_{kk}^* - \Sigma_{kI}^* - \Sigma_{Ik}^* + \Sigma_{II}^*, \quad \Sigma_{kI}^* = B_k^{-1}\Sigma_{kI}(B_I^{-1})^T,$$

$$\Sigma_{kI} = \sum_{i=1}^{2}\int_0^t h_{ki}(v,\theta_0)h_{Ii}(v,\theta_0)s_i^{(0)}(v,\theta_0)dA_0(u),$$

$$h_{ki}(v;\theta_0) = k(v)(z_i(v,\gamma_0) - e(v;\theta_0)) +$$

$$\frac{1}{y_i(v)}\int_v^t k(u)\frac{s^{(1)}(u,\theta_0)s_{*i}^{(0)}(u,\theta_0) - s_{*i}^{(1)}(u,\theta_0)s^{(0)}(u,\theta_0)}{s^{(0)}(u,\theta_0)}dA_0(u),$$

$$B_k = B_k(\tau), \quad ,B_k(t) = \sum_{i=1}^{2}\int_0^t k(u)\frac{\partial}{\partial\theta}(z_i(u,\gamma_0) - e(u,\theta_0))s_i^{(0)}(u,\theta_0)dA_0(u).$$

Sketch of the proof. Using method similar as in Bagdonavičius & Nikulin (1997) we show that the predictable covariation

$$< \frac{1}{\sqrt{n}}U_K, \frac{1}{\sqrt{n}}U_I > (t,\theta_0) = \frac{1}{n}\sum_{i=1}^{2}\int_0^t H_{Ki}(v,\theta_0)H_{Ii}(v,\theta_0)S_i^{(0)}(v,\theta_0)dA_0(u)+o_p(1),$$

uniformly on $[0,\tau]$, where

$$H_{Ki}(v;\theta_0) = K(u)(\tilde{z}_i(v,\gamma_0) - \tilde{E}(v;\theta_0)) +$$

$$\frac{1}{Y_i(v)}\int_v^t K(u)\frac{S^{(1)}(u,\theta_0)S_{*i}^{(0)}(u,\theta_0) - S_{*i}^{(1)}(u,\theta_0)S^{(0)}(u,\theta_0)}{S^{(0)}(u,\theta_0)}dA_0(u)$$

and

$$\hat{\theta}_K \xrightarrow{\mathbf{P}} \theta_0, \quad \hat{\theta}_I \xrightarrow{\mathbf{P}} \theta_0.$$

So

$$\mathbf{cov}(\sqrt{n}(\hat{\theta}_K - \theta_0), \sqrt{n}(\hat{\theta}_I - \theta_0)) \to B_k^{-1}\Sigma_{kI}(B_I^{-1})^T = \Sigma_{kI}^* \quad \text{as} \quad n \to \infty.$$

The statistic of chi-squared type test can be defined as

$$T_n = (\hat{\theta}_K - \hat{\theta}_I)^T(\hat{\Sigma}_{KI}^{**})^{-1}(\hat{\theta}_K - \hat{\theta}_I), \tag{12.61}$$

where

$$\hat{\Sigma}_{KI}^{**} = \hat{\Sigma}_{KK}^* - \hat{\Sigma}_{KI}^* - \hat{\Sigma}_{IK}^* + \hat{\Sigma}_{II}^*, \quad \hat{\Sigma}_{K_1K_2}^* = \hat{B}_{K_1}^{-1}\hat{\Sigma}_{K_1K_2}(\hat{B}_{K_2}^{-1})^T,$$

$$\hat{B}_K = \frac{1}{n}\int_0^\tau K(u)\frac{\partial}{\partial\theta}(\tilde{z}_i(u,\hat{\gamma}) - \tilde{E}(u,\hat{\theta}))dN(u),$$

$$\hat{\Sigma}_{K_1K_2} = \frac{1}{n}\sum_{i=1}^{s}\int_0^\tau H_{K_1i}(v,\hat{\theta})H_{K_2i}^T(v,\hat{\theta})\tilde{S}_i^{(0)}(v,\hat{\theta})d\hat{A}_0(v),$$

where $H_{Ki}(v,\hat{\theta})$ is obtained from $H_{Ki}(v,\theta_0)$ replacing θ_0 by $\hat{\theta}$ and $dA_0(u)$ by $d\hat{A}_0(u) = dN(u)/\tilde{S}^{(0)}(u,\hat{\theta})$. The distribution of the statistic T_n is approximated by the chi-square distribution with two degrees of freedom.

Example. Consider the hypothesis

$$H_{0g} : \alpha_i(t) = e^{\beta^T x_i(t)+\gamma A_i(t)}\alpha_0(t), \quad \gamma < 0,$$

i.e. the GPH_{GLL} model against alternatives, including the alternative

$$H_1 \; : \; \alpha_i(t) = e^{\beta^T x_i(t)} \frac{1}{1+\delta t} \alpha_0^*(t), \quad \delta > 0,$$

i.e. the IGF model.

The test statistic in this case is simple:

$$\hat{B}_K = -\frac{1}{n} \int_0^\tau K(u) \frac{\partial}{\partial \theta} \tilde{E}(u, \hat{\theta}) dN(u),$$

$$\hat{\Sigma}_{K_1 K_2} = \frac{1}{n} \int_0^\tau K_1(u) K_2(u) \frac{\partial}{\partial \theta} \tilde{E}(u, \hat{\theta}) dN(u),$$

and $\tilde{z}_i(u) = (x_i(u), \hat{A}_i(u))$ in the expression of $\tilde{E}(u, \hat{\theta})$.

12.6 Goodness-of-fit for parametric regression models

The most used parametric regression model in analysis of the FTR data is the AFT model. Note that if the explanatory variables are constant then, under specified survival distributions, this model coincides with some specified GPH1 model. For example, under the Weibull, loglogistic, lognormal, generalized Weibull, generalized loglogistic distribution under constant explanatory variables the AFT model is equivalent to the PH, logistic regression, generalized probit, GPH_{GW}, GPH_{GLL} model, respectively. In the case when these models do not coincide with the AFT model, the survival distributions under different constant explanatory variables are not from the same classes and parametric methods are not very attractive.

So let us consider model checking techniques for the parametric AFT model

$$S_{x(\cdot)}(t) = S_0 \left(\int_0^t e^{-\beta^T x(u)} du, \eta \right), \tag{12.62}$$

with specified parametric form of the baseline survival function $S_0(t, \eta)$. In terms of the hazard rates

$$\alpha_x(t, \theta) = \alpha_0 \left(\int_0^t e^{-\beta^T x(u)} du, \eta \right) e^{-\beta^T x(t)},$$

where α_0 are the baseline hazard rate.

12.6.1 Likelihood ratio tests

Let us consider the alternative: the AFT model with time-varying regression coefficients:

$$\hat{S}_{x(\cdot)}(t) = S_0 \left(\int_0^t e^{-\mu^T z(u)} du, \eta \right), \tag{12.63}$$

where

$$\mu = (\beta_1, \cdots, \beta_m, \gamma_1, \cdots, \gamma_r)^T.$$

So $\mu^T z(u) = \sum_{i=1}^{m}(\beta_j + \gamma_j\, g_j(u))x_j(u) = \beta^T x(u) + \gamma^T g(u)x(u)$, where $g_j(\cdot)$ are specified deterministic functions or paths of predictable processes and $g(u)$ is the diagonal matrix with the elements $g_j(u)$ on the diagonal. Examples of $g_j(t)$ are t; $\ln t$; $\mathbf{1}_{\{t \le t_0\}}$; $N(t-)$; $\hat{F}(t-)$, etc. Set $\nu = (\mu^T, \eta^T)^T$, $\theta = (\beta^T, \eta^T)^T$. If $\gamma = 0$, the model coincides with the AFT model.

Denote by $\hat{\nu} = (\hat{\theta}^T, \gamma^T)^T$ and $\tilde{\theta}$ the maximum likelihood estimators of the parameters ν and θ under the models (12.62) and (12.61), respectively. Note that if the baseline function is from classes considered in Chapter 5, the estimation in both the AFT and the alternative models is identical. In the alternative model the vector $z(\cdot)$ can be considered as a vector of explanatory variables. Denote by $L(\nu) = L(\theta, \eta)$ the likelihood function under the model (12.62).

The likelihood ratio statistic is:

$$2\left(\ln L(\hat{\theta}, \hat{\gamma}) - \ln L(\tilde{\theta}, 0)\right). \qquad (12.64)$$

If n is large then the distribution of the likelihood ratio statistic (cf.(4.40)) approximated by the chi-square law with m degrees of freedom.

12.6.2 Numerical methods for assessing goodness-of-fit

Let us consider numerical methods for assessing goodness-of-fit for parametric regression models. Such methods were developed by Lin and Spiekerman (1996).

Following Lin and Spiekerman, let us consider model checking techniques for the AFT model with constant explanatory variables.

Note that that for any x the survival function, cumulative hazard and hazard rate of the random variable

$$\varepsilon = \ln T_x - \beta^T x$$

are $S(t) = S_0(e^t)$, $A(t) = A_0(e^t)$ and $\alpha(t) = e^t\alpha_0(e^t)$, respectively.

The idea of the first goodness-of-fit test is to compare the parametric and semiparametric estimators of the cumulative hazard A.

Suppose that the explanatory variables are constant and the data are independent replicates $\{X_i, \delta_i, x^{(i)}\}$ of $\{X, \delta, x\}$ or, equivalently, independent replicates $\{N_i(\cdot), Y_i(\cdot), x^{(i)}\}$ of $\{N(\cdot), Y(\cdot), x\}$, where $N(t) = \mathbf{1}_{\{X \le t, \delta=1\}}$, $Y(t) = \mathbf{1}_{\{X \ge t\}}$.

The formula (6.14) implies that the semiparametric estimator of A has the form

$$\hat{A}^{(sem)}(t) = \tilde{A}_0(e^t, \hat{\beta}_s) = \sum_{i=1}^{n}\int_0^{e^t} \frac{dN_i(e^{\hat{\beta}_s^T x^{(i)}}u)}{\sum_{l=1}^{n}Y_l(e^{\hat{\beta}_s^T x^{(l)}}u)} =$$

$$\frac{\sum_{i=1}^{n}\delta_i \mathbf{1}_{\{\ln X_i - \hat{\beta}_s^T x^{(i)} \le t\}}}{\sum_{l=1}^{n}\mathbf{1}_{\{\ln X_l - \hat{\beta}_s^T x^{(l)} \ge \ln X_i - \hat{\beta}_s^T x^{(i)}\}}},$$

where $\hat{\beta}_s$ is the semiparametric estimator of β defined by (6.19).

Let $\hat{A}^{(par)}(t) = A(t, \hat{\eta})$ be the parametric estimator of $A(t, \eta)$; here $\hat{\eta}$ is the parametric maximum likelihood estimator of η defined by the score function of the form (4.16).

The first goodness-of-fit test is based on the statistic

$$W(t) = n^{1/2}\{\hat{A}^{(par)}(t) - \hat{A}^{(sem)}(t)\}.$$

This statistic can be written in the following way:

$$W(t) = n^{1/2}\{A(t,\hat{\eta}) - A(t,\eta_0)\} - n^{1/2}\{\tilde{A}(t,\hat{\beta}_s) - \tilde{A}(t,\beta_0)\} - n^{1/2}\{\tilde{A}(t,\beta_0) - A(t,\eta_0)\} =$$

$$W_1(t) - W_2(t) - W_3(t);$$

here $\tilde{A}(t,\beta) = \tilde{A}_0(e^t,\beta)$.

Denote by $\hat{\theta} = (\hat{\beta}^T, \hat{\eta}^T)^T$ the parametric estimator of θ. The delta method implies that

$$W_1(t) = \frac{\partial}{\partial\eta}A(t,\eta_0)n^{1/2}(\hat{\eta} - \eta_0) + o_p(1) = (0, \frac{\partial}{\partial\eta}A(t,\eta_0))n^{1/2}(\hat{\theta} - \theta_0) + o_p(1).$$

The formulas (4.21)-(4.24) imply that

$$n^{1/2}(\hat{\theta} - \theta_0) = (n^{-1}I(\theta_0))^{-1}n^{-1/2}U(\theta_0) + o_p(1),$$

$$n^{-1/2}U(\theta_0) = n^{-1/2}\sum_{i=1}^{n}\int_0^\infty \frac{\partial}{\partial\theta}\ln\alpha_i(v,\theta_0)dM_i(v,\theta_0) =$$

$$n^{-1/2}\sum_{i=1}^{n}\int_{-\infty}^{\infty} g_i(v,\eta_0)dM_i^*(v,\theta_0),$$

where

$$M_i^*(v,\theta_0) = M_i(e^{v+\beta_0^T x^{(i)}},\theta_0),$$

$$g_i(v,\eta_0) = (-(x^{(i)})^T(\ln\alpha)_1'(v,\eta_0), ((\ln\alpha)_2'(v,\eta_0)^T)^T.$$

So

$$W_1(t) = (0, \frac{\partial}{\partial\eta}A(t,\eta_0))(n^{-1}I(\theta_0))^{-1}n^{-1/2}\sum_{i=1}^{n}\int_{-\infty}^{\infty} g_i(v,\eta_0)dM_i^*(v,\theta_0)$$

The formula (6.14) and the Doob-Meier decomposition of N_i imply that

$$W_3(t) = n^{1/2}\left\{\sum_{i=1}^{n}\int_0^{e^t} \frac{dN_i(e^{\beta_0^T x^{(i)}}u)}{\sum_{l=1}^{n}Y_l(e^{\beta_0^T x^{(l)}}u)} - A(t,\eta_0)\right\} = n^{1/2}$$

$$\left\{\sum_{i=1}^{n}\int_{-\infty}^{t} \frac{Y_i(e^{\beta_0^T x^{(i)}+v})\alpha(v,\eta_0)dv + dM_i(e^{\beta_0^T x^{(i)}+v},\theta_0)}{\sum_{l=1}^{n}Y_l(e^{\beta_0^T x^{(l)}+v})} - \int_{-\infty}^{t}\alpha(v,\eta_0)dv\right\} =$$

$$n^{-1/2}\sum_{i=1}^{n}\int_{-\infty}^{t} \frac{dM_i^*(v,\theta_0)}{n^{-1}\sum_{l=1}^{n}\mathbf{1}_{\{\ln X_l - \beta_0^T x^{(l)} \geq v\}}}.$$

Set $\bar{y}(v,\beta) = \bar{x}(e^v, \beta)$, where $\bar{x}(v,\beta)$ is defined by (6.18). We have

$$W_2(t) = n^{1/2}\left\{\sum_{i=1}^{n}\int_0^{e^t}\frac{dN_i(e^{\hat{\beta}_s^T x^{(i)}}u)}{\sum_{l=1}^{n}Y_l(e^{\hat{\beta}_s^T x^{(l)}}u)} - \sum_{i=1}^{n}\int_0^{e^t}\frac{dN_i(e^{\beta_0^T x^{(i)}}u)}{\sum_{l=1}^{n}Y_l(e^{\beta_0^T x^{(l)}}u)}\right\} =$$

$$n^{1/2}\left\{\sum_{i=1}^{n}\int_{-\infty}^{t}\frac{Y_i(e^{\hat{\beta}^T x^{(i)}+v})\{\alpha(v+(\hat{\beta}_s-\beta_0)^T x^{(i)},\eta_0)-\alpha(v,\eta_0)\}dv}{\sum_{l=1}^{n}Y_l(e^{\hat{\beta}^T x^{(l)}+v})}\right.$$

$$\left.+\sum_{i=1}^{n}\int_{-\infty}^{t}\frac{dM_i(e^{\hat{\beta}_s^T x^{(i)}+v},\theta_0)}{\sum_{l=1}^{n}Y_l(e^{\hat{\beta}_s^T x^{(l)}+v})} - \sum_{i=1}^{n}\int_{-\infty}^{t}\frac{dM_i(e^{\beta_0^T x^{(i)}+v},\theta_0)}{\sum_{l=1}^{n}Y_l(e^{\beta_0^T x^{(l)}+v})}\right\} =$$

$$\int_{-\infty}^{t}\bar{y}^T(v,\beta_0)d\,\alpha(v,\eta_0)\,n^{1/2}(\hat{\beta}_s-\beta_0) + o_p(1).$$

By Chapter 6.1.4

$$n^{1/2}(\hat{\beta}_s-\beta_0) = \Sigma_*^{-1}n^{-1/2}\sum_{i=1}^{n}\int_{-\infty}^{\infty}\{x^{(i)}-\bar{y}(v,\beta_0)\}dM_i^*(v,\beta) + o_p(1).$$

So we have

$$W(t) = (0,\frac{\partial}{\partial\eta}A(t,\eta_0))(n^{-1}I(\theta_0))^{-1}n^{-1/2}\sum_{i=1}^{n}\int_{-\infty}^{\infty}g_i(v,\eta_0)dM_i^*(v,\beta_0)-$$

$$\int_{-\infty}^{t}\bar{y}^T(v,\beta_0)d\,\alpha(v,\eta_0)\,\Sigma_*^{-1}n^{-1/2}\sum_{i=1}^{n}\int_{-\infty}^{t}(x^{(i)}-\bar{y}(v,\beta_0)dM_i^*(v,\beta_0)-$$

$$n^{-1/2}\sum_{i=1}^{n}\int_{-\infty}^{t}\frac{dM_i^*(v,\beta_0)}{n^{-1}\sum_{l=1}^{n}\mathbf{1}_{\{\ln X_l-\beta_0^T x^{(l)}\geq v\}}}.$$

The null distribution of $W(u)$ is approximated by a zero-mean Gaussian process $\hat{W}(u)$ whose distribution can be generated through simulation. By replacing $M_i^*(u,\beta_0)$ with $N_i^*(u,\beta_0)V_i = N_i(e^{u+\beta_0^T x^{(i)}})V_i$, where $V_1,\cdots,V_n$ are independent standard normal variables which are independent of $Y_i, N_i, x^{(i)}$, and replacing the finite dimensional parameters by their respective estimators, we obtain

$$\hat{W}(t) = (0,\frac{\partial}{\partial\eta}A(t,\hat{\eta}))(n^{-1}I(\hat{\theta}))^{-1}n^{-1/2}\sum_{i=1}^{n}\int_{-\infty}^{\infty}g_i(v,\eta_0)V_i\,dN_i^*(v,\hat{\beta})-$$

$$\int_{-\infty}^{t}\bar{x}^T(v,\hat{\beta})d\,\alpha(v,\hat{\eta})\,\hat{\Sigma}^{-1}n^{-1/2}\sum_{i=1}^{n}\int_{-\infty}^{t}\{x^{(i)}-\bar{y}(v,\hat{\beta})\}V_i\,dN_i^*(v,\hat{\beta})-$$

$$n^{-1/2}\sum_{i=1}^{n}\int_{-\infty}^{t}\frac{V_i\,dN_i^*(v,\hat{\beta})}{n^{-1}\sum_{l=1}^{n}\mathbf{1}_{\{\ln X_l-\hat{\beta}^T x^{(l)}\geq v\}}},$$

where

$$\hat{\Sigma}_* = n^{-1}\sum_{i=1}^{n}\int_{-\infty}^{\infty}\mathbf{1}_{\{\ln X_i-\beta^T x^{(i)}\geq u\}}\{x^{(i)}-\bar{y}(u,\beta_0)\}^{\otimes 2}d\alpha(u,\hat{\eta}).$$

When approximating the distribution of W, the variables V_i are considered as random and the trajectories $\{N_i(\cdot), Y_i(\cdot), x^{(i)}\}$ as fixed. The conditional (given $\{N_i(\cdot), Y_i(\cdot), x^{(i)}\}$) distribution of $\hat{W}$ has the same limit as the unconditional distribution of W. So the distribution of W can be approximated by simulating samples $\{V_i; i = 1, \cdots, n\}$ while fixing the data $\{N_i(\cdot), Y_i(\cdot), x^{(i)}\}$.

The supremum test statistic is $Q = \sup_{t \in [0,\tau]} | W(t) |$. It is supposed that at the moment τ the failure and the censoring processes are stopped. Let q be the observed value of Q and let $\hat{Q} = \sup_{t \in [0,\tau]} | \hat{W}(t) |$. Then the P-value $\mathbf{P}(Q > q)$, can be approximated by $\mathbf{P}(\hat{Q} > q)$, the latter probability being approximated through simulation.

If the baseline hazard function is incorrectly parametrized but the AFT model is valid, then the semiparametric estimator $A^{(sem)}(t)$ converges to the true $A(t)$ whereas the parametric estimator $A^{(par)}(t)$ converges to a limit which is different from $A(t)$ at least for some t. Therefore the supremum test is consistent against any mis-specification of the baseline hazard function.

Similarly as in the semiparametric case let us consider the statistics

$$U^{(1)}(t,x) = \sum_{i=1}^{n} \int_0^t \mathbf{1}_{\{x^{(i)}(u) \leq x\}} d\hat{M}_i(u),$$

$$U^{(2)}(t,x) = \sum_{i=1}^{n} \int_0^t \mathbf{1}_{\{\hat{\beta}^T x^{(i)}(u) \leq x\}} d\hat{M}_i(u),$$

where $\hat{M}_i$ are the (parametric) martingale residuals

$$\hat{M}_i(t) = N_i(t) - \int_0^t Y_i(u)\, \alpha_i(u, \hat{\theta}) du.$$

Set $\hat{w}^{(1i)}(u,x) = \mathbf{1}_{\{x^{(i)}(u) \leq x\}}$ and $\hat{w}^{(2i)}(u,x) = \mathbf{1}_{\{\hat{\beta}^T x^{(i)}(u) \leq x\}}$. Similarly as for the above considered statistic W_1 (see also Chapter 9.5.3) the statistics $U^{(j)}(t,x)$ can be written

$$n^{-1/2}U^{(j)}(t,x) = n^{-1/2}\sum_{i=1}^{n} \hat{w}^{(ji)}(t,x)M_i(t) - J^{(j)}(t,x;\theta_0)n^{1/2}(\hat{\theta}-\theta_0) + o_p(1) =$$

$$n^{-1/2}\sum_{i=1}^{n} \hat{w}^{(ji)}(t,x)M_i(t) - J^{(j)}(t,x;\theta_0)(n^{-1}I(\theta_0))^{-1} \times$$

$$n^{-1/2}\sum_{i=1}^{n} \int_{-\infty}^{\infty} g_i(v,\eta_0)dM_i^*(v,\beta_0) + o_p(1),$$

where

$$J^{(j)}(t,x;\theta_0) = n^{-1}\sum_{i=1}^{n} \hat{w}^{(ji)}(t,x) \int_0^t Y_i(u) \frac{\partial}{\partial\theta}\alpha_i(u,\theta_0)du.$$

As in the case of the statistic $W(t)$, the distribution of $U^{(j)}(t,x)$ is approximated by the distribution of the statistic

$$n^{-1/2}\hat{U}^{(j)}(t,x) = n^{-1/2}\sum_{i=1}^{n}\hat{w}^{(ji)}(t,x)V_iN_i(t) - J^{(j)}(t,x;\hat{\theta})(n^{-1}I(\hat{\theta}))^{-1}\times$$

$$n^{-1/2}\sum_{i=1}^{n}\int_{-\infty}^{\infty}g_i(v,\hat{\eta})V_idN_i^*(v,\hat{\beta}).$$

The conditional (given $\{N_i(\cdot), Y_i(\cdot), x^{(i)}\}$) distribution of $n^{-1/2}U^{(j)}$ has the same limit as the unconditional distribution of $n^{-1/2}U(j)$.

The test statistic is defined as

$$T = \sup_{t,x} \mid n^{-1/2}\hat{U}^{(1)}(t,x) \mid .$$

The P-values of the statistic T can be estimated through simulation.

The test is consistent against any departure from the assumed parametric model (Lin and Spiekerman (1996)).

If one of the considered tests rejects the model, it means that there is misspecification for some aspects of the model. It can be bad choice of the AFT model, of the link function $r(x)$, of the parametrization of the baseline function A, etc.

CHAPTER 13

Estimation in degradation models

13.1 Introduction

Suppose that degradation under the explanatory variable $x(\cdot)$ is determined by a stochastic process (cf. (3.11))

$$Z_{x(\cdot)}(t) = Z \left(\int_0^t e^{\beta^T x(s)} \, ds \right).$$ (13.1)

Set

$$m(t) = \mathbf{E}(Z(t)), \quad m_{x(\cdot)}(t) = \mathbf{E}(Z_{x(\cdot)}(t)) = m \left(\int_0^t e^{\beta^T x(s)} \, ds \right).$$ (13.2)

The moment of the nontraumatic failure under explanatory variable $x(\cdot)$ is

$$T_{x(\cdot)} = \sup\{t : Z_{x(\cdot)}(t) < z_0\}.$$

It is the moment when degradation under $x(\cdot)$ attains a critical level z_0.

Let $C_{x(\cdot)}$ be the moment of the traumatic failure under $x(\cdot)$ and let us consider the model discussed in Chapter 3:

$$\mathbf{P}\{C_{x(\cdot)} > t \mid x(s), \ Z_{x(\cdot)}(s), \ 0 \le s \le t\} = \exp\{-\int_0^t \lambda \left(Z_{x(\cdot)}(s), x(s) \right) ds\}$$ (13.3)

In the particular case when the values of all explanatory variables are fixed, the model without explanatory variables

$$\mathbf{P}\{C > t \mid Z(s), \ 0 \le s \le t\} = \exp\{-\int_0^t \lambda \left(Z(s) \right) ds\}$$ (13.4)

may be considered. In this chapter we consider statistical estimation of reliability characteristics using degradation and failure time data with explanatory variables. In the case when traumatic failure times and degradation processes are dependent, this topic was considered by Bagdonavičius and Nikulin (2001) and Bagdonavičius et al. (2001).

13.2 Linear path models

Estimation procedures from degradation data using general path models are given in the book of Meeker and Escobar (1998) We consider only such topics which are not considered there.

The intensity of the traumatic failures is characterized by the function $\lambda(z, x)$ given in (13.3). In the preliminary stage of investigation, the form of this function is generally unknown. So the problem of its nonparametric estimation arises. The graph of obtained estimator gives an idea on the form of it.

We shall consider the most simple case of the linear path model when the intensity λ at any moment depends only on the value of degradation (and on the explanatory variable via degradation): $\lambda = \lambda(z)$. Generalizations for more general path models is one of the interesting directions of research.

One of the main applications of the linear path models is related with analysis of tire wear (degradation). Traumatic failures of tires are related with the production defects, mechanical damages, fatigue of the tire components, etc. So the traumatic failures can be of different modes.

The purpose of this book is to give analysis of failure time and degradadation data with explanatory variables. Nevertheless, for better understanding of further generalizations to the case of data with explanatory variables, we give at first analysis of failure time and degradation data without explanatory variables.

13.2.1 Model without explanatory variables

Suppose that under fixed values of the explanatory variables the wear process $Z(t)$ is modeled by the linear path model

$$Z(t) = t/A, \quad t \geq 0;$$

here A is a positive random variable with the distribution function π.

Suppose that a unit fails because of the natural cause (the degradation attains the critical level z_0) or because of traumatic events of one of s possible types (modes).

Denote by

$$T^{(0)} = z_0 A$$

the time of nontraumatic failure, and by $T^{(k)}$ ($k = 1, \cdots, s$) the failure time corresponding to the kth traumatic failure mode.

The reliability function $S^{(0)}(t)$ and the mean $E(T^{(0)})$ of the random variable $T^{(0)}$ are

$$S^{(0)}(t) = P(T^{(0)} > t) = 1 - \pi(t/z_0) \qquad E(T^{(0)}) = z_0 \int_0^\infty a \, d\pi(a). \quad (13.5)$$

We suppose that the random variables $T^{(1)}, \cdots, T^{(s)}$ are conditionally independent (given $A = a$) and for any k the model (13.4) with $C = T^{(k)}$ is true.

So the conditional survival function of $T^{(k)}$ is

$$S^{(k)}(t \mid a) = \mathbf{P}(T^{(k)} > t \mid A = a) = \mathbf{P}(T^{(k)} > t \mid Z(s) = s/a) =$$

$$\exp\left(-\int_0^t \lambda^{(k)}(s/a)ds\right) = \exp\left(-a\Lambda^{(k)}(t/a)\right), \qquad (13.6)$$

where

$$\Lambda^{(k)}(z) = \int_0^z \lambda^{(k)}(y)dy.$$

The conditional probability density function of $T^{(k)}$ is

$$p^{(k)}(t \mid a) = \lambda^{(k)}(t/a)e^{-a\Lambda^{(k)}(t/a)}. \qquad (13.7)$$

The failure time of an unit is the random variable

$$T = \min(T^{(0)}, T^{(1)}, \cdots, T^{(s)}).$$

Set

$$\Lambda(z) = \sum_{i=1}^s \Lambda^{(i)}(z).$$

The survival function and the mean of the random variable T are

$$S(t) = P(T > t) = \int_{t/z_0}^{\infty} e^{-a\Lambda(t/a)} d\pi(a), \qquad (13.8)$$

$$E(T) = E(T^{(0)}) - \int_0^{\infty} a^2 d\pi(a) \int_0^{z_0} (z_0 - y)e^{-a\Lambda(y)} d\Lambda(y). \qquad (13.9)$$

Set

$$V = \begin{cases} 0, & \text{if } T = T^{(0)}, \\ 1, & \text{if } T = T^{(1)}, \\ \cdots & \cdots \\ s, & \text{if } T = T^{(s)} \end{cases} \qquad (13.10)$$

The random variable V is the indicator of the failure mode. The 0 failure mode is nontraumatic. Other failure modes are traumatic.

13.2.2 Nonparametric estimation of the cumulative intensities

Suppose that the cumulative intensities $\Lambda^{(k)}$ are completely unknown. The purpose is to estimate

$$\Lambda^{(k)}(z), 0 \le z \le z_0.$$

Suppose that n units are on test and the failure moments T_i, the indicators of the failure modes V_i (cf. (13.10)) and the degradation values

$$Z_i = T_i/A_i \qquad (13.11)$$

at the failure moments T_i are observed. Thus, the data are:

$$(T_1, Z_1, V_1), \cdots, (T_n, Z_n, V_n). \qquad (13.12)$$

The formula (13.11) implies that the degradation rates A_i are known. Therefore the data can be defined as n independent copies of the vector (A, AT, V):

$$(A_1, Z_1, V_1), \cdots, (A_n, Z_n, V_n). \tag{13.13}$$

For $k = 1, \cdots, s$ and $0 \leq z \leq z_0$ set

$$N_n^{(k)}(z) = \sum_{i=1}^{n} \mathbf{1}_{\{Z_i \leq z, V_i = k\}}.$$

It is the number of units having a failure of the kth mode before the wear attains the level z.

Lemma 13.1. *Let $\mathcal{F}_z$ be the σ-algebra generated by the random variables $A_1, \cdots, A_n$ and $N_n^{(1)}(y), \cdots, N_n^{(s)}(y)$, $y \leq z$. Then the counting process $N_n^{(k)}(z)$ can be written as the sum*

$$N_n^{(k)}(z) = \int_0^z \lambda^{(k)}(y) Y_n(y) dy + M_n^{(k)}(z),$$

where

$$Y_n(z) = \sum_{i=1}^{n} A_i \mathbf{1}_{\{Z_i \geq z\}} = \sum_{Z_i \geq z} A_i, \tag{13.14}$$

and $(M_n^{(k)}(z), 0 \geq z \geq z_0)$ is a martingale with respect to the filtration $(\mathcal{F}_z, 0 \leq z \leq z_0)$.

Proof. Let $0 \leq y < z \leq z_0$. It is sufficient to prove that

$$\mathbf{E}\{N_n^{(k)}(z) - N_n^{(k)}(y) \mid \mathcal{F}_y\} = \mathbf{E}\{\int_y^z \lambda^{(k)}(u) Y_n(u) du \mid \mathcal{F}_y\}.$$

Additivity of mathematical expectation implies that it is sufficient to prove this for $n = 1$.

If $A_1 = a$ and $Z_1 \leq y$ then $N_1^{(k)}(z) = N_1^{(k)}(y)$. If $A_1 = a$ and $Z_1 > y$ then the random variable $N_1^{(k)}(z)$ takes two values, 0 and 1, and (we write $a \leq b, c$ for $a \leq b, a \leq c$)

$$\mathbf{P}\{N_1^{(k)}(z) = 1 \mid Z_1 > y, A_1 = a\} =$$

$$\mathbf{P}\{ay < T_1^{(k)} \leq az, T_1^{(1)}, \cdots, T_1^{(s)} \mid T_1 > ay, A_1 = a\} =$$

$$e^{a\Lambda(y)} \int_{ay}^{az} p^{(k)}(t \mid a) \prod_{l \neq k} S_l(t \mid a) dt = e^{a\Lambda(y)} a \int_y^z \lambda^{(k)}(u) e^{-a\Lambda(u)} du.$$

Thus,

$$\mathbf{E}\{N_1^{(k)}(z) - N_1^{(k)}(y) \mid \mathcal{F}_y\} = \mathbf{1}_{\{Z_1 > y\}} A_1 \int_y^z \lambda^{(k)}(u) e^{-A_1\{\Lambda(u) - \Lambda(y)\}} du.$$

Analogously, for $u > y$

$$\mathbf{E}\{\mathbf{1}_{\{Z_1 > u\}} \mid \mathcal{F}_y\} = \mathbf{1}_{\{Z_1 > y\}} e^{-A_1\{\Lambda(u) - \Lambda(y)\}},$$

thus

$$\mathbf{E}\{\int_y^z \lambda^{(k)}(u)Y_1(u)du \mid \mathcal{F}_y\} = A_1 \int_y^z \lambda^{(k)}(u)\mathbf{E}\{1_{\{Z_1>u\}} \mid \mathcal{F}_y\}du =$$

$$1_{\{Z_1>y\}}A_1 \int_y^z \lambda^{(k)}(u)e^{-A_1\{\Lambda(u)-\Lambda(y)\}}du.$$

The lemma implies that optimal estimators of the cumulative intensities $\Lambda^{(k)}(z)$ are Nelson-Aalen type (see Andersen et al (1993),p.p. 177-178):

$$\hat{\Lambda}^{(k)}(z) = \int_0^z Y_n^{-1}(y)\, dN_n^{(k)}(y)$$

$$= \sum_{Z_i \le z, V_i=k} Y_n^{-1}(Z_i) = \sum_{Z_i \le z, V_i=k} \left(\sum_{Z_j \ge Z_i} A_j\right)^{-1}.$$

Thus

$$\hat{\Lambda}_n^{(k)}(z) = \sum_{Z_i \le z, V_i=k} \left(\sum_{Z_j \ge Z_i} \frac{T_j}{Z_j}\right)^{-1}. \tag{13.15}$$

The estimator is correctly defined if $Z_i \ge z$ and $V_i = k$ for some i. If such i do not exist then the estimator is defined as 0.

Note the remarkable fact that for nonparametric estimation of the cumulative intensities we do not need to specify the form of distribution π.

The estimators $\hat{\Lambda}_n^{(k)}(z)$ are piecewise-constant right-continuous functions, i.e. they are random elements of the Skorokhod space $D[0, z_0]$ (see Appendix, Section A12). The properties of Nelson-Aalen type estimators are well known (see Andersen et al (1993)). In our special case the following result holds.

Theorem 13.1. *If $EA < \infty$, then the estimator $\hat{\Lambda}_n^{(k)}(z)$ is consistent, and the random vector function*

$$\sqrt{n}\big(\hat{\Lambda}_n^{(1)}(z) - \Lambda_1(z), \cdots, \hat{\Lambda}_n^{(s)}(z) - \Lambda_s(z)\big)$$

converges in distribution on the space $D[0, z_0] \times \cdots \times D[0, z_0]$ to the random vector

$$\big(W_1(\sigma_1^2(z)), \cdots, W_s(\sigma_s^2(z))\big),$$

where

$$\sigma_k^2(z) = \int_0^z \frac{\lambda^{(k)}(y)}{b(y)}dy, \quad b(z) = E\left(Ae^{-A\Lambda(z)}\right)$$

and $W_1(z), \cdots, W_s(z)$ are independent standard Wiener processes.

13.2.3 Nonparametric estimation of unconditional reliability characteristics

Under the given plan of experiments estimation of the unconditional reliability characteristics is trivial.

The survival function $S(t)$ (cf. (13.8)) is estimated by the empirical survival function

$$\hat{S}(t) = \frac{1}{n} \sum_{i=1}^{n} \mathbf{1}_{\{T_i > t\}}.$$

The mean failure time is estimated by the empirical mean:

$$\widehat{ET} = \overline{T} = \frac{1}{n} \sum_{i=1}^{n} T_i.$$

The probability $p^{(k)}(t)$ of the failure of the kth mode in the interval $[0, t]$, and the probability $p^{(k)}$ of the failure of the kth mode in the interval $[0, \infty)$ are estimated by:

$$\hat{p}^{(k)}(t) = \frac{1}{n} \sum_{i=1}^{n} \mathbf{1}_{\{V_i = k, T_i \leq t\}}, \quad \hat{p}^{(k)} = \frac{1}{n} \sum_{i=1}^{n} \mathbf{1}_{\{V_i = k\}}.$$

The probability $p^{(tr)}(t)$ of a traumatic failure in the interval $[0, t]$ and the probability $p^{(tr)}$ of the a traumatic failure during the experiment are estimated by

$$\hat{p}^{(tr)}(t) = \frac{1}{n} \sum_{i=1}^{n} \mathbf{1}_{\{V_i \neq 0, T_i \leq t\}}, \quad \hat{p}^{(tr)} = \frac{1}{n} \sum_{i=1}^{n} \mathbf{1}_{\{V_i \neq 0\}}.$$

13.2.4 Prediction of the residual reliability characteristics

Suppose that at the moment t the degradation level is measured to be z. Using the estimators of the cumulative intensities obtained from the above considered experiment, the residual reliability characteristics can be predicted.

For $z \in [0, z_0]$ denote by

$$Q(\Delta; t, z) = P(T \leq t + \Delta \mid T > t, Z(t) = z) \qquad (13.16)$$

the conditional probability to fail in the interval $(t, t + \Delta]$ given that at the moment t a unit is functioning and its degradation value is z. This probability can be written in terms of the cumulative intensity Λ: for $\Delta < t(z_0/z - 1)$

$$Q(\Delta; t, z) = 1 - \exp\left\{ -\frac{t}{z} \left(\Lambda(z(1 + \Delta/t)) - \Lambda(z) \right) \right\}; \qquad (13.17)$$

if $\Delta \geq t(z_0/z - 1)$ then $Q(\Delta, t, z) = 1$.

The probability $Q(\Delta; t, z)$ is estimated by: for $\Delta < t(z_0/z - 1)$

$$\hat{Q}(\Delta; t, z) = 1 - \exp\left\{ -\frac{t}{z} \left(\hat{\Lambda}_n(z(1 + \Delta/t)) - \hat{\Lambda}_n(z) \right) \right\};$$

if $\Delta \geq t(z_0/z - 1)$ then $\hat{Q}(\Delta, t, z) = 1$.

For $z \in [0, z_0]$ denote by

$$Q^{(k)}(\Delta; t, z) = P(T \leq t + \Delta, V = k \mid T > t, Z(t) = z) \qquad (13.18)$$

the conditional probability to have a traumatic failure of the kth mode in
the interval $(t, t + \Delta]$ given that at the moment t an unit is functioning and
its degradation value is z. This probability can be written in terms of the
cumulative intensities $\Lambda^{(i)}$: for $k = 1, \cdots, s$

$$Q^{(k)}(\Delta; t, z) = \frac{t}{z} \int_z^{\min\{z(1+\Delta/t), z_0\}} \exp\left\{ -\frac{t}{z} (\Lambda(y) - \Lambda(z)) \right\} d\Lambda^{(k)}(y).$$

(13.19)

The probability $Q^{(k)}(\Delta; t, z)$ is estimated by:

$$\hat{Q}^{(k)}(\Delta; t, z) = \frac{t}{z} \int_z^{\min\{z(1+\Delta/t), z_0\}} \exp\left\{ -\frac{t}{z} \left(\hat{\Lambda}(y) - \hat{\Lambda}(z) \right) \right\} d\hat{\Lambda}^{(k)}(y)$$

$$= \frac{t}{z} \exp\left\{ \frac{t}{z} \hat{\Lambda}(z) \right\} \sum_{z < Z_i < \min\{z(1+\Delta/t), z_0\}, V_i = k} \exp\left\{ -\frac{t}{z} \hat{\Lambda}(Z_i) \right\} Y_n^{-1}(Z_i).$$

Analogously, the conditional probability to have a nontraumatic failure in the
interval $(t, t + \Delta]$ given that at the moment t an unit is functioning and its
degradation value is z, has the form: for $\Delta < t(z_0/z - 1)$

$$Q^{(0)}(\Delta; t, z) = P(T \le t + \Delta, V = 0 \mid T > t, Z(t) = z) = 0, \qquad (13.20)$$

and for $\Delta \ge t(z_0/z - 1)$

$$Q^{(0)}(\Delta; t, z) = \exp\left\{ -\frac{t}{z} (\Lambda(z_0) - \Lambda(z)) \right\}. \qquad (13.21)$$

The probability $Q^{(0)}(\Delta; t, z)$ is estimated by: for $\Delta < t(z_0/z - 1)$

$$\hat{Q}^{(0)}(\Delta; t, z) = 0,$$

and for $\Delta \ge t(z_0/z - 1)$

$$\hat{Q}^{(0)}(\Delta; t, z) = \exp\left\{ -\frac{t}{z} \left(\hat{\Lambda}(z_0) - \hat{\Lambda}(z) \right) \right\}.$$

For $z \in [0, z_0]$ denote by

$$Q^{(tr)}(\Delta; t, z) = P(T \le t + \Delta, V \ne 0 \mid T > t, Z(t) = z) \qquad (13.22)$$

the conditional probability to have a traumatic failure in the interval $(t, t + \Delta]$
given that at the moment t an unit is functioning and its degradation value is
z. This probability can be written in terms of the cumulative intensities $\Lambda^{(i)}$:
for $\Delta < t(z_0/z - 1)$

$$Q^{(tr)}(\Delta; t, z) = 1 - \exp\left\{ -\frac{t}{z} (\Lambda(z(1 + \Delta/t)) - \Lambda(z)) \right\}; \qquad (13.23)$$

if $\Delta \ge t(z_0/z - 1)$ then

$$Q^{(tr)}(\Delta; t, z) = 1 - \exp\left\{ -\frac{t}{z} (\Lambda(z_0) - \Lambda(z)) \right\}. \qquad (13.24)$$

The probability $Q^{(tr)}(\Delta; t, z)$ is estimated by: for $\Delta < t(z_0/z - 1)$

$$\hat{Q}^{(tr)}(\Delta; t, z) = 1 - \exp\left\{-\frac{t}{z}\left(\hat{\Lambda}(z(1 + \Delta/t)) - \hat{\Lambda}(z)\right)\right\};$$

if $\Delta \geq t(z_0/z - 1)$ then

$$\hat{Q}^{(tr)}(\Delta; t, z) = 1 - \exp\left\{-\frac{t}{z}\left(\hat{\Lambda}(z_0) - \hat{\Lambda}(z)\right)\right\}.$$

13.2.5 Prediction of the ideal reliability characteristics

The random variable $T^{(0)}$ means the failure time of a tire when all traumatic failure modes are eliminated. The survival function and the mean of $T^{(0)}$ are:

$$S^{(0)}(t) = P(T^{(0)} > t) = 1 - \pi(t/z_0), \quad E(T^{(0)}) = z_0\, E(A). \qquad (13.25)$$

The mean $E(T^{(0)})$ shall be called the *ideal resource of the tire*, corresponding to the nontraumatic failure.

The survival function $S^{(0)}(t)$ is estimated by

$$\hat{S}^{(0)}(t) = 1 - \hat{\pi}_n(t/z_0) = \frac{1}{n}\sum_{i=1}^{n}\mathbf{1}_{\{z_0 A_i > t\}}$$

The ideal resource $E(T^{(0)})$ is estimated by

$$\widehat{ET^{(0)}} = z_0\int_0^\infty a\, d\hat{\pi}_n(a) = \frac{z_0}{n}\sum_{i=1}^{n} A_i.$$

The random variable $T^{(0k)} = \min\left(T^{(0)}, T^{(k)}\right)$ is the failure time of an unit when all traumatic failure modes, with exception of the kth, are eliminated.

Note that if $t < az_0$ then

$$P(T^{(0k)} > t \mid A = a) = P(T^{(k)} > t \mid A = a) = e^{-a\Lambda^{(k)}(t/a)},$$

and if $t \geq az_0$ then $P(T^{(0k)} > t \mid A = a) = 0$. It implies that the survival function $S^{(0k)}(t)$ of the random variable $T^{(0k)}$ is

$$S^{(0k)}(t) = P(T^{(0k)} > t) = \int_{t/z_0}^\infty e^{-a\Lambda^{(k)}(t/a)} d\pi(a), \qquad (13.26)$$

and the mean $E(T^{(0k)})$, called the *ideal resource of a unit*, corresponding to the kth failure mode, is

$$E(T^{(0k)}) = \int_0^\infty S^{(0k)}(t)dt = \int_0^\infty a\, d\pi(a)\int_0^{z_0} e^{-a\Lambda^{(k)}(z)}dz.$$

It may be written in the form

$$E(T^{(0k)}) = E(T^{(0)}) - \int_0^\infty a^2 d\pi(a)\int_0^{z_0}(z_0 - y)e^{-a\Lambda^{(k)}(y)}d\Lambda^{(k)}(y). \qquad (13.27)$$

The estimator of the survival function $S^{(0k)}(t)$ is

$$\hat{S}^{(0k)}(t) = \int_{t/z_0}^{\infty} e^{-a\hat{\Lambda}_n^{(k)}(t/a)}d\hat{\pi}_n(a) = \frac{1}{n}\sum_{z_0A_i>t} e^{-A_i\hat{\Lambda}_n^{(k)}(t/A_i)}$$

The estimator of the ideal resource is

$$\widehat{ET}^{(0k)} = \widehat{ET}^{(0)} - \int_0^{\infty} a^2 d\hat{\pi}(a)\int_0^{z_0}(z_0-y)e^{-a\hat{\Lambda}_n^{(k)}(y-)}d\hat{\Lambda}_n^{(k)}(y) =$$

$$= \widehat{ET}^{(0)} - \frac{1}{n}\sum_{i=1}^{n}A_i^2\sum_{V_j=k}(z_0-Z_j)e^{-A_i\hat{\Lambda}_n^{(k)}(Z_j-)}Y_n^{-1}(Z_j).$$

13.2.6 Prediction of the reliability characteristics related with elimination of particular failure modes

Suppose that the cause of a particular traumatic failure mode is supposed to be eliminated. Note that elimination of a failure mode may increase the number of failures of other modes. Indeed, if, say, the kth failure mode is present then some failures of, say, the lth mode may be not observed if these failures are preceded by a failure of the kth mode.

Denote by $T^{(-l)} = \min(T^{(0)}, T^{(1)}, \cdots, T^{(l-1)}, T^{(l+1)}, \cdots, T^{(s)})$ the failure time of an unit when the lth failure mode is absent.

Set

$$\Lambda^{(-l)}(z) = \sum_{i\neq l}\Lambda^{(i)}(z)$$

The survival function and the mean of the random variable $T^{(-l)}$ are

$$S^{(-l)}(t) = \int_{t/z_0}^{\infty} e^{-a\Lambda^{(-l)}(t/a)}d\pi(a), \tag{13.28}$$

$$E(T^{(-l)}) = E(T^{(0)}) - \int_0^{\infty} a^2 d\pi(a)\int_0^{z_0}(z_0-y)e^{-a\Lambda^{(-l)}(y)}d\Lambda^{(-l)}(y). \tag{13.29}$$

The estimator of the survival function $S^{(-l)}(t)$ is

$$\hat{S}^{(-l)}(t) = \int_{t/z_0}^{\infty} e^{-a\hat{\Lambda}_n^{(-l)}(t/a)}d\hat{\pi}_n(a) = \frac{1}{n}\sum_{z_0A_i>t} e^{-A_i\hat{\Lambda}_n^{(-l)}(t/A_i)}.$$

The estimator of $E(T^{(-l)})$ is

$$\widehat{ET}^{(-l)} = \widehat{ET}^{(0)} - \frac{1}{n}\sum_{i=1}^{n}A_i^2\sum_{V_j\neq l}(z_0-Z_j)e^{-A_i\hat{\Lambda}_n^{(-l)}(Z_j-)}Y_n^{-1}(Z_j).$$

Denote by $p_{(-l)}^{(k)}$ the probability of the failure of the kth mode when the cause of, say, the lth traumatic failure mode is eliminated. For $k = 1, \cdots, s, k \neq l$, this probability is

$$p_{(-l)}^{(k)} = \int_0^{\infty} a\,d\pi(a)\int_0^{z_0} e^{-a\Lambda^{(-l)}(x)}d\Lambda^{(k)}(x), \tag{13.30}$$

and for $k = 0$

$$p_{(-l)}^{(0)} = \int_0^\infty e^{-a\Lambda^{(-l)}(z_0)} d\pi(a). \tag{13.31}$$

The estimator of the probability $p_{(-l)}^{(k)}$:

$$\hat{p}_{(-l)}^{(k)} = \int_0^\infty a\, d\hat{\pi}(a) \int_0^{z_0} e^{-a\hat{\Lambda}_n^{(-l)}(x)} d\hat{\Lambda}_n^{(k)}(x) =$$

$$\frac{1}{n} \sum_{i=1}^n A_i \sum_{V_j=k} e^{-A_i \hat{\Lambda}_n^{(-l)}(Z_j-)} Y_n^{-1}(Z_j),$$

$$\hat{p}_{(-l)}^{(0)} = \int_0^\infty e^{-a\hat{\Lambda}_n^{(-l)}(z_0)} d\hat{\pi}(a) = \frac{1}{n} \sum_{i=1}^n e^{-A_i \hat{\Lambda}_n^{(-l)}(z_0)}.$$

The probability of the failure of the kth mode in the interval $[0, t]$ when the cause of the lth traumatic failure mode is eliminated, is

$$p_{(-l)}^{(k)}(t) = \int_0^\infty a\, d\pi(a) \int_0^{\min\{t/a, z_0\}} e^{-a\Lambda^{(-l)}(x)}(x) d\Lambda^{(k)}(x) \quad (k, l \neq 0, \ k \neq l), \tag{13.32}$$

$$p_{(-l)}^{(0)}(t) = \int_0^{t/z_0} e^{-a\Lambda^{(-l)}(z_0)} d\pi(a). \tag{13.33}$$

The estimator of the probability $p_{(-l)}^{(k)}(t)$:

$$\hat{p}_{(-l)}^{(k)}(t) = \int_0^\infty a\, d\hat{\pi}(a) \int_0^{\min\{t/a, z_0\}} e^{-a\hat{\Lambda}_n^{(-l)}(x)} d\hat{\Lambda}_n^{(k)}(x) =$$

$$\frac{1}{n} \sum_{i=1}^n A_i \sum_{V_j=k, Z_j \leq \min\{t/A_i, z_0\}} e^{-A_i \hat{\Lambda}_n^{(-l)}(Z_j-)} Y_n^{-1}(Z_j) \quad (k = 1, \cdots, s),$$

and

$$\hat{p}_{(-l)}^{(0)}(t) = \int_0^{t/z_0} e^{-a\hat{\Lambda}_n^{(-l)}(z_0)} d\hat{\pi}(a) = \frac{1}{n} \sum_{i=1, A_i \leq t/z_0}^n e^{-A_i \hat{\Lambda}_n^{(-l)}(z_0)}.$$

The probability of a traumatic failure when the cause of the lth traumatic failure mode is eliminated, is

$$p_{(-l)}^{(tr)} = \int_0^\infty a\, d\pi(a) \int_0^{z_0} e^{-a\Lambda^{(-l)}(x)} d\Lambda^{(-l)}(x), \tag{13.34}$$

and the probability of a traumatic failure in the interval $[0, t]$ when the cause of the lth traumatic failure mode is eliminated, is

$$p_{(-l)}^{(tr)}(t) = \int_0^\infty a\, d\pi(a) \int_0^{\min\{t/a, z_0\}} e^{-a\Lambda^{(-l)}(x)}(x) d\Lambda^{(-l)}(x). \tag{13.35}$$

The probability $p^{(tr)}_{(-l)}$ is estimated by

$$p^{(tr)}_{(-l)} = 1 - \frac{1}{n} \sum_{i=1}^{n} e^{-A_i \hat{\Lambda}_n^{(-l)}(z_0)}.$$

The probability $p^{(tr)}_{(-l)}(t)$ is estimated by

$$p^{(tr)}_{(-l)}(t) = 1 - \frac{1}{n} \sum_{i=1}^{n} e^{-A_i \hat{\Lambda}_n^{(-l)}(\min\{t/A_i, z_0\})}.$$

13.2.7 Semiparametric and parametric estimation

The graphs of the estimators $\hat{\Lambda}_n^{(k)}(z)$ give an idea of the form of the cumulative intensity functions $\Lambda^{(k)}(z)$. So the functions $\lambda^{(k)}(z)$ may be chosen from specified classes. Then semiparametric or parametric estimation of the reliability characteristics can be done. Semiparametric estimation is used when the distribution of the random variable A is completely unknown. Parametric estimation is used when the distribution of A is taken from a specified family of distributions.

Analysis of the tire failure time and wear data by nonparametric methods (see Bagdonavičius et al (2000, 2001)) shows that the intensities $\lambda^{(k)}(z)$ typically have one of the following forms:

$$\alpha_k z^{\nu_k}, \quad \alpha_k (z - u_k)^{\nu_k} \tag{13.36}$$

(production defects and defects caused by fatigue of tire components) or

$$\beta_k + \alpha_k z^{\nu_k} \tag{13.37}$$

(failures caused by the mechanical damages).

Suppose that the function $\lambda^{(k)}(z)$ is from a class of functions

$$\lambda^{(k)}(z) = \lambda^{(k)}(z, \gamma_k),$$

where γ_k is a possibly multidimensional parameter. For example, in the case of the classes (13.36)-(13.37) the parameter γ_k is (α_k, ν_k), (α_k, ν_k, u_k) or $(\alpha_k, \beta_k, \nu_k)$.

Suppose that the data are of the form (13.12).

If $V_i = k \ (k = 1, \ldots, s)$ then $T_i = T_i^{(k)}$ and A_i are observed and and it is known that $T_i^{(l)} > T_i^{(k)}, l \neq k$. The term of likelihood function corresponding to the ith unit is

$$p^{(k)}(T_i \mid A_i) \prod_{l \neq k} S^{(l)}(T_i \mid A_i) \, p_A(A_i) = \lambda^{(k)}(Z_i) \prod_{l=1}^{s} S^{(l)}(T_i \mid A_i) \, p_A(A_i),$$

where $p_A(a)$ is the density function of A. In the case of semiparametric estimation this term is absent.

If $V_i = 0$ then A_i are observed and it is known that $T_i^{(k)} > T_i^{(0)} = z_0 A_i, k \neq$

0. The term of likelihood function corresponding to the ith unit is

$$\prod_{l=1}^{s} S^{(l)}(T_i) \mid A_i)\, p_A(A_i).$$

Set

$$\delta_i = \begin{cases} 1, & \text{if } V_i = k,\ k = 1, \ldots, s, \\ 0, & \text{if } V_i = 0. \end{cases}$$

The likelihood function is

$$L = \prod_{i=1}^{n} \left\{ \lambda^{(V_i)}(Z_i) \right\}^{\delta_i} \prod_{l=1}^{s} S_i^{(l)}(T_i \mid A_i)\, p_A(A_i),$$

We write $B^0 = 1$ even when B is not defined. Note that

$$\lambda^{(V_i)}(Z_i) = \sum_{k=1}^{s} \lambda^{(k)}(Z_i)\, \mathbf{1}_{\{V_i=k\}}, \quad S^{(l)}(T_i \mid A_i) = \exp\{-A_i\, \Lambda^{(k)}(Z_i)\}.$$

If $V_i = 0$ then $\delta_i = 0$ and $\{\lambda^{(V_i)}(Z_i)\}^{\delta_i} = 1$.
  So

$$\ln L = \sum_{i=1}^{n} \sum_{k=1}^{s} \ln\{\lambda^{(k)}(Z_i)\}\, \mathbf{1}_{\{V_i=k\}} - \sum_{i=1}^{n} \sum_{k=1}^{s} A_i\, \Lambda^{(k)}(Z_i) + \ln p_A(A_i).$$

Estimators $\hat{\gamma}_k$ verify the equations

$$\frac{\partial \ln L}{\partial \gamma_k} = \sum_{i=1}^{n} \frac{\partial}{\partial \gamma_k} \ln\{\lambda^{(k)}(Z_i; \hat{\gamma}_k)\}\, \mathbf{1}_{\{V_i=k\}} - \sum_{i=1}^{n} A_i \frac{\partial}{\partial \gamma_k} \Lambda^{(k)}(Z_i; \hat{\gamma}_k) = 0.$$

$$(13.38)$$

Example 13.1.

$$\lambda^{(k)}(z; \alpha_k, \nu_k) = \alpha_k z^{\nu_k}.$$

The solution of the equations (13.38) is

$$\hat{\alpha}_k = \frac{n_k\, (\hat{\nu}_k + 1)}{\sum_{i=1}^{n} T_i\, Z_i^{\hat{\nu}_k}},$$

and $\hat{\nu}_k$ verifies the equation

$$\frac{1}{n_k} \sum_{i=1}^{n} \mathbf{1}_{\{V_i=k\}}\, \ln Z_i - \frac{\sum_{i=1}^{n} T_i\, Z_i^{\hat{\nu}_k} \ln Z_i}{\sum_{i=1}^{n} T_i\, Z_i^{\hat{\nu}_k}} + \frac{1}{\hat{\nu}_k + 1} = 0;$$

here

$$n_k = \sum_{i=1}^{n} \mathbf{1}_{\{V_i=k\}}.$$

Example 13.2.

$$\lambda^{(k)}(z; \alpha_k, \beta_k, \nu_k) = \beta_k + \alpha_k z^{\nu_k}.$$

The estimators $\hat{\alpha}_k, \hat{\beta}_k, \hat{\nu}_k$ verify the equations

$$\sum_{i=1}^{n} \frac{\mathbf{1}_{\{V_i=k\}}}{\hat{\beta}_k + \hat{\alpha}_k Z_i^{\hat{\nu}_k}} - \sum_{i=1}^{n} T_i = 0,$$

$$(\hat{\nu}_k + 1) \sum_{i=1}^{n} \frac{Z_i^{\hat{\nu}_k}}{\hat{\beta}_k + \hat{\alpha}_k Z_i^{\hat{\nu}_k}} \mathbf{1}_{\{V_i=k\}} - \sum_{i=1}^{n} T_i Z_i^{\hat{\nu}_k} = 0,$$

$$(\hat{\nu}_k + 1) \sum_{i=1}^{n} \frac{Z_i^{\hat{\nu}_k} \ln Z_i}{\hat{\beta}_k + \hat{\alpha}_k Z_i^{\hat{\nu}_k}} \mathbf{1}_{\{V_i=k\}} - \sum_{i=1}^{n} T_i Z_i^{\hat{\nu}_k} \left\{ \ln Z_i - \frac{1}{\hat{\nu}_k + 1} \right\} = 0.$$

After estimation of the parameters γ_k of the functions $\Lambda^{(k)}(z) = \Lambda^{(k)}(z, \gamma_k)$, these functions are estimated by

$$\hat{\Lambda}^{(k)}(z) = \Lambda^{(k)}(z, \hat{\gamma}_k)$$

Semiparametric predictors of the conditional, ideal, etc. reliability characteristics are obtained evidently: in the expressions (13.16)-(13.35) of these characteristics all functions $\Lambda^{(k)}(t)$ are replaced by their estimators $\hat{\Lambda}^{(k)}(t)$, and the distribution function $\pi(a)$ by its estimator

$$\hat{\pi}(a) = \sum_{i=1}^{n} \mathbf{1}_{\{A_i \leq a\}}. \tag{13.39}$$

The main unconditional reliability characteristics (see Chapter 13.2.3.) can be estimated using the formulas (13.8), (13.9),

$$p^{(k)} = P(V = k) =$$

$$\int_0^\infty d\pi(a) \int_{t^{(k)} \leq z_0 a, t_1, \cdots, t_{k-1}, t_{k+1}, \cdots, t_m} p_1(t_1 \mid a) \cdots p_m(t_m \mid a) dt_1 \cdots dt_m$$

$$= \int_0^\infty a d\pi(a) \int_0^{z_0} e^{-a\Lambda(x)} d\Lambda^{(k)}(x), \tag{13.40}$$

$$p^{(k)}(t) = P(V = k, T \leq t) = \int_0^\infty a d\pi(a) \int_0^{\min\{t/a, z_0\}} e^{-a\Lambda(x)} d\Lambda^{(k)}(x). \tag{13.41}$$

$$p^{(0)} = P(V = 0) = \int_0^\infty d\pi(a) \int_{z_0 a \leq t_1, \cdots, t_m} f_1(t_1 \mid a) \cdots f_m(t_m \mid a) dt_1 \cdots dt_m$$

$$= \int_0^\infty e^{-a\Lambda(z_0)} d\pi(a) = 1 - \sum_{i=1}^{s} p_i, \tag{13.42}$$

$$p^{(0)}(t) = P(V = 0, T \leq t) = \int_0^{t/z_0} e^{-a\Lambda(z_0)} d\pi(a), \tag{13.43}$$

$$p^{(tr)} = P(V \neq 0) = \int_0^\infty a d\pi(a) \int_0^{z_0} e^{-a\Lambda(x)} d\Lambda(x), \tag{13.44}$$

$$p^{(tr)}(t) = P(V \neq 0, T \leq t) = \int_0^\infty a d\pi(a) \int_0^{\min\{t/a, z_0\}} e^{-a\Lambda(x)} d\Lambda(x). \tag{13.45}$$

Semiparametric estimators of these reliability characteristics are obtained by replacing all functions $\Lambda^{(k)}(t)$ by their estimators $\hat{\Lambda}^{(k)}(t)$, and the distribution function $\pi(a)$ by it's estimator $\hat{\pi}$.

In the case of parametric estimation the distribution function π is taken from a specified family of distributions

$$\pi(a) = \pi(a, \eta),$$

and the estimators $\hat{\eta}$ of the unknown parameters η of this distribution are estimated by the method of maximum likelihood using the complete data

$$A_1, \cdots, A_n.$$

The real data show that the families of the gamma and Weibull distributions are the good choices. Applicability of a specified family is verified using the standard goodness-of-fit tests.

Parametric estimators and predictors of the main reliability characteristics are obtained as follows: in the expressions (13.40)-(13.45) and (13.16)-(13.35) all functions $\Lambda^{(k)}(t)$ are replaced by their estimators $\hat{\Lambda}^{(k)}(t)$, and the distribution function $\pi(a)$ by its estimator $\hat{\pi}(a) = \pi(a, \hat{\eta})$.

13.2.8 Right censored data

Semiparametric and parametric estimation can be evidently modified when units are observed not necessary until the failure (nontraumatic or traumatic). Observation of the ith unit may be censored at the moment C_i (random or nonrandom). We suppose that at the moment C_i the value of the degradation Z_i is measured. So in the case of censoring $A_i = C_i/Z_i$ and V_i is not observed.

The likelihood function (supposing that the censoring moments C_i do not depend on the degradation and traumatic failure processes) is

$$L = \prod_{i=1}^{n} \left\{ \lambda^{(V_i)}(Z_i) \right\}^{\delta_i \varepsilon_i} \prod_{l=1}^{s} S_i^{(l)}(T_i \wedge C_i \mid A_i) \, p_A(A_i),$$

where $\varepsilon_i = \mathbf{1}_{\{T_i \le C_i\}}$ and

$$\delta_i = \begin{cases} 1, & \text{if } V_i = k, \; k = 1, \ldots, s, \\ 0, & \text{if } V_i = 0. \end{cases}$$

Then

$$\ln L = \sum_{i=1}^{n} \varepsilon_i \sum_{k=1}^{s} \ln\{\lambda^{(k)}(Z_i)\} \, \mathbf{1}_{\{V_i = k\}} - \sum_{i=1}^{n} \sum_{k=1}^{s} A_i \, \Lambda^{(k)}(Z_i) + \ln p_A(A_i).$$

After the estimation of the parameters γ_k the functions $\Lambda^{(k)}(z)$ are estimated by $\hat{\Lambda}^{(k)}(z) = \Lambda^{(k)}(z, \hat{\gamma}_k)$. Semiparametric and parametric estimators and predictors of the main reliability characteristics are obtained using the same formulas as in the uncensored case. See, for example, Huber (2000).

13.2.9 Model with explanatory variables

Suppose that the degradation process $Z_{x(\cdot)}(t)$ under the explanatory variable $x(\cdot)$ is modeled by

$$Z_{x(\cdot)}(t) = A^{-1} \int_0^t e^{\beta^T x(s)} \, ds, \quad t \geq 0,$$

where A is a positive random variable with the distribution π.

Denote by $T_{x(\cdot)}^{(0)}$ the moment of the nontraumatic failure under $x(\cdot)$, and by $T_{x(\cdot)}^{(k)}$ $(k = 1, \cdots, s)$ the moment of the traumatic failure of the kth mode under $x(\cdot)$.

We suppose that the random variables $T_{x(\cdot)}^{(1)}, \cdots, T_{x(\cdot)}^{(s)}$ are conditionally independent (given $A = a$) but depend on the degradation level.

Suppose that the model (13.3) is true for each k. Then the conditional survival function of $T_{x(\cdot)}^{(k)}$ is

$$S_{x(\cdot)}^{(k)}(t \mid a) = \mathbf{P}(T_{x(\cdot)}^{(k)} > t | A = a) = \exp\left\{ -\int_0^t \lambda^{(k)} (a^{-1} \int_0^s e^{\beta^T x(u)} \, du) ds \right\}$$

$$= \exp\left\{ -a \int_0^{a^{-1} f_{x(\cdot)}(t,\beta)} e^{-\beta^T x(g_{x(\cdot)}(av,\beta))} \, d\Lambda^{(k)}(v) \right\},$$

where

$$\Lambda^{(k)}(z) = \int_0^z \lambda^{(k)}(y) dy,$$

and

$$g_{x(\cdot)}(t,\beta) = f_{x(\cdot)}^{-1}(t,\beta)$$

is the inverse function of

$$f_{x(\cdot)}(t,\beta) = \int_0^t e^{\beta^T x(s)} \, ds.$$

with respect to the first argument.

If $x = $const then

$$f_x(t,\beta) = e^{\beta^T x} t, \quad g_x(t,\beta) = e^{-\beta^T x} t$$

and

$$S_x^{(k)}(t \mid a) = \exp\left\{ -ae^{-\beta^T x} \Lambda^{(k)}(a^{-1} e^{\beta^T x} t) \right\}.$$

13.2.10 Nonparametric estimation of the cumulative intensities

Suppose that the cumulative intensities Λ_k are completely unknown.

The purpose is to estimate $\Lambda_k(z), 0 \leq z \leq z_0$.

Suppose that n units are on test. The ith unit is tested under explanatory variable $x^{(i)}(\cdot)$, and the failure moments T_i, failure modes V_i and the degradation levels

$$Z_i = A_i^{-1} f_{x^{(i)}(\cdot)}(T_i,\beta)$$

at the failure moments T_i are observed. So the data has the form

$$(T_1, Z_1, V_1, x^{(1)}), \cdots, (T_n, Z_n, V_n, x^{(n)}). \tag{13.46}$$

For $k = 1, \cdots, s$ and $0 \le z \le z_0$ set

$$N_n^{(k)}(z) = \sum_{i=1}^{n} N_{in}^{(k)}(z), \quad N_{in}^{(k)}(z) = \mathbf{1}_{\{Z_i \le z, V_i = k\}}.$$

The parameter β shows the influence of the explanatory variables on the degradation. The intensities of the traumatic events $\lambda^{(k)}(z)$ do not depend on β and the distribution of the traumatic events depend on β only via degradation. So for estimation of the parameter β the data $(T_1, Z_1, x^{(1)}), \cdots, (T_n, Z_n, x^{(n)})$ is used.

Note that the random variables

$$A_i(\beta) = Z_i^{-1} f_{x^{(i)}(\cdot)}(T_i, \beta)$$

are independent identically distributed with the mean, say m, which does not depend on β. So the parameter β is estimated by the method of least squares, minimizing the sum

$$\sum_{i=1}^{n} (A_i(\beta) - m)^2,$$

which gives the system of equations

$$n \sum_{i=1}^{n} Z_i^{-2} \int_0^{T_i} x^{(i)}(u) e^{\beta^T x^{(i)}(u)} du \int_0^{T_i} e^{\beta^T x^{(i)}(u)} du -$$

$$\sum_{i=1}^{n} Z_i^{-1} \int_0^{T_i} x^{(i)}(u) e^{\beta^T x^{(i)}(u)} du \sum_{i=1}^{n} Z_i^{-1} \int_0^{T_i} e^{\beta^T x^{(i)}(u)} du = 0. \tag{13.47}$$

If $x^{(i)}$ are constant then this system has the form

$$n \sum_{i=1}^{n} x^{(i)} \{Z_i^{-1} e^{\beta^T x^{(i)}} T_i\}^2 - \sum_{i=1}^{n} Z_i^{-1} x^{(i)} e^{\beta^T x^{(i)}} T_i \sum_{i=1}^{n} Z_i^{-1} e^{\beta^T x^{(i)}} T_i = 0.$$

$$\tag{13.48}$$

For estimation of the cumulative intensities suppose at first that the parameters β are known.

Note that the data $(T_i, V_i, Z_i, x^{(i)})$ is equivalent to the data $(T_i, N_i^{(k)}(z), k = 1, \ldots, s, 0 \le z \le z_0, x^{(i)})$. Indeed, the random variables Z_i and V_i define the stochastic processes $N_i^{(k)}(z)$, $0 \le z \le z_0$. Vice versa, if $N^{(k)}(z) = 0$ for all $0 \le z \le z_0$ and $k = 1, \ldots, s$, then $V_i = 0$ and $Z_i = z_0$. If there exist $k \ne 0$ and z_i such that $N_i^{(k)}(z_i-) = 0$, $N_i^{(k)}(z_i) = 1$, then $V_i = k$ and $Z_i = z_i$.

If β is known then the data $(T_i, V_i, Z_i, x^{(i)})$ is equivalent to the data $(A_i, V_i, Z_i, x^{(i)})$ and hence, to the data $(A_i, N_i^{(k)}(z), k = 1, \ldots, s, 0 \le z \le z_0)$, because the random variables $Z_i, T_i,$ and A_i have the following relations: $A_i = Z_i / f_{x_i}(T_i, \beta)$.

Let $\mathcal{F}_z$ be the σ-algebra generated by the random variables $A_1, \cdots, A_n$ and $N_n^{(1)}(y), \cdots, N_n^{(s)}(y)$, $y \leq z$.

Set $f_i = f_{x^{(i)}}$, $g_i = g_{x^{(i)}}$, $S_i^{(k)} = S_{x^{(i)}}^{(k)}$, $S_i^{(k)} = S_{x^{(i)}}^{(k)}$.

Lemma 13.2. *The counting process* $N_n^{(k)}(z)$ *can be written as the sum*

$$N_n^{(k)}(z) = \int_0^z Y_n(y, \beta) \lambda^{(k)}(y) \, dy + M_n^{(k)}(z, \beta),$$

where

$$Y_n(z, \beta) = \sum_{i=1}^n Y_{in}(z, \beta),$$

$$Y_{in}(z, \beta) = A_i \exp\left\{ -\beta^T x^{(i)}(g_i(A_i z, \beta)) \right\} 1_{\{Z_i \geq z\}},$$

and $(M_n^{(k)}(z), 0 \geq z \geq z_0)$ *is a martingale with respect to the filtration* $(\mathcal{F}_z, 0 \leq z \leq z_0)$.

Proof. Let $0 \leq y < z \leq z_0$. It is sufficient to prove that

$$\mathbf{E}\{N_1^{(k)}(z) - N_1^{(k)}(y) \mid \mathcal{F}_y\} = \mathbf{E}\{\int_y^z \lambda^{(k)}(u) Y_1^{(k)}(u) du \mid \mathcal{F}_y\}.$$

If $A_1 = a$ and $Z_1 \leq y$ then $N_1^{(k)}(z) = N_1^{(k)}(y)$. If $A_1 = a$ and $Z_1 > y$ then the random variable $N_1^{(k)}(z)$ takes two values, 0 and 1, and

$$\mathbf{P}\{N_1^{(k)}(z) = 1 \mid Z_1 > y, \ A_1 = a\} == P\{N_1^{(k)}(z) = 1 \mid A_1 = a, T_1 > g_1(ay, \beta)\} =$$

$$= P\{g_1(ay, \beta) < T_1^{(k)} \leq g_1(az, \beta), T_1^{(1)}, \cdots, T_1^{(s)} \mid A_1 = a, T_1 > g_1(ay, \beta)\}$$

$$= \frac{1}{\prod_{l=1}^s S_1^{(l)}(g_1(ay, \beta) \mid a)} \int_{g_1(ay,\beta)}^{g_1(az,\beta)} p_1^{(k)}(t \mid a) \prod_{l \neq k} S_1^{(l)}(t \mid a) \, dt =$$

$$= \exp\left\{ a \int_0^y e^{-\beta^T x^{(1)}(g_1(au,\beta))} d\Lambda(u) \right\} a \int_y^z \lambda^{(k)}(v) \times$$

$$\exp\left\{ -a \int_0^v e^{-\beta^T x^{(1)}(g_1(au,\beta))} d\Lambda(u) \right\} e^{-\beta^T x^{(1)}(g_1(av,\beta))} \, dv.$$

So

$$E\{N_1^{(k)}(z) - N_1^{(k)}(y) \mid \mathcal{F}_y\} = 1_{\{Z_1 > y\}} \exp\left\{ A_1 \int_0^y e^{-\beta^T x^{(1)}(g_1(A_1 u,\beta))} d\Lambda(u) \right\} \times$$

$$A_1 \int\limits_y^z \lambda^{(k)}(v) \exp\left\{ -A_1 \int\limits_0^v e^{-\beta^T x^{(1)}(g_1(A_1 u,\beta))}d\Lambda(u) \right\} e^{-\beta^T x^{(1)}(g_1(A_1 v,\beta))}\, dv.$$

(13.49)

If $A_1 = a$ and $Z_1 \leq y$ then $\mathbf{1}_{\{Z_i \geq z\}} = 0$. If $A_1 = a$, $Z_1 > y$ then for $v > y$

$$P\{\mathbf{1}_{\{Z_1 \geq v\}} = 1 \mid A_1 = a, Z_1 > y\} = P\{Z_1 \geq v \mid A_1 = a, Z_1 > y\}$$

$$= \exp\left\{ a \int\limits_0^y e^{-\beta^T x^{(1)}(g_1(au,\beta))}d\Lambda(u) \right\} \exp\left\{ -a \int\limits_0^v e^{-\beta^T x^{(1)}(g_1(au,\beta))}d\Lambda(u) \right\}.$$

$$E\left\{ \int\limits_y^z Y_1^{(k)}(u,\beta)\lambda^{(k)}(u)\, du \mid \mathcal{F}_y \right\} = \mathbf{1}_{\{Z_1 > y\}} \exp\left\{ A_1 \int\limits_0^y e^{-\beta^T x^{(1)}(g_1(A_1 u,\beta))}d\Lambda(u) \right\} \times$$

$$A_1 \int\limits_y^z \lambda_k(v) \exp\left\{ -A_1 \int\limits_0^v e^{-\beta^T x^{(1)}(g_1(A_1 u,\beta))}d\Lambda(u) \right\} e^{-\beta^T x^{(1)}(g_1(A_1 v,\beta))}\, dv.$$

(13.50)

The result of the lemma follows from the equalities (13.49) and (13.50).

Corollary 13.1. *If β is known the optimal estimators of the cumulative intensities are*

$$\tilde\Lambda_n^{(k)}(z,\beta) = \sum_{i:Z_i \leq z, V_i = k} (Y_n^{(k)}(Z_i,\beta))^{-1}$$

$$= \sum_{i:Z_i \leq z, V_i = k} \left(\sum_{j:Z_j \geq Z_i} \frac{f_j(T_j,\beta)}{Z_j} \exp\left\{ -\beta^T x^{(j)}(T_j) \right\} \right)^{-1}.$$ (13.51)

Remark 13.1. *If $x_i(t) \equiv x_i = \text{const}$ $(i = 1, \ldots, n)$ then the estimator (13.51) does not depend on β and is:*

$$\hat\Lambda_n^{(k)}(z) = \int\limits_0^z \frac{dN_n^{(k)}(y)}{Y_n(y)} = \sum_{i:Z_i \leq z, V_i = k} \left(\sum_{j:Z_j \geq Z_i} \frac{T_j}{Z_j} \right)^{-1}.$$ (13.52)

Remark 13.2. *If the degradation values are measured at the moments of switch-ups from one constant value to another then under the piecewise-constant explanatory variables the estimators (13.51) can be modified to obtain estimators which do not depend on β.*

Indeed, suppose that the explanatory variables $x^{(i)}(\cdot)$ have the form

$$x^{(i)}(t) = x_l^{(i)}, \qquad t \in [t_{i,l-1}, t_{il}),$$

where $0 \leq t_{i0} < t_{i1} < \ldots < t_{im_i} = \infty$ and $x_1^{(i)}, \ldots, x_{m_i}^{(i)}$ are constant explanatory variables. Lemma 13.2 implies that

$$Y_{in}^{(k)}(z) = A_i e^{-\beta^T x^{(i)}(g_i(A_i z,\beta))} \mathbf{1}_{\{Z_i \geq z\}}.$$

Note that

$$t_{i,l-1} \le g_i(A_i z, \beta) \le t_{il} \iff f_i(t_{i,l-1}, \beta) \le A_i z \le f_i(t_{il}, \beta).$$

So

$$Y_{in}^{(k)}(z) = A_i e^{-\beta^T x_l^{(i)}} \mathbf{1}_{\{Z_i \ge z\}}$$

when

$$A_i^{-1} f_i(t_{i,l-1}, \beta) \le z \le A_i^{-1} f_i(t_{il}, \beta).$$

Suppose that at the moments $t_{il} \le T_i$ the values of degradation are measured. Set

$$Z_{il} = Z(t_{il}) = A_i^{-1} f_i(t_{il}, \beta).$$

Then

$$\Delta Z_{il} = Z_{il} - Z_{i,l-1} = A_i^{-1} \int_{t_{i,l-1}}^{t_{il}} e^{\beta^T x_l^{(i)}} du = A_i^{-1} e^{\beta^T x_l^{(i)}} \Delta t_{il},$$

which implies

$$A_i^{-1} e^{\beta^T x_l^{(i)}} = \frac{\Delta Z_{il}}{\Delta t_{il}};$$

here $\Delta t_{il} = t_{il} - t_{i,l-1}$.

So the function $Y_{in}^{(k)}(z)$ can be written in the form

$$Y_{in}^{(k)}(z) = \frac{\Delta t_{il}}{\Delta Z_{il}} \mathbf{1}_{\{Z_i \ge z\}},$$

if $Z_{i,l-1} \le z \le Z_{il}, l-1, \ldots, \mu_i$, where

$$\mu_i = \max\{l : t_{i,l-1} \le T_i\} \qquad \text{and} \qquad Z_{i\mu_i} = Z_i.$$

We have

$$Y_{in}^{(k)}(z) = \sum_{l=1}^{\mu_i} \frac{\Delta t_{il}}{\Delta Z_{il}} \mathbf{1}_{\{Z_i \ge z, Z_{i,l-1} \le z \le Z_{il}\}}.$$

The estimator $\tilde{\Lambda}_n^{(k)}(z, \beta)$ does not depend on β and can be written in the following form

$$\hat{\Lambda}_n^{(k)}(z) = \int_0^z \frac{dN_n^{(k)}(y)}{Y_n(y)} = \sum_{i: Z_i \le z, V_i = k} \left(\sum_{j: Z_j \ge Z_i} \sum_{l: Z_{j,l-1} \le Z_i \le Z_{jl}} \frac{\Delta t_{jl}}{\Delta Z_{jl}} \right)^{-1}.$$

$$(13.53)$$

Note that under known β the triplets (T_i, V_i, Z_i) are equivalent to the triplets $(T_i, V_i, Z_{il}, l = 1, \ldots, \mu_i)$, because

$$\Delta Z_{il} = \frac{Z_i}{f_i(T_i, \beta)} e^{\beta^T x_{il}} \Delta t_{il}, \qquad Z_{il} = \sum_{s=1}^{l} \Delta Z_{is}.$$

So the adding of the measurements $Z_{il}, l \ne \mu_i$ does not change the filtration considered in Lemma 13.2.

If the explanatory variables are more complicated time-varying and the data are (13.46) then the estimators of the cumulative intensities are obtained from replacing β by $\hat{\beta}$ in (13.51):

$$\hat{\Lambda}_n^{(k)}(z) = \tilde{\Lambda}_n^{(k)}(z,\hat{\beta}) = \int_0^z \frac{dN_n^{(k)}(y)}{Y_n^{(k)}(y,\hat{\beta})} = \sum_{i:Z_i \leq z, V_i = k} (Y_n^{(k)}(Z_i,\hat{\beta}))^{-1}$$

$$= \sum_{i:Z_i \leq z, V_i = k} \left(\sum_{j:Z_j \geq Z_i} \frac{f_j(T_j,\beta)}{Z_j} \exp\left\{ -\hat{\beta}^T x^{(j)}(g_i(\frac{Z_i}{Z_j}f_j(T_j,\beta),\beta))\right\} \right)^{-1}.$$

$$(13.54)$$

13.2.11 Estimation and prediction of the reliability under given explanatory variables

Estimation and prediction of the reliability under given explanatory variables is done similarly as in the case when the explanatory variables are absent.

Indeed, all reliability characteristics (13.8)-(13.9), (13.16)-(13.35), (13.40)-(13.45) can be rewritten to this more general case. For example, the survival function $S_{x(\cdot)}$ under the explanatory variable $T_{x(\cdot)}$ generalizes the survival function $S(t)$ (cf.(13.8)) and is

$$S_{x(\cdot)}(t) = \int_{t/z_0}^{\infty} \exp\left\{ -a \int_0^{a^{-1}f_{x(\cdot)}(t,\beta)} e^{-\beta^T x(g_{x(\cdot)}(av,\beta))}d\Lambda(v) \right\} d\pi(a),$$

$$(13.55)$$

the probability $p_{x(\cdot)}^{(k)}$ of a failure of the kth mode under $x(\cdot)$ generalizes the probability $p^{(k)}$ (cf. (13.40)) and is

$$p_{x(\cdot)}^{(k)} = \int_0^{\infty} a d\pi(a) \int_0^{z_0} e^{-\beta^T x(g_{x(\cdot)}(az,\beta))} \times$$

$$\exp\left\{ -a \int_0^z e^{-\beta^T x(g_{x(\cdot)}(au,\beta))}d\Lambda(u) \right\} d\Lambda^{(k)}(z),$$

the conditional probability

$$Q_{x(\cdot)}(\Delta;t,z) = P(T_{x(\cdot)} \leq t+\Delta \mid T_{x(\cdot)} > t, Z_{x(\cdot)}(t) = z) \qquad (13.56)$$

to fail in the interval $(t, t+\Delta]$ under $x(\cdot)$ given that at the moment t a unit is functioning and its degradation value is z generalizes the probability $Q(\Delta;t,z)$ (cf. (13.17)) and is:

$$Q_{x(\cdot)}(\Delta;t,z) = 1 - \exp\left\{ -\frac{f_{x(\cdot)}(t,\beta)}{z} \int_z^{z\frac{f_{x(\cdot)}(t+\Delta,\beta)}{f_{x(\cdot)}(t,\beta)}} e^{-\beta^T x(g_{x(\cdot)}(av,\beta))}d\Lambda(v) \right\},$$

$$\text{if} \quad \Delta < g_{x(\cdot)}(\frac{z_0}{z}f_{x(\cdot)}(t,\beta),\beta) - t;$$

otherwise $Q(\Delta,t,z) = 1$.

And so on. Expressions of other reliability characteristics can be found in Bagdonavičius et al (2001). Estimation of them is evident: the unknown parameters β, cumulative intensities $\Lambda^{(k)}$, and the distribution function π are replaced by their estimators in all expressions. The estimator of β verifies the equations (13.47) or (13.48), the estimators of $\Lambda^{(k)}$ are given by (13.51), (13.52) or (13.54), and the estimator of π is

$$\hat{\pi}(a) = \sum_{i=1}^{n} \mathbf{1}_{\{f_i(T_i,\hat{\beta}) \leq a Z_i\}}.$$

Semiparametric and parametric estimation procedures considered in 13.2.7 and 13.2.8. can be generalized (see Bagdonavičius et al (2001)).

13.3 Gamma and shock processes

In general path models the finite-dimensional distributions of the degradation process are degenerate. It is not so in the case of most other processes such as gamma and shock processes. Now we shall follow Bagdonavičius and Nikulin (2001). Suppose that the model (13.3) with degradation process defined by (13.1) is considered. Assume that finite-dimensional distributions of the stochastic process $Z(t)$ are not degenerated.

The function $\lambda(z, x)$ is supposed to have a specified form, for example, one of the forms (3.19), (3.21), (3.23), (3.28), (3.31).

13.3.1 Data and likelihood structure when the form of the mean degradation form is specified

Suppose that n items are observed. The ith item is observed under the vector of explanatory variables $x^{(i)}(\cdot) = \left(x_1^{(i)}(\cdot), \ldots, x_s^{(i)}(\cdot)\right)^T$, and at the moments

$$0 < t_{i1} < t_{i2} < \ldots < t_{im_i}$$

the values $Z_{x^{(i)}(\cdot)}(t_{ij})$ of the degradation level are supposed to be measured without errors.

The values of explanatory variables are supposed to be observed during the experiment. Most often they should be constant in time or step-functions.

Set

$$Z_i(t) = Z_{x^{(i)}(\cdot)}(t), \quad Z_{ij} = Z_i(t_{ij}).$$

The value Z_{ij} is measured if a traumatic event does not occur in the interval $[0, t_{ij}]$ and $Z_{i,j-1} < z_0$. The value Z_{ij} is not measured if a traumatic event occurs before the moment t_{ij} or $Z_{i,j-1} \geq z_0$.

Denote by C_i the moment of the traumatic failure of the ith unit. Set $\Delta Z_{ij} = Z_{ij} - Z_{i,j-1}$, $Z_{i0} = 0$, $t_{i,m_i+1} = \infty$.

The structure and the size of the data is determined by the following non-intersecting events:

1) A_{i0}: a traumatic failure occurs before the moment t_{i1}. Then only the moment of the traumatic event C_i is observed;

2) A_{ij} $(j = 1, \cdots, m_i-1)$: a traumatic event occurs in the interval $(t_{ij}, t_{i,j+1}]$ and $Z_{ij} < z_0$. Then the values of degradation $Z_{i1}, \cdots, Z_{ij}$ and the moment of the traumatic event C_i are observed.

3) B_{ij} $(j = 1, \cdots, m_i)$: a traumatic event occurs after the moment t_{ij}, $Z_{i,j-1} < z_0$, $Z_{ij} \geq z_0$. Then the values of degradation $Z_{i1}, \cdots, Z_{ij}$ are observed and the moment of traumatic event is censored at the moment t_{ij}, i.e. it is known that $C_i > t_{ij}$.

Denote by η_i the *observed number of degradation measurements* of the ith item. Set

$$\delta_i = \begin{cases} 1, & \text{if the traumatic event of the ith item is observed,} \\ 0, & \text{otherwise.} \end{cases}$$

Take notice that

$$A_{ij} = \{\eta_i = j, \delta_i = 1\}, \quad (j = 1, \cdots, m_i - 1),$$

$$B_{ij} = \{\eta_i = j, \delta_i = 0\} \quad (j = 1, \cdots, m_i),$$

$$A_{i0} = \{\eta_i = 0, \delta_i = 0\} = \{\eta_i = 0\}.$$

Suppose that the function $m(t)$ is parametrized via parameters $\gamma = (\gamma_0, ..., \gamma_m)^T$:

$$m = m(t; \gamma),$$

and it is supposed to be continuously differentiable. Then the mean degradation under $x(\cdot)$ has the form

$$m_{x(\cdot)}(t; \beta, \gamma) = m\left(\int_0^t e^{\beta^T x(s)}\, ds; \gamma\right).$$

For example, in the case of the model (3.1)

$$\Delta m_{ij}(\beta, \gamma) = \gamma_0 \int_{t_{i,j-1}}^{t_{ij}} e^{\beta^T x^{(i)}(s)} ds + \frac{\gamma_1}{\gamma_2} \exp\{-\gamma_2 \int_0^{t_{i,j-1}} e^{\beta^T x^{(i)}(s)} ds\}$$

$$\left(1 - \exp\{-\gamma_2 \int_{t_{i,j-1}}^{t_{ij}} e^{\beta^T x^{(i)}(s)} ds\}\right).$$

Denote by $\tau_{x(\cdot)} = \mathbf{E}\{T_{x(\cdot)} \mid x(\cdot)\}$ the mean time needed to attain the level z_0 under $x(\cdot)$.

Our purpose is to estimate reliability characteristics

$$m_{x(\cdot)}(t), \quad S_{x(\cdot)}(t), \quad \tau_{x(\cdot)}, \quad Q_{x(\cdot)}(t), \quad G_{x(\cdot)}(t),$$

and the coefficients β_j, where $x(\cdot)$ is possibly different from $x^{(1)}(\cdot), \cdots, x^{(n)}(\cdot)$. The coefficient β_j shows the influence of component $x_j(\cdot)$ of the vector $x(\cdot) = \left(x_0(\cdot), \ldots x_s(\cdot)\right)^T$ on the degradation.

In what follows we denote by p_X the probability density function and by F_X the cumulative distribution function of any random variable X. Set

$$p_{C_i > t, Z_{i1}, \dots, Z_{ij}}(z_{i1}, \dots, z_{ij}) = \int_t^\infty p_{C_i, z_{i1}, \dots, z_{ij}}(v, z_{i1}, \dots, z_{ij}) dv.$$

On the sets A_{i0}, A_{ij} $(j = 1, \cdots, m_i - 1)$ and B_{ij} $(j = 1, \cdots, m_i)$ the likelihood function has respectively the following forms

$$L = \prod_{i=1}^n p_{C_i}(C_i);$$

$$L = \prod_{i=1}^n p_{C_i, Z_{i1}, \dots, Z_{ij}}(C_i, Z_{i1}, \dots, Z_{ij}),$$

and

$$L = \prod_{i=1}^n p_{C_i > t_{ij}, Z_{i1}, \dots, Z_{ij}}(Z_{i1}, \dots, Z_{ij}).$$

Thus, the likelihood function given $x^{(1)}(\cdot), \cdots, x^{(n)}(\cdot)$ is

$$L = \prod_{i=1}^n p_{C_i}^{k_i}(C_i)\, p_{C_i, Z_{i1}, \dots, Z_{ij}}^{\delta_i(1-k_i)}(C_i, Z_{i1}, \dots, Z_{ij})|_{j=\eta_i} \times$$

$$p_{C_i > t_{ij}, Z_{i1}, \dots, Z_{ij}}^{(1-\delta_i)(1-k_i)}(Z_{i1}, \dots, Z_{ij})|_{j=\eta_i}. \tag{13.57}$$

where $k_i = \mathbf{1}_{\{\eta_i = 0\}}$ and $A^0 = 1$, when A is not defined. The parameters to be estimated are the parameters of the function $\lambda(z, x)$ and the parameters of the process $Z_{x(\cdot)}(t)$.

13.3.2 Estimation of degradation characteristics and reliability when traumatic events do not depend on degradation

Suppose that the traumatic event intensity does not depend on degradation. Then the likelihood function (13.57) has the form

$$L(\theta^*) = \prod_{i=1}^n p_{C_i}^{k_i + (1-k_i)\delta_i}(C_i)\, S_{C_i}^{1-k_i)}(t_{i\eta_i})\, p_{Z_{i1}, \dots, Z_{ij}}^{(1-k_i)}(Z_{i1}, \dots, Z_{ij})|_{j=\eta_i}. \tag{13.58}$$

Example 13.3. Gamma process.

The density function of degradation measurements has the form:

$$p_{Z_{i1}, \dots, Z_{ij}}(z_{i1}, \dots, z_{ij}) = \prod_{l=1}^j p_{\Delta Z_{il}}(\Delta z_{il}),$$

where

$$p_{\Delta Z_{ij}}(s; \theta) = \frac{1}{\Gamma(\nu_{ij}(\theta))\sigma^{2\nu_{ij}(\theta)}} s^{\nu_{ij}(\theta)-1} e^{-s/\sigma^2}, \quad s \geq 0,$$

$$\nu_{ij}(\beta,\gamma,\sigma^2) = \frac{\Delta m_{ij}(\beta,\gamma)}{\sigma^2}$$

and

$$\Delta m_{ij}(\beta,\gamma) = m_{ij}(\beta,\gamma) - m_{i,j-1}(\beta,\gamma) =$$

$$m\left(\int_0^{t_{ij}} e^{\beta^T x^{(i)}(s)}\,ds;\gamma\right) - m\left(\int_0^{t_{i,j-1}} e^{\beta^T x^{(i)}(s)}\,ds;\gamma\right).$$

Set $\theta = (\beta^T,\gamma^T,\sigma^2)^T$. If the model (3.19) is considered then the likelihood function depends on the parameters $\theta^* = (\theta,\alpha_0)$. In the case of the model (3.21) the likelihood function depends on $\theta^* = (\theta,\beta^*)$.

Example 13.4. Shock process

The density function of degradation measurements has the form:

$$p_{Z_{i1},\dots,Z_{ij}}(z_{i1},\dots,z_{ij}) = p_{\Delta Z_{i1},\dots,\Delta Z_{ij}}(\Delta z_{i1},\dots,\Delta z_{ij}).$$

Set

$$s_{ij}(\beta) = \int_0^{t_{ij}} e^{\beta^T x^{(i)}(u)}\,du, \quad s_{i0}(\beta) = 0.$$

For $\Delta z_{i1} = \dots = \Delta z_{ij} = 0$:

$$p_{\Delta Z_{i1},\dots,\Delta Z_{ij}}(\Delta z_{i1},\dots,\Delta z_{ij}) = \mathbf{P}\{N(s_{ij}(\beta)) = 0\};$$

for $\Delta z_{ik_1},\dots,\Delta z_{ik_s} > 0,\ \Delta z_{il} = 0,\ l \neq k_1,\dots,k_s,\ 1 \le k_1 < \dots, k_s \le j$;

$$p_{\Delta Z_{i1},\dots,\Delta Z_{ij}}(\Delta z_{i1},\dots,\Delta z_{ij}) = \sum_{l_1=1}^{\infty}\cdots\sum_{l_s=1}^{\infty} g_{l_1}(z_{ik_1})\cdots g_{l_s}(z_{ik_s})\times$$

$$\mathbf{P}\{N(s_{ik_1}(\beta)) - N(s_{ik_1-1}(\beta)) = l_1, \cdots, N(s_{ik_s}(\beta)) - N(s_{ik_s-1}(\beta)) = l_s,$$
$$N(s_{il}(\beta)) - N(s_{i,l-1}(\beta)) = 0, l \neq k_1, \cdots k_s\};$$

for $\Delta z_{i1},\dots,\Delta z_{ij} > 0$:

$$p_{\Delta Z_{i1},\dots,\Delta Z_{ij}}(\Delta z_{i1},\dots,\Delta z_{ij} = \sum_{l_1=1}^{\infty}\cdots\sum_{l_j=1}^{\infty} g_{l_1}(z_{i1})\cdots g_{l_j}(z_{ij})\times$$

$$\mathbf{P}\{N(s_{i1}(\beta)) - N(s_{i0}(\beta)) = l_1, \cdots, N(s_{ij}(\beta)) - N(s_{ij-1}(\beta)) = l_j\}.$$

In the particular case when $N(t)$ is an homogenous Poisson process with the parameter λ, and X_n have exponential distribution with the parameter μ we have: for $\Delta z_{i1} = \dots = \Delta z_{ij} = 0$:

$$p_{\Delta Z_{i1},\dots,\Delta Z_{ij}}(\Delta z_{i1},\dots,\Delta z_{ij}) = \exp\{-\lambda s_{ij}(\beta)\};$$

for $\Delta z_{ik_1},\dots,\Delta z_{ik_s} > 0,\ \Delta z_{il} = 0,\ l \neq k_1,\dots,k_s,\ 1 \le k_1 < \dots, k_s \le j$;

$$p_{\Delta Z_{i1},\dots,\Delta Z_{ij}}(\Delta z_{i1},\dots,\Delta z_{ij})$$

$$= \exp\{-\lambda s_{ij}(\beta) - \mu(\Delta z_{ik_1} + \dots + \Delta z_{ik_s}\}\prod_{r=1}^{s}\sum_{l_r=1}^{\infty}\frac{\{\lambda\mu(s_{ik_r}(\beta) - s_{i,k_r-1}(\beta))\}^{l_r}}{l_r!(l_r-1)!}$$

for $\Delta z_{i1}, ..., \Delta z_{ij} > 0$:

$$p_{\Delta Z_{i1}, ..., \Delta Z_{ij}}(\Delta z_{i1}, ..., \Delta z_{ij})$$

$$= \exp\{-\lambda s_{ij}(\beta) - \mu(\Delta z_{i1} + ... + \Delta z_{ij}\} \prod_{r=1}^{j} \sum_{l_r=1}^{\infty} \frac{\{\lambda\mu(s_{ir}(\beta) - s_{i,r-1}(\beta))\}^{l_r}}{l_r!(l_r-1)!}.$$

Denote by $\widehat{\theta}^*$ the maximum likelihood estimator of θ^*. Then the estimators of $m_{x(\cdot)}(t)$, $S_{x(\cdot)}(t)$ and $\tau_{x(\cdot)}$ are

$$\widehat{m}_{x(\cdot)}(t) = m\left(\int_0^t e^{\hat{\beta}^T x(s)}\, ds; \hat{\gamma}\right),$$

$$\widehat{\tau}_{x(\cdot)} = \sup\{t : \widehat{m}_{x(\cdot)}(t) < z_0\}, \quad \hat{S}_{x(\cdot)}(t) = S_{x(\cdot)}(t, \hat{\theta}).$$

For example, in the case of the gamma-process

$$\hat{S}_{x(\cdot)}(t) = \frac{1}{\Gamma(\hat{\nu}_{x(\cdot)}(t))}\sigma^{2\hat{\nu}_{x(\cdot)}(t)} \int_0^{z_0} x^{\hat{\nu}_{x(\cdot)}(t)-1}e^{-x/\hat{\sigma}^2}\, dx,$$

where $\hat{\nu}_{x(\cdot)}(t) = \widehat{m}_{x(\cdot)}(t)/\hat{\sigma}^2$.

In the case of shock models

$$S_{x(\cdot)}(t) = \sum_{i=0}^{\infty} G_i(z_0)\mathbf{P}\{N(f_{x(\cdot)}(t)) = i\},$$

where $G_0(z_0) = 1$ and an estimator of this survival function is obtained by replacing unknown parameters by their maximum likelihood estimators. For example, in the particular case when $N(t)$ is an homogenous Poisson process with the parameter λ, and X_n have exponential distribution with the parameter μ, we have:

$$\hat{S}_{x(\cdot)}(t) = e^{-\hat{\lambda}\int_0^t e^{\hat{\beta}^T x(u)}du}\left\{1 + \sum_{l=1}^{\infty} \frac{\{\hat{\lambda}\hat{\mu}\int_0^t e^{\hat{\beta}^T x(u)}du\}^l}{l!(l-1)!}\int_0^{z_0} v^{l-1}e^{-\hat{\mu}v}dv\right\}.$$

For the model (3.19) the estimators of Q and $G_{x(\cdot)}$ are

$$\hat{Q}(t) = e^{-\hat{\alpha}_0 t}, \quad \hat{G}_{x(\cdot)}(t) = \hat{Q}(t)\,\hat{S}_{x(\cdot)}(t)$$

and for the model (3.21) the estimators of $Q_{x(\cdot)}$ and $G_{x(\cdot)}$ are

$$\hat{Q}_{x(\cdot)}(t) = \exp\left\{-\int_0^t e^{\hat{\beta}^{*T} x(s)}ds\right\}, \quad \hat{G}_{x(\cdot)}(t) = \hat{Q}_{x(\cdot)}(t)\,\hat{S}_{x(\cdot)}(t).$$

Asymptotic properties and confidence intervals

Asymptotic properties of the estimators are obtained using standard likelihood theory and the Fisher information matrix.

Set

$$I(\widehat{\theta}^*) = -\frac{\partial^2 \ln L(\widehat{\theta}^*)}{\partial \theta^{*2}}.$$

When n is large, an approximate $(1 - \alpha)$-confidence interval for $m_{x(\cdot)}(t)$ is:

$$\left(\widehat{m}_{x(\cdot)}(t) \exp\{-\frac{\widehat{\sigma}_t}{\widehat{m}_{x(\cdot)}(t)\sqrt{n}}w_{1-\alpha/2}\}, \quad \widehat{m}_{x(\cdot)}(t) \exp\{\frac{\widehat{\sigma}_t}{\widehat{m}_{x(\cdot)}(t)\sqrt{n}}w_{1-\alpha/2}\}\right),$$

where $w_{1-\alpha/2}$ is the $(1 - \alpha/2)$ – quantile of the standard normal distribution and

$$\widehat{\sigma}_t^2 = \left(\frac{\partial \widehat{m}_{x(\cdot)}(t)}{\partial \widehat{\theta}^*}\right)^T \widehat{I}^{-1}(\widehat{\theta}^*) \frac{\partial \widehat{m}_{x(\cdot)}(t)}{\partial \widehat{\theta}^*}.$$

An approximate $(1 - \alpha)$-confidence interval for $\tau_{x(\cdot)}$ is:

$$\left(\widehat{\tau}_{x(\cdot)} \exp\{-\frac{\widehat{u}_{\tau_{x(\cdot)}}}{\widehat{\tau}_{x(\cdot)}\sqrt{n}}y_{1-\alpha/2}\}, \quad \widehat{\tau}_{x(\cdot)} \exp\{\frac{\widehat{u}_{\tau_{x(\cdot)}}}{\widehat{\tau}_{x(\cdot)}\sqrt{n}}y_{1-\alpha/2}\}\right),$$

where

$$\widehat{u}_{\tau_{x(\cdot)}} = \widehat{\sigma}_t / \widehat{m}'_{x(\cdot)}(\widehat{\tau}_{x(\cdot)}).$$

13.3.3 Estimation of reliability and degradation characteristics when traumatic events depend on degradation

At first we suppose that the intensity of traumatic events depends linearly on degradation and does not depend on explanatory variables:

$$G_{x(\cdot)}(t) = \mathbf{E}\left\{\exp\{-\alpha_0 t - \alpha_1 \int_0^t Z_{x(\cdot)}(s)ds\} \mathbf{1}_{\{Z_{x(\cdot)}(t)<z_0\}} \mid x(s), 0 \le s \le t\right\},$$

is considered. We assume also that $Z_{x(\cdot)}(0) = 0$.

The parameter α_1 characterizes the influence of degradation on the intensity of traumatic events.

Set $\theta^* = (\theta^T, \alpha_0, \alpha_1)^T$ and $\Delta\tau_i = \max_{1 \le l \le m_i} \Delta t_{il}$. The formula (3.24) implies that for small $\Delta\tau_i$, for all $t \in [t_{ij}, t_{i,j+1})$ and

$$Z_{i1} \le ... \le Z_{ij} < z_0, \quad Z_{i0} = 0, \quad (j = 1, ..., m_i - 1),$$

we have

$$\mathbf{P}\{C_i > t \mid x^{(i)}(s), 0 \le s \le t, Z_{i1}, ..., Z_{ij}\}$$

$$\approx \exp\left\{-\alpha_0 t - \alpha_1\left(S_{ij} + Z_{ij}(t - t_{ij})\right)\right\},$$

where

$$S_{ij} = \frac{1}{2}\sum_{k=1}^{j}(Z_{i,k-1} + Z_{ik})\Delta t_{ik}, \quad S_{i0} = 0, \quad Z_{i0} = 0.$$

This approximation is used because of the fact that the σ-algebra generated by $x(s), Z_{x(s)}, 0 \le s \le t$ is near to the σ-algebra generated by $x(s), Z_{i1}, ..., Z_{ij}, 0 \le s \le t$ and the integral $\int_0^t Z_{x(\cdot)}(s)ds$ is near to the sum $S_{ij} + Z_{ij}(t - t_{ij})$ when $\Delta\tau_i$ is small. For $t = t_{im_i}$

$$\mathbf{P}\{C_i > t \mid x^{(i)}(s), 0 \le s \le t, Z_{i1}, ..., Z_{im_i}\} \approx \exp\left\{-\alpha_0 t - S_{im_i}\right\}.$$

The formula (3.27) implies that for all $t \in [t_{ij}, t_{i,j+1})$ $(j = 0, \cdots, m_i - 1)$ the conditional density of C_i is

$$p_{C_i | x^{(i)}(s),\, 0 \le s \le t,\, Z_{i1},...,Z_{ij}}(t)$$

$$\approx (\alpha_0 + \alpha_1 Z_{ij}) \exp \left\{ -\alpha_0 t - \alpha_1 \left(S_{ij} + Z_{ij}(t - t_{ij}) \right) \right\}$$

and for $t \in (0, t_{i1})$,

$$p_{C_i | x^{(i)}(s),\, 0 \le s \le t}(t) \approx \alpha_0\, e^{-\alpha_0 t}.$$

Then the likelihood function is approximated by the function:

$$\prod_{i=1}^{n} p_{Z_{i1},...,Z_{ij}}^{1-k_i}(Z_{i1}, ..., Z_{ij}; \theta)\big|_{j=\eta_i}\, e^{-\alpha_0 V_i - \alpha_1 \left(S_{i\eta_i} + Z_{i\eta_i}(V_i - t_{i\eta_i}) \right)} \times$$

$$(\alpha_0 + \alpha_1 Z_{i\eta_i})^{k_i + \delta_i(1-k_i)}, \qquad (13.59)$$

where

$$V_i = \begin{cases} C_i, & \text{si} \quad \delta_i = 1 \\ t_{i\eta_i}, & \text{si} \quad \delta_i = 0. \end{cases}$$

Let $\hat{\theta}^* = (\hat{\theta}, \hat{\alpha})$ be the maximum likelihood estimators. The estimators $\hat{m}_{x(\cdot)}(t)$, $\hat{S}_{x(\cdot)}(t)$ and $\hat{\tau}_{x(\cdot)}(t)$ have the the same forms as in the previous section; the expressions for $\hat{\theta}^*$ are evidently different.

The survival function of the time to failure $G_{x(\cdot)}(t)$ is estimated by the statistic

$$\hat{G}_{x(\cdot)}(t) = \exp \left\{ -\hat{\alpha}_0 t - \hat{\alpha}_1 \int_0^t \hat{m}_{x(\cdot)}(s)ds \right\} \mathbf{1}\{\hat{m}_{x(\cdot)}(t) < z_0\}.$$

For example, if $x(t) \equiv x = \text{const}$ and $m_1'(t) = \gamma_0 + \gamma_1 e^{-\gamma_2 t}$, then

$$m(t) = \gamma_0 t + \frac{\gamma_1}{\gamma_2}(1 - e^{-\gamma_2 t})$$

and

$$\hat{G}_{x(\cdot)}(t) = \exp \left\{ -\hat{\alpha}_0 t - \hat{\alpha}_1 [\hat{\gamma}_0 e^{\hat{\beta}^T x} \frac{t^2}{2} + \frac{\hat{\gamma}_1}{\hat{\gamma}_2}t + \frac{\hat{\gamma}_1}{\hat{\gamma}_2^2} e^{-\hat{\beta}^T x} (e^{-\hat{\gamma}_2 e^{\hat{\beta}^T x} t} - 1)] \right\} \times$$

$$\mathbf{1}\{\hat{\gamma}_0 e^{\hat{\beta}^T x} t + \frac{\hat{\gamma}_1}{\hat{\gamma}_2}(1 - e^{-\hat{\gamma}_2 e^{-\hat{\beta}^T x} t}) < z_0\}.$$

Now we suppose that the intensity of traumatic events at any moment t depends not only on the level of degradation but also on the values of the explanatory variables at this moment. So, we consider the model (3.30) and suppose that

$$G_{x(\cdot)}(t) = \mathbf{E} \left\{ \exp\{ -\int_0^t e^{\beta^{*T} x(s)}(1 + \alpha Z_{x(\cdot)}(s)ds\} \times \right.$$

$$\mathbf{1}_{\{Z_{x(\cdot)}(t) < z_0\}} \mid x(s),\, 0 \le s \le t \right\},$$

where the parameters $\beta^* = (\beta_0^*, \cdots, \beta_s^*)^T$ and α show the influence of explanatory variables and degradation on the intensity of traumatic moments, respectively. The intensity at the moment t would be $e^{\lambda^T x(t)}$, if degradation were absent.

The approximation of the likelihood function (13.59) is generalized simply to this more general case. Set $\theta^* = (\theta, \alpha, \beta^*)$. If the $\Delta\tau_i$ are small, then the likelihood function can be approximated by the function

$$L(\theta^*) = \prod_{i=1}^{n} p_{Z_{i1},\ldots,Z_{ij}}^{1-k_i}(Z_{i1},\ldots,Z_{ij};\theta)|_{j=\eta_i} \exp\{\int_0^{V_i} e^{\beta^{*T}x^{(i)}(v)}dv-$$

$$\alpha[\frac{1}{2}\sum_{k=1}^{\eta_i}(Z_{i,k-1}e^{\beta^{*T}x^{(i)}(t_{i,k-1})} + Z_{ik}e^{\beta^{*T}x^{(i)}(t_{ik})})+$$

$$Z_{i\eta_i}(e^{\beta^{*T}x^{(i)}(V_i)} - e^{\beta^{*T}x^{(i)}(t_{i\eta_i})})]\} \left(e^{\beta^{*T}x^{(i)}(t_{i\eta_i})}(1+\alpha Z_{i\eta_i})\right)^{k_i+\delta_i(1-k_i)}.$$

$$(13.60)$$

The estimators $\hat{m}_{x(\cdot)}(t)$, $\hat{S}_{x(\cdot)}(t)$ and $\hat{\tau}_{x(\cdot)}(t)$ have the the same forms as in previous section.

The survival function $G_{x(\cdot)}(\cdot)$ is estimated by the statistic

$$\hat{G}_{x(\cdot)}(t) = \exp\{-\int_0^t e^{\{\hat{\beta}^*\}^T x(s)}(1+\hat{\alpha}\hat{m}_{x(\cdot)}(s))ds\}\mathbf{1}\{\hat{m}_{x(\cdot)}(t) < z_0\}.$$

The influence of increasing degradation on intensity of the traumatic process can be nonlinear. The methodology of estimation is the same. For example, if the model is

$$G_{x(\cdot)}(t) = \mathbf{E}\left\{\exp\{-\int_0^t \lambda\left(Z_{x(\cdot)}(s), x(s)\right)ds\}\mathbf{1}_{\{Z_{x(\cdot)}(t)<z_0\}} \mid x(s), 0 \le s \le t\right\},$$

with the intensity

$$\lambda(z, x) = e^{\beta^{*T}x}(1+\alpha_1 z^{\alpha_2}),$$

i.e. with power rule influence of degradation on the intensity of traumatic events, then, in constructing the likelihood function, the integrals in the conditional survival functions are approximated in similar manner to the linear case.

13.3.4 Estimation when the form of the mean degradation is unknown

Suppose that the mean degradation $m_{x(\cdot)}(t)$ is completely unknown.

Denote by $g_{x(\cdot)}(t, \beta) = f_{x(\cdot)}^{-1}(t, \beta)$ the inverse of the increasing continuous function (given $x(\cdot)$)

$$f_{x(\cdot)}(t, \beta) = \int_0^t e^{\beta^T x(u)}du.$$

For all $x(\cdot)$ the process

$$Z_{x(\cdot)}(g_{x(\cdot)}(t)) = \sigma^2 \gamma(t)$$

is a gamma-process which does not depend on $x(\cdot)$ and

$$\mathbf{E}\left\{Z_{x(\cdot)}(g_{x(\cdot)}(t))\right\} = m(t), \quad \mathbf{Var}\left\{Z_{x(\cdot)}(g_{x(\cdot)}(t))\right\} = \sigma m(t), \qquad (13.61)$$

where $m(t)$ is a completely unknown function.

As a rule the explanatory variable x is constant in time or a step function. If $x(t) \equiv x = const$, then

$$f_x(t; \beta) = e^{\beta^T x} t, \quad g_x(t; \beta) = e^{-\beta^T x} t.$$

If $x(\cdot)$ is a step function of the form

$$x(t) = x_l, \quad t \in [s_{l-1}, s_l] \quad (l = 1, ..., r_i; s_0 = 0),$$

then

$$f_{x(\cdot)}(t; \beta) = s_{l-1}^*(\beta) + e^{\beta^T x_l}(t - s_{l-1}), \quad t \in [s_{l-1}, s_l],$$

$$g_{x(\cdot)}(t; \beta) = s_{l-1} + e^{-\beta^T x_l}(t - s_{l-1}^*(\beta)), \quad t \in [s_{l-1}^*(\beta), s_l^*(\beta)],$$

where

$$s_l^*(\beta) = \sum_{j=1}^{l} e^{\beta^T x_j}(s_j - s_{j-1}), \quad l = 1, ..., r_i,; \quad s_0^*(\beta) = 0.$$

Suppose that the data are the same as in the previous section. The process $Z_i(t) = Z_{x^{(i)}(\cdot)}(t)$ is defined on $[0, t_{i\eta_i}]$. The process

$$\tilde{Z}_i(t) = \sum_{j=1}^{m_i} \left\{ Z_i(t_{i,j-1}) + \frac{t - t_{i,j-1}}{t_{ij} - t_{i,j-1}}(Z_i(t_{ij}) - Z_i(t_{i,j-1})) \right\} \times$$

$$\mathbf{1}_{[t_{i,j-1}, t_{ij}]}(t)\mathbf{1}\{t_{ij} \leq U_i\},$$

is a piecewise-linear approximation of the process $Z_i(t)$ and is also defined on $[0, t_{i\eta_i}]$.

The equalities (13.61) imply that for all $i = 1, ..., n$

$$\mathbf{E}\left\{\tilde{Z}_i(g_i(t, \beta))\right\} \approx m(t), \quad \mathbf{Var}\left\{\tilde{Z}_i(g_i(t, \beta))\right\} \approx \sigma m(t),$$

if the differences Δt_{ij} are small; here $g_i := g_{x^{(i)}}$. Set $f_i := f_{x^{(i)}}$.

The processes $\tilde{Z}_i(g_i(t, \beta))$ are censored at the moments $t_i^*(\beta) = g_i(t_{i\eta_i}, \beta)$. Order these censoring moments:

$$t_{(1)}^*(\beta) < \cdots < t_{(n)}^*(\beta).$$

For $t \in [0, t_{(1)}^*(\beta))$ set

$$\tilde{m}(t, \beta) = \frac{1}{n} \sum_{i=1}^{n} \tilde{Z}_i(g_i(t, \beta))$$

and for $t \in [t_{(j-1)}^*(\beta), t_{(j)}^*(\beta))$ set

$$\tilde{m}(t, \beta) =$$

$$\tilde{m}(t^*_{(j-1)}(\beta)) + \frac{1}{n-j+1} \sum_{i:t^*_i(\beta)>t^*_{(j-1)}(\beta)} \left(\tilde{Z}_i(g_i(t,\beta)) - \tilde{Z}_i(g_i(t^*_{(j-1)}(\beta),\beta)) \right).$$

Obvious modifications can be made if there are *ex aequo* among these censoring moments.

Let us consider, for example, the case when the degradation process is gamma. Set $\Delta Z_{ij} = Z_{ij} - Z_{i,j-1}$. Then $\Delta Z_{ij} = \sigma \Delta \gamma_{ij}$

$$\Delta \gamma_{ij} \sim G\left(1, \frac{\Delta m_{ij}(\beta)}{\sigma} \right) = G(1, \nu_{ij}(\beta,\sigma)),$$

where

$$\Delta m_{ij}(\beta) = m(f_i(t_{ij};\beta);\beta) - m(f_i(t_{i,j-1};\beta);\beta).$$

For the models (3.20) and (3.22) the likelihood function is written in the form (13.58) and for the models (3.23) and (3.28) in the forms (12.59) and (12.60), respectively, with only the difference that $\nu_{ij}(\theta)$ (defined in Example 13.3) are replaced by

$$\nu_{ij}(\theta) = \Delta \tilde{m}_{ij}(\beta)/\sigma,$$

where

$$\Delta \tilde{m}_{ij}(\beta) = \tilde{m}(f_i(t_{ij};\beta);\beta) - \tilde{m}(f_i(t_{i,j-1};\beta);\beta).$$

Denote by $\hat{\theta}^*$, the maximum likelihood estimator of the parameter θ^*. The function $m(t)$ is estimated by the statistic $\hat{m}(t) = \tilde{m}(t,\hat{\beta})$ and is defined for all t, such that

$$\sum_{i=1}^{n} \mathbf{1}\{t \le \int_0^{t_{i\eta_i}} e^{\hat{\beta}^T x^{(i)}(u)} du\} > 0.$$

The estimator of the mean degradation $m_{x(\cdot)}(t)$ under any explanatory variable $x(\cdot)$ is estimated by the statistic

$$\hat{m}_{x(\cdot)}(t) = \hat{m}(\int_0^t e^{\hat{\beta}^T x(u)} du)$$

and is defined for all t, such that

$$\sum_{i=1}^{n} \mathbf{1}\{\int_0^t e^{\hat{\beta}^T x(u)} du \le \int_0^{t_{i\eta_i}} e^{\hat{\beta}^T x^{(i)}(u)} du\} > 0.$$

Some results from stochastic process theory

A.1 Stochastic process. Filtration

Definition A1. *A collection of random variables $X = X(t), t \geq 0$ defined on the same probability space $(\Omega, \mathcal{F}, \mathbf{P})$ is called a stochastic process.*

Definition A2. *For any fixed $\omega \in \Omega$ the real function $\{X(\cdot, \omega)\}$ is called the path of a stochastic process X.*

An event occurs almost surely (a.s.) if it occurs with the probability one.

Definition A.3. *A stochastic process is*
1. Cadlag, if its trajectories are right-continuous with finite left-hand limits;
2. Caglad, if its trajectories are left-continuous with finite right-hand limits;
3. Square integrable, if $\sup_{0 \leq t < \infty} \mathbf{E}\{X(t)\}^2 < \infty$;
4. Bounded, if $\sup_{0 \leq t < \infty} | X(t) | < C = const$ a.s.

In Chapter 4 it was noted that right censored data may be presented in the form

$$(X_1, \delta_1), \cdots, (X_n, \delta_n). \tag{A.1}$$

or, equivalently,

$$(N_1(t), Y_1(t), t \geq 0), \cdots, (N_n(t), Y_n(t), t \geq 0), \tag{A.2}$$

where

$$N_i(t) = \mathbf{1}_{\{X_i \leq t, \delta_i = 1\}}, \quad Y_i(t) = \mathbf{1}_{\{X_i \geq t\}}, \quad X_i = T_i \wedge C_i, \quad \delta_i = \mathbf{1}\{T_i \leq C_i\}, \tag{A.3}$$

T_i and C_i being failure and censoring times of the ith unit ($i = 1, \cdots, n$). The random variables T_i and C_i are supposed to be defined on the probability space $(\Omega, \mathcal{F}, \mathbf{P})$.

Note that $N(t) = \sum_{i=1}^{n} N_i(t)$ and $Y(t) = \sum_{i=1}^{n} Y_i(t)$ are the number of observed failures in the interval $[0, t]$ and the number of units at risk just prior the moment t, respectively.

The stochastic processes N_i and Y_i show dynamics of failures and censorings over time. If the values of $\{N_i(s), Y_i(s), 0 \leq s \leq t, i = 1, \cdots, n\}$ are known, the *history* of failures and censorings in the interval $[0, t]$ is known. The data (1.2) gives all history of failures and censorings during the experiment. The notion of the history is formalized by the following notion of the *filtration*.

Denote by
$$\mathcal{F}_t = \sigma\{N_i(s), Y_i(s), 0 \le s \le t\}$$
the σ-algebra generated by $N_i(s), Y_i(s), 0 \le s \le t$.

The σ-algebra $\mathcal{F}_t$ contains all events related with failure and censoring processes which can occur until the moment t.

It is clear that $\mathcal{F}_s \subset \mathcal{F}_t \subset \mathcal{F}, \quad 0 \le s \le t$.

Definition A4. *The family of σ-algebras* $\mathbf{F} = \{\mathcal{F}_t, t \ge 0\}$ *is called the filtration (or history) generated by the data (1.2).*

Definition A5. *A stochastic process is adapted to the filtration* $\mathbf{F}$ *if for any fixed $t \ge 0$ the random variable $X(t)$ is $\mathcal{F}_t$-measurable, i.e. for any Borel set B of* $\mathbf{R}$ *the event* $\{X(t) \in B\} \in \mathcal{F}_t$.

If a process is $\mathbf{F}$-adapted then its value is known at the moment t given the history in the interval $[0, t]$.

A.2 Counting process

Definition A.6. *An adapted to a filtration* $\mathbf{F}$ *process* $\{X(t), t \ge 0\}$ *is called a counting process if $X(0) = 0$, $X(t) < \infty$ a.s. for all $t > 0$, and its trajectories are right continuous nondecreasing piecewise constant with jumps of size 1.*

The processes N_i and N are counting processes.

Definition A.7. *A multivariate process* $(X_1(t), ..., X_m(t), t \ge 0)$ *is called a multivariate counting process if each X_j is a counting process and no two component processes jump at the same time with probability one.*

We suppose that the failure times T_i are absolutely continuous random variables, so $(N_1(t), ..., N_n(t))$ is a multivariate counting process.

A.3 Stochastic integral

Let $Y = Y(t, \omega), t \ge 0, \omega \in \Omega$ be a cadlag *finite variation* process, i.e. with cadlag paths and such that for all $\omega \in \Omega, t \ge 0$ the supremum
$$\sup \sum_{i=1}^{m} | Y(t_i) - Y(t_{i-1}) |$$
is finite; the supremum is taken with respect to all partitions $0 = t_0 \le t_1 \le \cdots \le t_m = t$ of the interval $[0, t]$.

In this book the stochastic integral of the stochastic process X with respect to the stochastic integral Y:
$$\int_0^t X(u)dY(u) = \int_{[0,t]} X(u)dY(u),$$

is understood as a *pathwise* Lebesque-Stieltjes integral. By convention, $Y(0-) = 0$, so

$$\int_0^0 X(u)dY(u) = X(0)Y(0).$$

If the paths of the process Y are right-continuous step-functions (in particular, if Y is a counting process) then

$$\int_0^t X(u)dY(u) = \sum_{i:\tau_i \leq t} X(\tau_i)\Delta Y(\tau_i); \qquad (A.4)$$

here $\tau_1 < \cdots < \tau_m$ are jump points of Y, $\Delta Y(\tau_i) = Y(\tau_i) - Y(\tau_{i-1})$.

If a cadlag finite variation process

$$Y(u) = \int_0^t Z(u)du$$

is a pathwise Lebesque integral of a process $Z(u)$ then

$$\int_0^t X(u)dY(u) = \int_0^t X(u)Z(u)du. \qquad (A.5)$$

Integration by parts formula. *If both X and Y are cadlag finite variation processes then*

$$\int_0^t X(u-)dY(u) = X(t)Y(t) - X(0)Y(0) - \int_0^t Y(u)dX(u) \qquad (A.6)$$

$$= X(t)Y(t) - X(0)Y(0) - \int_0^t Y(u-)dX(u) - \sum_{0<u\leq t} \Delta X(u)\Delta Y(u).$$

Estimating parameters and testing hypothesis we need properties of processes of the type

$$U(t) = (U_1(t), ..., U_m(t))^T,$$

where

$$U_j(t) = \sum_{i=1}^n U_{ij}(t), \quad U_{ij}(t) = \int_0^t H_{ij}(u)\,dM_i(u) \ (i = 1, ..., n), \qquad (A.7)$$

H_{ij} are left-continuous processes, and

$$M_i(t) = N_i(t) - \int_0^t Y_i(u)\,\alpha_i(u)du. \qquad (A.8)$$

Properties of the process $U(t)$ can be studied using the fact that $U_j(t)$ are (local) *martingales*. To define the notion of a martingale we need the notion of the *conditional expectation*.

A.4 Conditional expectation

Suppose that the random variable X has the finite mean $\mathbf{E}(X)$.

Definition A.8. *The conditional expectation of the random variable X with respect to the σ-algebra $\mathcal{F}_t \subset \mathcal{F}$ is a random variable Y, denoted by $\mathbf{E}(X \mid \mathcal{F}_t)$, such that*

1. Y *is $\mathcal{F}_t$ measurable;*
2. $\mathbf{E}(X\mathbf{1}_B) = \mathbf{E}(Y\mathbf{1}_B)$ *for all $B \in \mathcal{F}_t$.*

The conditional expectation exists and is a.s. unique. It minimizes the distance $\mathbf{E}(X - \tilde{X})^2$ in the class $\mathcal{H}_t$ of all $\mathcal{F}_t$ - measurable random variables $\tilde{X}$.

The random variable X is not necessary $\mathcal{F}_t$ - measurable. The conditional expectation $Y = \mathbf{E}(X \mid \mathcal{F}_t)$ is the nearest to X random variable which is $\mathcal{F}_t$ - measurable.

Properties of the conditional expectation:

a) $\mathbf{E}\{\mathbf{E}(X \mid \mathcal{F}_t)\} = \mathbf{E}(X)$;

b) *for $s \leq t$*

$$\mathbf{E}\{\mathbf{E}(X \mid \mathcal{F}_s) \mid \mathcal{F}_t\} = \mathbf{E}\{\mathbf{E}(X \mid \mathcal{F}_t) \mid \mathcal{F}_s\} = \mathbf{E}(X \mid \mathcal{F}_s) \quad a.s.$$

c) *if Y is $\mathcal{F}_t$ - measurable, then*

$$\mathbf{E}(XY \mid \mathcal{F}_t) = Y\,\mathbf{E}(X \mid \mathcal{F}_t) \quad a.s.,$$

in particular

$$\mathbf{E}(Y \mid \mathcal{F}_t) = Y \quad a.s.$$

d) *if $a, b \in \mathbf{R}$, then*

$$\mathbf{E}(aX_1 + bX_2 \mid \mathcal{F}_t) = a\mathbf{E}(X_1 \mid \mathcal{F}_t) + b\mathbf{E}(X_2 \mid \mathcal{F}_t);$$

e) *(Jensen's inequality). If g is a convex real-valued function then*

$$\mathbf{E}(g(X) \mid \mathcal{F}_t) \geq g(\mathbf{E}(X \mid \mathcal{F}_t));$$

f) *if $X \leq Y$ then $\mathbf{E}(X \mid \mathcal{F}_t) \leq \mathbf{E}(Y \mid \mathcal{F}_t)$;*

g) *if $A \in \mathcal{F}_t$ is an atom of $\mathcal{F}_t$, (i.e. $\mathbf{P}(A) > 0$ and $\mathbf{P}(B) = \mathbf{P}(A)$ or $\mathbf{P}(B) = 0$ for any $B \subset A$, $B \in \mathcal{F}_t$) then for almost all $\omega \in A$*

$$\mathbf{E}(X \mid \mathcal{F}_t)(\omega) = \frac{1}{\mathbf{P}(A)}\,\mathbf{E}(X\mathbf{1}_A).$$

A.5 Martingale

Definition A.9. *A cadlag process $M(t)$, $t \geq 0$ is called a martingale (submartingale) with respect to the filtration $\mathbf{F}$ if*

1. M *is adapted;*
2. $\mathbf{E}\mid M(t)\mid < \infty$ *for all $0 \leq t < \infty$;*
3. $\mathbf{E}(M(t) \mid \mathcal{F}_s) = M(s)$ $(\mathbf{E}(M(t) \mid \mathcal{F}_s) \geq M(s))$ *a.s. for all $t \geq s \geq 0$.*

So the process M is a $\mathbf{F}$-martingale if, for any $t \geq s \geq 0$, $\mathcal{F}_s$ - measurable random variable which is nearest to $M(t)$, is $M(s)$.

The conditional mean $\mathbf{E}(M(t) - M(s) \mid \mathcal{F}_s)$ does not depend on the history up to the moment s and is equal to 0.

The counting processes N_i are nonnegative right-continuous $\mathbf{F}$-submartingales because $N_i(t) - N_i(s) \geq 0$ for any $t \geq s \geq 0$ and hence $\mathbf{E}\{N_i(t) - N_i(s) \geq 0 \mid \mathcal{F}_s\} \geq 0$.

A.6 Predictable process and Doob-Meyer decomposition

Definition A.10. *A process $\{H(t,\omega), t \geq 0, \omega \in \Omega\}$ is predictable if it is measurable with respect to the σ-algebra on $[0,\infty) \times \Omega$ generated by adapted left-continuous processes.*

In particular, *left continuous adapted processes and deterministic measurable functions are predictable* (see Fleming and Harrington (1991), Andersen et al (1993)). Namely such predictable processes are usually used in this book.

If H is a predictable process then the random variable $H(t)$ is $\mathcal{F}_{t-}$-measurable; here

$$\mathcal{F}_{t-} = \sigma\{N_i(s), Y_i(s), 0 \leq s < t, i = 1, \cdots, n\}.$$

So the value of a predictable process is known at the moment t if the history in the interval $[0, t)$ is known.

Suppose that the process H has the form:

$$H(t) = Z\, \mathbf{1}_{(a,b]}(t),$$

where Z is a $\mathcal{F}_a$-measurable random variable and $0 \leq a < b$ are fixed numbers. It is left-continuous and $\mathbf{F}$-adapted, hence $\mathbf{F}$-predictable. Predictable processes which are sums of finite number of such processes are called *simple predictable processes*. It can be shown that any bounded predictable processes can be written as the limit of increasing sequences of simple predictable processes.

We shall use the famous Doob-Meyer decomposition of submartingales as sums of predictable processes and martingales.

Theorem A.1. (Doob-Meyer decomposition). *Let X be a right continuous nonnegative $\mathbf{F}$-submartingale. Then there exists a right-continuous martingale M and an non-decreasing right-continuous predictable process Λ such that $\mathbf{E}(\Lambda(t)) < \infty$ and*

$$X(t) = M(t) + \Lambda(t) \quad a.s.$$

for any $t \geq 0$. If $\Lambda(0) = 0$ a.s. then this decomposition is a.s. unique, i.e. if $X^*(t) = M^*(t) + \Lambda^*(t)$ for any $t \geq 0$ with $\Lambda^*(0) = 0$, then for any $t \geq 0$,

$$\mathbf{P}\{M^*(t) \neq M(t)\} = \mathbf{P}\{\Lambda^*(t) \neq \Lambda(t)\}.$$

The process Λ is called the *compensator* of the submartingale X.

Let us consider the data (A.2). The counting processes N_i and N are sub-martingales. Let us find their compensators with respect to the filtration $\mathbf{F}$ generated by the data.

Theorem A.2. *Suppose that*
1) the failure times $T_1, \cdots, T_n$ are absolutely continuous random variables;
2) the failure times $T_1, \cdots, T_n$ and the censoring times $C_1, \cdots, C_n$ are mutually independent.
Then the compensator of the counting process N_i is

$$\Lambda_i(t) = \int_0^t Y_i(u)\,\alpha_i(u)du. \tag{A.9}$$

Proof. Let us consider the process $M_i(t)$. We skip the index i in all expressions. Let us show that $M(t), t \geq 0$ verifies the three axioms of a martingale.
$N(t)$ is evidently $\mathcal{F}_t$ - measurable. Let us consider the integral

$$\Lambda(t) = \int_0^t Y(u)\alpha(u)du = A(X \wedge t);$$

here

$$A(t) = \int_0^t \alpha(u)du.$$

The random variable $X \wedge t$ is $\mathcal{F}_t$ - measurable because for any $t \leq x$

$$\{X \wedge t \leq x\} = \Omega \in \mathcal{F}_t,$$

and for any $t > x$

$$\{X \wedge t \leq x\} = \{X > x\}^c = \{N(u) = 0, Y(u) = 1, 0 \leq u \leq x\}^c \in \mathcal{F}_x \subset \mathcal{F}_t.$$

The function A is continuous, so $A(X \wedge t)$ is also $\mathcal{F}_t$ - measurable.
The mean $\mathbf{E}\,|\,M(t)\,|$ is finite for any $t \geq 0$:

$$\mathbf{E}|M(t)| \leq \mathbf{E}N(t) + \mathbf{E}\int_0^t \mathbf{1}\{X \geq u\}\alpha(u)du \leq 1 + \int_0^t \mathbf{P}(T \geq u, C \geq u)\alpha(u)du \leq$$

$$1 + \int_0^t \mathbf{P}(T \geq u)\alpha(u)du = 2 - S_T(t) \leq 2.$$

Independence of the pairs $(X_1, \delta_1), \cdots, (X_n, \delta_n)$ implies that the conditional expectation of $M_i(t)$ with respect to the σ-algebra generated by $N_j(u), Y_j(u), 0 \leq u \leq s, j = 1, \cdots, n$ is the same as the conditional expectation of $M_i(t)$ with respect to the σ-algebra generated by $N_i(u), Y_i(u), 0 \leq u \leq s$. So we can suppose that $\mathcal{F}_s$ is the last one. As previously, we skip the index i in what follows.
Take $0 \leq s \leq t$. On the event $\{X \leq s\}$ we have :

$$N(t) - N(s) = \mathbf{1}_{\{s < X \leq t, T \leq C\}} = 0,$$

$$\int_0^t Y(u)\alpha(u)du - \int_0^s Y(u)\alpha(u)du = \int_{(s,t]} \mathbf{1}\{X \geq u\}\alpha(u)du = 0.$$

So $M(t) = M(s)$ on $\{X \leq s\}$. If $\mathbf{P}\{X > s\} = 0$ then $M(t) = M(s)$ a.s., which implies

$$\mathbf{E}(M(t)|\mathcal{F}_s) = \mathbf{E}(M(s)|\mathcal{F}_s) = M(s) \text{ a.s.}$$

Suppose that $\mathbf{P}(X > s) > 0$. Then the event $\{X > s\} = \{N(u) = 0, Y(u) = 1, 0 \leq u \leq s\}$ is an atom of $\mathcal{F}_s$ and on $\{X > s\}$ a.s. (the property g) of the conditional expectation)

$$\mathbf{E}(M(t) - M(s)|\mathcal{F}_s) = \mathbf{E}\left(\mathbf{1}_{\{s < X \leq t, T \leq C\}} - \int_{(s,t]} \mathbf{1}_{\{X \geq u\}}\alpha(u)du|\mathcal{F}_s\right) =$$

$$\frac{1}{\mathbf{P}(X > s)}\mathbf{E}\left(\mathbf{1}_{\{s < X \leq t, T \leq C\}} - \int_{(s,t]} \mathbf{1}_{\{X \geq u\}}\alpha(u)du\right) =$$

$$\frac{1}{\mathbf{P}(X > s)}\left\{\int_{(s,t]} \mathbf{P}(C \geq u)p_T(u)du - \int_{(s,t]} \mathbf{P}(C \geq u, T \geq u)\,\alpha(u)du\right\} = 0.$$

We denoted by p_T the probability density of T.

The process $\Lambda_i(t) = \Lambda_i(t \wedge X_i)$ is predictable because it is continuous and adapted.

A.7 Predictable variation and predictable covariation

Jensen's inequality for the conditional expectation implies that the square M^2 of a square-integrable $\mathbf{F}$-martingale is a $\mathbf{F}$-submartingale:

$$\mathbf{E}\{M^2(t) \mid \mathcal{F}_s\} \geq (\mathbf{E}\{M(t) \mid \mathcal{F}_s\})^2 = M^2(s).$$

Doob-Meyer theorem implies that this submartingale has a compensator.

Definition A.11. *The compensator $< M, M >$ of the $\mathbf{F}$-submartingale M^2 is called the predictable variation of the $\mathbf{F}$-martingale M.*

If M_1 and M_2 are two square-integrable $\mathbf{F}$-martingales then the equality

$$M_1 M_2 = \frac{1}{4}(M_1 + M_2)^2 - \frac{1}{4}(M_1 - M_2)^2$$

implies that the product $M_1 M_2$ is a difference of two submartingales with predictable variations

$$\frac{1}{4} < M_1 + M_2, M_1 + M_2 > \quad \text{and} \quad \frac{1}{4} < M_1 - M_2, M_1 - M_2 >$$

Definition A.12. *The process*

$$< M_1, M_2 >= \frac{1}{4}(< M_1 + M_2, M_1 + M_2 > - < M_1 - M_2, M_1 - M_2 >)$$

is called the predictable covariation of martingales M_1 and M_2.

The definition implies that the difference

$$M_{12}^* = M_1 M_2 - <M_1, M_2>$$

is a **F**-martingale.

Theorem A.3. *Suppose that the conditions of Theorem A.2 are verified. Then the predictable variations and covariations of the counting process N_i martingales $M_i = N_i - \Lambda_i$ are:*

$$<M_i, M_i>(t) = \Lambda_i(t), \quad <M_i(t), M_{i'}(t)>= 0 \ (i \neq i') \qquad (A.10)$$

Proof. Let us consider the process $M_{ii'}^* = M_i(t) M_{i'}(t) \ (i \neq i')$.
Take $0 \leq s \leq t$. On the event $\{X_i \leq s\}$ we have : $M_i(t) = M_i(s)$ and hence

$$\mathbf{E}\{M_i(t) M_{i'}(t) \mid \mathcal{F}_s\} = M_i(s) \, \mathbf{E}\{M_{i'}(t) \mid \mathcal{F}_s\} = M_i(s) M_{i'}(s).$$

The same on the event $\{X_{i'} \leq s\}$. Independence of the martingales M_i and $M_{i'}$ implies that on the event $\{X_i > s, X_{i'} > s\}$

$$\mathbf{E}\{M_i(t) M_{i'}(t) \mid \mathcal{F}_s\} = \frac{1}{\mathbf{P}(X_i > s, X_{i'} > s)} \mathbf{E}\{M_i(t)\mathbf{1}_{\{X_i>s\}} M_{i'}(t)\mathbf{1}_{\{X_{i'}>s\}}\}$$

$$= \mathbf{E}\{M_i(t) \mid \mathcal{F}_s\}\mathbf{E}\{M_{i'}(t) \mid \mathcal{F}_s\} = M_i(s) M_{i'}(s).$$

Let us consider the process $M_i^* = M_i^2 - \Lambda_i$. We skip indices in what follows.
Note that $N^2(t) = N(t)$. We have

$$\Delta = M^*(t) - M^*(s) = N(t) - N(s) - \int_{(s,t]} Y(u)\,\alpha(u)du - 2N(t)\int_0^t Y(u)\,\alpha(u)du+$$

$$2N(s)\int_0^s Y(u)\alpha(u)du + 2\int_s^t \left(\int_0^v Y(u)\,\alpha(u)du\right) Y(v)\,\alpha(v)dv =$$

$$\Delta_1 + \Delta_2 + \Delta_3 + \Delta_4 + \Delta_5.$$

From the part a) of the proof we have that $\mathbf{E}\{\Delta_1 + \Delta_2 \mid \mathcal{F}_s\} = 0$.
If $X \leq s$ then $\Delta = 0$.
If $\mathbf{P}\{X > s\} = 0$ then $\Delta = 0$ and $\mathbf{E}\{\Delta \mid \mathcal{F}_s\} = 0$ a.s.
Suppose that $\mathbf{P}\{X > s\} > 0$. Then on the set $\{X > s\}$ we have $N(s) = 0$ and $\Delta_4 = 0$. So on this set

$$\mathbf{E}\{\Delta \mid \mathcal{F}_s\} = \frac{1}{\mathbf{P}\{X > s\}}\mathbf{E}\left\{-2\int_0^t \mathbf{1}_{\{s<T\leq t, T\leq C, T\geq u, C\geq u\}}\alpha(u)du+\right.$$

$$\left. 2\int_s^t \left(\int_0^v \alpha(u)du\right)\mathbf{1}\{X \geq v\}\alpha(v)dv\right\} =$$

$$\frac{1}{\mathbf{P}\{X > s\}}\left\{-2\int_0^t \left(\int_{s\vee u}^t \mathbf{P}\{X \geq v\}\alpha(v)dv\right)\alpha(u)du+\right.$$

$$\left. 2\int_s^t \left(\int_0^v \alpha(u)du\right)\mathbf{P}\{X \geq v\}\alpha(v)dv\right\} = 0.$$

Corollary A.1. Under the assumptions of the theorem for any $t_1, t_2 \geq s$

$$\mathbf{E}\{M_i(t_1)M_{i'}(t_2) - M_i(s)M_{i'}(s) - \mathbf{1}_{\{i=i'\}} \int_s^{t_1 \wedge t_2} Y_i(u)\,\alpha_i(u)du \mid \mathcal{F}_s\} = 0$$

$$(A.11)$$

Proof. Suppose that $t_2 \geq t_1 \geq s$. Then

$$\mathbf{E}\{M_i(t_1)M_{i'}(t_2) - M_i(s)M_{i'}(s) - \mathbf{1}_{\{i=i'\}} \int_s^{t_1} Y_i(u)\,\alpha_i(u)du \mid \mathcal{F}_s\} =$$

$$\mathbf{E}\{M_i(t_1)M_{i'}(t_1) - M_i(s)M_{i'}(s) - \mathbf{1}_{\{i=i'\}} \int_s^{t_1} Y_i(u)\,\alpha_i(u)du + M_i(t_1)\times$$

$$(M_{i'}(t_2) - M_{i'}(t_1)) \mid \mathcal{F}_s\} = \mathbf{E}\{M_i(t_1)\mathbf{E}\{M_{i'}(t_2) - M_{i'}(t_1) \mid \mathcal{F}_{t_1}\} \mid \mathcal{F}_s\} = 0.$$

A.8 Stochastic integrals with respect to martingales

Let us consider integrals of the form (A.7). Under some assumptions these integrals are martingales and their predictable variations and covariations can be found.

Theorem A.4. *Suppose that M_i are counting process N_i martingales, and H_{ij} are predictable processes such that for all $t \geq 0, i = 1, ..., m, j = 1, ..., n$*

$$\mathbf{E} \mid \int_0^t H_{ij}(u)dM_i(u) \mid < \infty. \qquad (A.12)$$

Then U_{ij} are $\mathbf{F}$-martingales and

$$< U_{ij}, U_{i'j'} > (t) = \int_0^t H_{ij}(u)H_{i'j'}(u)d < M_i, M_{i'} > (u) =$$

$$\begin{cases} \int_0^t H_{ij}(u)H_{ij'}(u)Y_i(u)\alpha_i(u)du, & i = i', \\ 0, & i \neq i'. \end{cases} \qquad (A.13)$$

In particular,

$$< U_{ij}, U_{ij} > (t) = \int_0^t H_{ij}^2(u)d < M_i, M_i > (u) = \int_0^t H_{ij}^2(u)Y_i(u)\alpha_i(u)du,$$

$$(A.14)$$

Proof. The result is implied by Theorem T6 given in Brémaud (1981, Ch.1). We shall consider the theorem under stronger condition that H_{ij} are bounded predictable processes and N_i are the counting processes given in Theorem A2.

1) Suppose that H_{ij} are simple predictable processes of the form:

$$H_{ij}(t) = Z_{ij}\mathbf{1}_{(a_{ij}, b_{ij}]}, \qquad (A.15)$$

where Z_{ij} is a bounded $\mathcal{F}_{a_{ij}}$-measurable random variables.

Let us show that U_{ij} are $\mathbf{F}$-martingales. We skip the indices.

The process

$$U(t) = Z \left(M(t \wedge b) - M(t \wedge a) \right)$$

is $\mathcal{F}_t$-measurable and $E \mid U(t) \mid < \infty$ for all $t > 0$.

Fix $t > s \geq 0$. If $(a, b] \cap (s, t] = \emptyset$ then $U(t) = U(s)$ and $\mathbf{E}\{U(t) - U(s) \mid \mathcal{F}_s\} = 0$.

If $(a, b] \cap (s, t] \neq \emptyset$ then

$$\mathbf{E}\{U(t) - U(s) \mid \mathcal{F}_s\} = \mathbf{E}\{Z(M(t \wedge b) - M(s \wedge a)) \mid \mathcal{F}_s\} =$$

$$\mathbf{E}\{Z \, \mathbf{E}\{M(t \wedge b) - M(s \vee a) \mid \mathcal{F}_{s \wedge a}\} \mid \mathcal{F}_s\} = 0.$$

Let us show (1.13). Fix i, j, i', j'. Ordered $0, a_{ij}, b_{ij}, a_{i'j'}, b_{i'j'}, +\infty$ can take from 2 to 6 values. Denote these values by $t_0 < \cdots < t_m$. Set $H_{ij}^{(k)} = Z_{ij} \mathbf{1}_{\{(t_{k-1}, t_k] \subset (a_{ij}, b_{ij}]\}}$. Then for $s \in [t_{k-1}, t_k)$

$$\mathbf{E}\{U_{ij}(t) U_{i'j'}(t) - U_{ij}(s) U_{i'j'}(s) \mid \mathcal{F}_s\} =$$

$$\mathbf{E}\{U_{ij}(t \wedge t_k) U_{i'j'}(t \wedge t_k) - U_{ij}(s) U_{i'j'}(s) \mid \mathcal{F}_s\} +$$

$$\sum_{l=k+1}^{m} \mathbf{E}\{\mathbf{E}\{U_{ij}(t \wedge t_l) U_{i'j'}(t \wedge t_l) - U_{ij}(t \wedge t_{l-1}) U_{i'j'}(t_{l-1}) \mid \mathcal{F}_{t \wedge t_{l-1}}\} \mid \mathcal{F}_s\}$$

$$= \mathbf{E}\{(U_{ij}(t \wedge t_k) - U_{ij}(s))(U_{i'j'}(t \wedge t_k) - U_{i'j'}(s)) \mid \mathcal{F}_s\} +$$

$$\sum_{l=k+1}^{m} \mathbf{E}\{\mathbf{E}\{(U_{ij}(t \wedge t_l) - U_{ij}(t \wedge t_{l-1}))(U_{i'j'}(t \wedge t_l) - U_{i'j'}(t \wedge t_{l-1})) \mid \mathcal{F}_{t \wedge t_{l-1}}\} \mid \mathcal{F}_s\}$$

$$= H_{ij}^{(k)} H_{i'j'}^{(k)} \mathbf{E}\{(M_i(t \wedge t_k) - M_i(s))(M_{i'}(t \wedge t_k) - M_{i'}(s)) \mid \mathcal{F}_s\} +$$

$$\sum_{l=k+1}^{m} \mathbf{E}\{H_{ij}^{(l)} H_{i'j'}^{(l)} \mathbf{E}\{(M_i(t \wedge t_l) - M_i(t \wedge t_{l-1}))(M_{i'}(t \wedge t_l) - M_{i'}(t \wedge t_{l-1})) \mid \mathcal{F}_{t_{l-1}}\}$$

$$\mid \mathcal{F}_s\} = H_{ij}^{(k)} H_{i'j'}^{(k)} \mathbf{E}\{< M_i, M_{i'} > (t \wedge t_k) - < M_i, M_{i'} > (s) \mid \mathcal{F}_s\} +$$

$$\sum_{l=k+1}^{m} \mathbf{E}\{H_{ij}^{(l)} H_{i'j'}^{(l)} \mathbf{E}\{< M_i, M_{i'} > (t \wedge t_l) - < M_i, M_{i'} > (t \wedge t_{l-1}) \mid \mathcal{F}_{t_{l-1}}\} \mid \mathcal{F}_s\} =$$

$$\mathbf{E}\{\int_s^{t \wedge t_k} H_{ij}^{(k)} H_{i'j'}^{(k)} \, d < M_i, M_{i'} > (u) \mid \mathcal{F}_s\} +$$

$$\sum_{l=k+1}^{m} \mathbf{E}\{\mathbf{E}\{\int_{t \wedge t_{l-1}}^{t \wedge t_l} H_{ij}^{(l)} H_{i'j'}^{(l)} \, d < M_i, M_{i'} > (u) \mid \mathcal{F}_{t_{l-1}}\} \mid \mathcal{F}_s\} =$$

$$\mathbf{E}\{\int_s^t H_{ij}(u) H_{i'j'}(u) \, d < M_i, M_{i'} > (u) \mid \mathcal{F}_s\}.$$

2) If H_{ij} are finite sums of simple predictable processes: $H_{ij} = \sum_{l=1}^{m_{ij}} H_{ijl}$, then

$$\mathbf{E}\{U_{ij}(t) - U_{ij}(s) \mid \mathcal{F}_s\} = \int_s^t H_{ij}(u)\,dM_i(u) \mid \mathcal{F}_s\} = \sum_{l=1}^{m_{ij}} \int_s^t H_{ijl}(u)\,dM_i(u) \mid \mathcal{F}_s\} = 0,$$

and

$$\mathbf{E}\{U_{ij}(t)U_{i'j'}(t) - U_{ij}(s)U_{i'j'}(s) - \mathbf{1}_{\{i=i'\}} \int_s^t H_{ij}(u)H_{ij'}(u)Y_i(u)\,\alpha_i(u)du \mid \mathcal{F}_s\} =$$

$$\sum_{l=1}^{m_{ij}} \sum_{l'=1}^{m_{i'j'}} \mathbf{E}\{\int_0^t H_{ijl}(u)\,dM_i(u) \int_0^t H_{i'j'l'}(u)\,dM_{i'}(u) - \int_0^s H_{ijl}(u)\,dM_i(u) \int_0^s H_{i'j'l'}(u)\,dM_i$$

$$- \mathbf{1}_{\{i=i'\}} \int_s^t H_{ijl}(u)H_{ij'l'}(u)Y_i(u)\,\alpha_i(u)du \mid \mathcal{F}_s\} = 0.$$

3) If H_{ij} is any bounded predictable process then it can be written as the limit of increasing sequence of simple predictable processes. Monotone class arguments can be used to get the result of the theorem (see Fleming and Harrington (1991)).

Corollary A.2. *The processes $U_j = \sum_{i=1}^n U_{ij}$ are $\mathbf{F}$-martingales and*

$$< U_j, U_{j'} > (t) = \sum_{i=1}^n \int_0^t H_{ij}(u)H_{ij'}(u)d < M_i, M_i > (u)$$

$$= \sum_{i=1}^n \int_0^t H_{ij}(u)H_{ij'}(u)Y_i(u)\alpha_i(u)du, \qquad (A.16)$$

Corollary A.3. *The means and the covariances of the statistics U_j have the form*

$$\mathbf{E}(U_j(t)) = 0, \qquad (A.17)$$

$$\mathbf{Cov}(U_j(s), U_{j'}(t)) = \mathbf{E}(< U_j, U_{j'} > (s \wedge t)) =$$

$$\mathbf{E}\left(\sum_{i=1}^n \int_0^{s \wedge t} H_{ij}(u)H_{ij'}(u)Y_i(u)\alpha_i(u)du\right); \qquad (A.18)$$

in particular

$$\mathbf{Var}(U_j(t)) = \mathbf{E}(< U_j, U_j > (t)) = \mathbf{E}\left(\sum_{i=1}^n \int_0^t H_{ij}^2(u)Y_i(u)\alpha_i(u)du\right). \qquad (A.19)$$

Proof. U_j is a martingale, so $\mathbf{E}(U_j(t)) = \mathbf{E}(U_j(0)) = 0$.
$M_{jj'} = U_j U_{j'} - < U_j, U_{j'} >$ is a martingale, $M_{jj'}(0) = 0$, so $\mathbf{E}\{U_j(t)U_{j'(t)}\} = \mathbf{E}\{< U_j, U_{j'} > (t)\}$. If $s \leq t$ then the last equality implies that

$$\mathbf{Cov}(U_j(s), U_{j'}(t)) - \mathbf{E}(< U_j, U_{j'} > (s)) =$$

$$\mathbf{E}\{U_j(s)U_{j'}(s) - < U_j, U_{j'} > (s)\} + \mathbf{E}\{U_j(s)(U_{j'(t)} - U_{j'(s)})\}$$

$$= \mathbf{E}\{U_j(s)\mathbf{E}\{U_{j'(t)} - U_{j'(s)} \mid \mathcal{F}_s\}\} = 0.$$

The condition (1.12) may be hard to verify in practice or not even true. By chance, application of the central limit theorem for martingales may be done if U_i are not necessary martingales. It is sufficient if they are the *local martingales*. In such a case weaker and easier verified conditions are needed. So we need the notion of *localization*.

A.9 Localization

Definition A.13. *A random variable T is called the stopping time if for all $t \geq 0$*

$$\{T \leq t\} \in \mathcal{F}_t.$$

As a rule the stopping time is the moment when a certain event occurs. For example, it can be the moment when the ith failure occurs. If the history up to the moment t is known then it is known whether or not such event has already occurred.

Definition A.14. *A sequence of stopping times $\{T_n\}$ is called a localizing sequence if $T_n \xrightarrow{\mathbf{P}} \infty$.*

Definition A.15. *A process X has a certain property locally if there exists a localizing sequence $\{T_n\}$ such that for each n the process $X(t \wedge T_n)\mathbf{1}_{\{T_n > 0\}}$ has the property.*

So a process X is *locally bounded* if there exist constants c_n and a localizing sequence $\{T_n\}$ such that

$$\sup_{t \leq T_n} \mid X(t) \mid \leq c_n \text{ a.s. on } \{T_n > 0\}.$$

It can be shown that any *left-continuous adapted process is locally bounded.*

A process M is a *local martingale* if there exists a localizing sequence $\{T_n\}$ such that for each n the process

$$M(t \wedge T_n)\mathbf{1}_{\{T_n > 0\}}$$

is a martingale.

Theorem A.5. (Optional stopping theorem). *If $M(t), t \geq 0$ is a right-continuous martingale and T is a stopping time then $X(t \wedge T)$ is a martingale.*

See the proof in Fleming and Harrington (1991).

A.10 Stochastic integrals with respect to martingales (continuation)

Theorem A.6. *Suppose that assumptions of Theorem A.1 hold, H_{ij} are left-*

continuous adapted processes. Then U_{ij} and U_i are local square-integrable **F**-*martingales*

Proof. We skip indexes i and j in what follows. We need to show that there exists a localizing sequence $\{T_n\}$ such that for all n the process

$$\int_0^{t \wedge T_n} H(u)dM(u)\,\mathbf{1}_{\{T_n>0\}}$$

is a **F**-martingale.

H is a locally bounded process. So there exist constants c_n and a localizing sequence $\{T_n\}$ such that

$$\mid H(t \wedge T_n) \mid \le c_n \text{ a.s. on } \{T_n > 0\}.$$

Then

$$\int_0^t \tilde{H}_n(u)d\tilde{M}_n(u) = \int_0^{t \wedge T_n} H(u)dM(u)\,\mathbf{1}_{\{T_n>0\}}, \qquad (A.20)$$

where

$$\tilde{H}_n(t) = H(t \wedge T_n)\mathbf{1}_{\{T_n>0\}}, \quad \tilde{M}_n(t) = M(t \wedge T_n).$$

For fixed n the process $\tilde{H}_n(t)$ is left-continuous adapted and bounded. The optional stopping theorem implies that $\tilde{M}_n(t)$ is a **F**-martingale. For all $t \in [0, \infty]$ and any fixed n

$$\mid \int_0^t \tilde{H}_n(u)d\tilde{M}_n(u) \mid = \mid \int_0^{t \wedge T_n} H(u)dN(u) -$$

$$\int_0^{t \wedge T_n} H(u)\alpha(u)Y(u)du \mid \le c_n + c_n A(T_n). \qquad (A.21)$$

So the integral in the left side of (1.20) verifies the conditions of Theorem A.3. It implies that this integral is a **F**-martingale.

A.11 Weak convergence

Let us consider the space D, where $D = D[0, \tau]$ is a space of cadlag functions on $[0, \tau]$ or $D = \mathbf{R}$. Denote by $D^p = D_1 \times \cdots \times D_p$ the product of p such spaces.

Let $\rho(x, y)$ be the distance between the functions x and y, $x, y \in D$. It can be defined in various ways. In the space $D = D[0, \tau]$ the most used are Skorokhod and supremum norm distances. Let $\Lambda = \{\lambda(\cdot)\}$ be the set of strictly increasing continuous functions on $[0, \tau]$ such that $\lambda(0) = 0, \lambda(\tau) = \tau$.

Skorohod metric:

$$\rho(x, y) =$$

$$\inf\{\varepsilon > 0 : \text{exists } \lambda \in \Lambda : \sup_{0 \le t \le \tau} \mid \lambda(t) - t \mid \le \varepsilon, \ \sup_{0 \le t \le \tau} \mid x(t) - y(\lambda(t)) \mid \le \varepsilon\};$$

$$(A.22)$$

Supremum norm metric:

$$\rho(x,y) = \sup_{0 \leq t \leq \tau} | x(t) - y(t) | . \qquad (A.23)$$

In the space $D = \mathbf{R}$ the most used is

Euclidean metric

$$\rho(x,y) = | x - y | . \qquad (A.24)$$

An open ε-ball with center x is defined as $B_\varepsilon(x) = \{y : \rho(x,y) < \varepsilon\}$. A set $G \subset D$ is open if for any $x \in G$ there exists an open ball $B_\varepsilon(x) \subset G$. The boundary ∂G of the set G is a set of points $x \in G$ such that in any open ball $B_\varepsilon(x)$ there exist points $y, z \in D : y \in G$, $z \notin G$.

The smallest σ-algebra containing all open sets of D is called the Borel σ-algebra of D and is denoted by $\mathcal{B}(D)$. The smallest σ-algebra containing all sets of the form $A = A_1 \times \cdots A_s$, where $A_1, \cdots, A_s$ are open sets of $D_1, \cdots, D_s$, respectively, is called the *Borel σ-algebra* of D^s and is denoted by $\mathcal{B}^s(D^s)$.

We use capital letters for random elements and small letters for non-random elements. So, if we write $X \in D$ when $D = \mathbf{R}$, it means that $X = X(\omega)$ is a random variable defined on the probability space $(\Omega, \mathcal{F}, \mathbf{P})$. If we write $x \in D$, it means that x is a real number.

If we write $X \in D$ when $D = D[0, \tau]$, it means that $X = X(t, \omega)$ is a cadlag stochastic process on $[0, \tau]$, i.e. for any fixed $t \in [0, t]$ $X(t, \cdot)$ is a random variable defined on the probability space $(\Omega, \mathcal{F}, \mathbf{P})$ and for any fixed $\omega \in \Omega$ the trajectory $X(\cdot, \omega)$ is a real cadlag function on $[0, \tau]$. If we write $x \in D$, it means that x is a real cadlag function on $[0, \tau]$.

If we write $X \in D^s$, it means that $X = (X_1, \cdots, X_s)^T$, where $X_i \in D_i$, $(i = 1, \cdots, s)$.

Any $X \in D^s$ generates a probability measure $\mathbf{P}^X$ on $(D^s, \mathcal{B}^s(D^s))$:

$$\mathbf{P}^X(B) = \mathbf{P}\{X \in B\}, \quad \text{for any} \quad B \in \mathcal{B}^s(D^s).$$

Definition A.16. The sequence $\{X^{(n)}\}$, $X^{(n)} \in D^s$, weakly converges to $X \in D^s$ if for any $B \in \mathcal{B}^s(D^s)$ such that $\mathbf{P}^X(\partial B) = 0$ we have:

$$\mathbf{P}^{X^{(n)}}(B) \to \mathbf{P}^X(B).$$

Weak convergence is denoted by $X^{(n)} \xrightarrow{\mathcal{D}} X$.

The Borel σ-algebra generated by the Skorohod metric is wider then the σ-algebra generated by the supremum norm metric. So weak convergence in the sense of Skorohod metric implies weak convergence in the sense of supremum norm metric. If the limit process is continuous, both convergences are equivalent. In all applications considered here the limit processes are continuous Gaussian processes.

A.12 Central limit theorem for martingales

The limit distribution of integrals (1.4) can be found by using the central limit theorem for martingales, Rebolledo (1980). This theorem is based on the following property of Gaussian processes with independent increments.

Characterization of Gaussian processes with independent increments. *Suppose that for all $t \geq 0$ the matrix $\Sigma(t) = (\sigma_{jj'}(t))_{m \times m}$ is deterministic positively definite, $\sigma_{jj'}(0) = 0$.*

A m-variate stochastic process $V = (V_1, \cdots, V_m)$ is a Gaussian process having components with independent increments and

$$\mathbf{E}(V_j(t)) = 0 \quad and \quad \mathbf{Cov}(V_j(s), V_{j'}(t)) = \sigma_{jj'}(s \wedge t) \quad (s, t \geq 0)$$

if and only if V_i are locally square integrable martingales with continuous paths such that for all $t \geq 0$

$$V_j(0) = 0, \quad < V_j, V_{j'} > (t) = \sigma_{jj'}(t) \quad (j, j' = 1, \cdots, m).$$

So if the jumps of the martingales $U_j^{(n)}$ on $[0, \tau]$ tend to zero (Lindenberg condition) and
$$< U_j^{(n)}, U_{j'}^{(n)} > (t) \xrightarrow{\mathbf{P}} \sigma_{jj'}(t) \text{ for all } t \in [0, \tau] \text{ as } n \to \infty \text{ then the following}$$
convergence should be expected:

$$U^{(n)} = (U_1^{(n)}, ..., U_m^{(n)})^T \xrightarrow{\mathcal{D}} V = (V_1, ..., V_m)^T \text{ on } (D[0, \tau])^m. \quad (A.25)$$

Exact formulation of the expression *if the jumps of the martingales $U_j^{(n)}$ on $[0, \tau]$ tend to zero* is given in the following theorem.

Fix $k \in \mathbf{N}$ and set

$$U_j^{(n)}(t) = \sum_{i=1}^{n} \int_0^t H_{ij}^{(n)}(v) \, dM_i(v)$$

or

$$U_j^{(n)}(t) = \sum_{i=1}^{k} \int_0^t H_{ij}^{(n)}(v) \, dM_i(v),$$

and

$$U_{j\varepsilon}^{(n)}(t) = \sum_{i=1}^{n} \int_0^t H_{ij}^{(n)}(v) \, \mathbf{1}\{|H_{ij}^{(n)}(v)| \geq \varepsilon\} \, dM_i(v)$$

or

$$U_{j\varepsilon}^{(n)}(t) = \sum_{i=1}^{k} \int_0^t H_{ij}^{(n)}(v) \, \mathbf{1}\{|H_{ij}^{(n)}(v)| \geq \varepsilon\} \, dM_i(v),$$

respectively $(j = 1, \cdots, m)$. Fix a moment τ.

Theorem A.7. *Suppose that the multiplicative intensity model holds and a) the integrals*

$$\Lambda_i(t) = \int_0^t Y_i(u) \, \alpha_i(u) du$$

are continuous on $[0, \tau]$;

b) $H_{ij}^{(n)}(v)$ are left-continuous with finite right limits adapted processes on $[0, \tau]$;

c) $< U_j^{(n)}, U_{j'}^{(n)} > (t) \xrightarrow{\mathbf{P}} \sigma_{jj'}(t)$ for all $t \in [0, \tau]$ as $n \to \infty$;

d) $< U_{j\varepsilon}^{(n)}, U_{j\varepsilon}^{(n)} > (t) \xrightarrow{\mathbf{P}} 0$ for all $t \in [0, \tau]$ as $n \to \infty$;

e) the matrix $\Sigma(t) = (\sigma_{jj'}(t))_{m \times m}$ is positively definite for all $t \in [0, \tau]$.
Then

$$U^{(n)} = (U_1^{(n)}, ..., U_m^{(n)})^T \xrightarrow{D} V = (V_1, ..., V_m)^T \quad on \ (D[0, \tau])^m,$$

where V is a m-variate Gaussian process having components with independent increments, $V_j(0) = 0$ a.s. and for all $0 \le s \le t \le \tau$:

$$\mathbf{cov}(V_j(s), V_{j'}(t)) = \sigma_{jj'}(s).$$

Corollary A.4. *Under the assumptions of the theorem*

$$U^{(n)}(\tau) \xrightarrow{D} N(0, \Sigma(\tau)) \quad as \ n \to \infty.$$

Corollary A.5. *Suppose that $a(t)$ is a $m \times 1$ non-random vector. If the assumptions of Theorem A.7 are verified for $j, j' = 1, \cdots, m+1$,*

$$< U_j^{(n)}(\tau), U_{m+1}^{(n)}(t) > = 0 \quad (j = 1, \cdots m),$$

and a univariate process $Z^{(n)}$ is the linear combination

$$Z^{(n)}(t) = a^T(t) U^{(n)}(\tau) + U_{m+1}^{(n)}(t),$$

then

$$Z^{(n)} \xrightarrow{D} Z,$$

where Z is a zero mean Gaussian process with independent increments and

$$\mathbf{Cov}(Z(s), Z(t)) = a^T(s) \Sigma_{m \times m}(\tau) a(t) + \sigma_{m+1, m+1}(s);$$

Set $M^{(n)}(t) = \sum_{i=1}^{n} M_i(t)$.

Theorem A.8. (Lenglart's inequality, Lenglart (1977)). *Suppose that H is an adapted caglad process. Then for all $\varepsilon, \eta, \tau > 0$*

$$\mathbf{P}\left\{\sup_{[0, \tau]}\{\int_0^t H(s) dM^{(n)}(s)\}^2 \ge \varepsilon\right\} \le \frac{\eta}{\varepsilon} + \mathbf{P}\left\{\int_0^\tau H^2(s) d < M^{(n)} > (s) \ge \eta\right\}.$$

Corollary A.6. *The convergence*

$$\int_0^\tau (H^{(n)}(s))^2 d < M^{(n)} > (s) \xrightarrow{\mathbf{P}} 0$$

implies the convergence

$$\sup_{[0,\tau]} \int_0^t H^{(n)}(s)dM^{(n)}(s) \xrightarrow{\text{P}} 0.$$

A.13 Nonparametric estimators of the cumulative hazard and the survival function

Let us consider nonparametric estimation of the cumulative hazard and the survival function from the data (1.2) when the failure times $T_1, \cdots, T_n$ are identically distributed. Theorem A.2 implies that

$$\mathbf{E}\{N(t)\} = \mathbf{E}\{\int_0^t Y(u)\, dA(u)\}, \qquad (A.26)$$

where

$$A(t) = \int_0^t \alpha(u)\, du$$

is the cumulative hazard function. The equality holds even when the function A is not necessary continuous, i.e; *ex aequo* of failures are possible. This equality implies that an estimator of the cumulative hazard function can be defined as a solution of the integral equation

$$\hat{A}(t) = \int_0^t \frac{dN(u)}{Y(u)} = \sum_{j:\delta_j=1, X_j \le t} \frac{1}{n_j}, \qquad (A.27)$$

where

$$n_j = Y(X_j) = \sum_{l=1}^n \mathbf{1}_{\{X_l \ge X_j\}}$$

is the number of units at risk just prior to X_j. It is the Nelson-Aalen estimator of the cumulative hazard A.

The equality

$$S(t) = 1 + \int_0^t dS(u) = 1 - \int_0^t S(u)dA(u)$$

implies the following equation for an estimator of the survival function S:

$$\hat{S}(t) = 1 - \int_0^t \hat{S}(u-)d\hat{A}(u)$$

or

$$\hat{S}(t) = 1 - \int_0^t \hat{S}(u-)\frac{dN(u)}{Y(u)}. \qquad (A.28)$$

Hence

$$\hat{S}(t) = \hat{S}(t-)\left(1 - \frac{\Delta N(t)}{Y(t)}\right),$$

where $\Delta N(t) = N(t) - N(t-)$. It gives

$$\hat{S}(t) = \prod_{s:s\leq t}\left(1 - \frac{\Delta N(s)}{Y(s)}\right) = \prod_{j:\delta_j=1,X_j\leq t}\left(1 - \frac{1}{n_j}\right). \qquad (A.29)$$

If there are *ex aequo*, and $T_1^0 < \cdots < T_m^0$ are the distinct moments of observed failures, d_i is a number of failures at the moment T_i^0, and $n_i = Y(T_i^0)$ is the number of units at risk just prior to T_i^0 then

$$\hat{A}(t) = \sum_{i:T_i^0\leq t}\frac{d_i}{n_i}, \quad \hat{S}(t) = \prod_{i:T_i^0\leq t}\left(1 - \frac{d_i}{n_i}\right). \qquad (A.30)$$

Let G_i be the survival function of the censoring time C_i.

Theorem A.9. *If the failure times $T_1, \cdots, T_n$ are absolutely continuous random variables, the failure times $T_1, \cdots, T_n$ and the censoring times $C_1, \cdots, C_n$ are mutually independent, and there exists a survival function G with $G(\tau-) > 0$ such that*

$$\sup_{s\in[0,\tau]} \mid n^{-1}\sum_{i=1}^{n} G_i(s) - G(s) \mid \to 0 \text{ as } n \to \infty,$$

then

$$n^{1/2}(\hat{A} - A) \xrightarrow{D} U, \quad n^{1/2}(\hat{S} - S) \xrightarrow{D} -SU \text{ as } n \to \infty, \qquad (A.31)$$

on $D[0,\tau]$, where U is a Gaussian martingale with $U(0) = 0$ and

$$\mathbf{Cov}\,(U(s), U(t)) = \sigma^2(s \wedge t) = \int_0^{s\wedge t}\frac{dA(u)}{y(u)}, \qquad (A.32)$$

where $y(u) = S(u)G(u-)$.

A.14 Product-integral

Let $X(t), t \geq 0$ be a cadlag stochastic process. Denote by $t_1, \cdots, t_{m(t)}$ the jump points of X in the interval $[0,t]$ and $\Delta X(t_i) = X(t_i) - X(t_{i-1})$. Note that $m(t)$ is random.

Definition A.17 *The stochastic process*

$$Z(t) = \prod_{0\leq s\leq t}(1 + dX(s)) = \prod_{i=1}^{m(t)}(1 + \Delta X(t_i))\exp\{X(t) - \sum_{i=1}^{m(t)}\Delta X(t_i)\} \qquad (A.33)$$

is called the product-integral of the process X.

If the paths of X are continuous then

$$Z(t) = \prod_{0\leq s\leq t}(1 + dX(s)) = e^{X(t)}. \qquad (A.34)$$

If the paths of X are step-functions, then

$$Z(t) = \prod_{0 \leq s \leq t}(1 + dX(s)) = \prod_{i=1}^{m(t)}(1 + \Delta X(t_i)). \qquad (A.35)$$

We set

$$\prod_{0 \leq s \leq t}(1 - dX(s)) = \prod_{0 \leq s \leq t}(1 + d(-X(s))). \qquad (A.36)$$

If the cumulative hazard function $A(\cdot)$ is continuous then the survival function can be written as the product-integral:

$$S(t) = e^{-A(t)} = \prod_{0 \leq s \leq t}(1 - dA(s)). \qquad (A.37)$$

The Nelson-Aalen estimator of the function $A(t)$ is a step-function with jumps $1/n_j$ at the points $X_j : \delta_j = 1$. So

$$\prod_{0 \leq s \leq t}(1 - d\hat{A}(s)) = \prod_{j:\delta_j=1, X_j \leq t}\left(1 - \Delta\hat{A}(X_j)\right) =$$

$$\prod_{j:\delta_j=1, X_j \leq t}\left(1 - \frac{1}{n_j}\right) = \hat{S}(t), \qquad (A.38)$$

where $\hat{S}(t)$ is the Kaplan-Meier estimator. So

$$\hat{S}(t) = \prod_{0 \leq s \leq t}(1 - d\hat{A}(s)). \qquad (A.39)$$

The same relation holds if there are ex aequo: the estimator $\hat{A}$ has jumps d_i/n_i at the points T_i^0, so

$$\prod_{0 \leq s \leq t}(1 - d\hat{A}(s)) = \prod_{j:T_j^0 \leq t}\left(1 - \frac{d_j}{n_j}\right) = \hat{S}(t). \qquad (A.40)$$

A.15 Delta method

The most of the estimators and test statistics used here are functionals of cadlag stochastic processes. In many situations weak convergence of such functionals can be obtained by using the functional delta method. We refer to the book of Andersen et al. (1993), where this method is described and give here several results implied by this method and used in this book. Convergence is considered in the sense of supremum norm.

Let $\{a_n\}$ be a sequence of real numbers.

Theorem A.10. Let $g = (g_1, \cdots, g_q) : \mathbf{R}^p \to \mathbf{R}^q$ be a differentiable vector-function, and

$$J_g(x) = \| \frac{\partial g_i(x)}{\partial x_j} \|_{q \times p}$$

be the Jacobi matrix of partial derivatives of coordinate functions g_i.
  If

$$a_n(X^{(n)} - x) \xrightarrow{\mathcal{D}} Z \quad as \ a_n \to \infty$$

on $\mathbf{R}^p$, then

$$a_n(g(X^{(n)}) - g(x)) \xrightarrow{\mathcal{D}} J_g(x)Z \quad \text{as } a_n \to \infty. \qquad (A.41)$$

on $\mathbf{R}^q$.

Theorem A.11. *Suppose that*

1) $\{X_1^n \in D[0,\tau]\}$ and $\{X_2^n \in D[0,\tau]\}$ are sequences of cadlag stochastic processes, the second being of bounded variation and bounded by a positive constant M;

2) $X_1, X_2 \in D[0,\tau]$ are cadlag stochastic processes of bounded variation the second being bounded by M such that

$$(a_n(X_1^n - X_1), a_n(X_2^n - X_2)) \xrightarrow{\mathcal{D}} (Z_1, Z_2),$$

on $D[0,\tau] \times D[0,\tau]$; here $Z_1, Z_2 \in D[0,\tau]$.

Then

$$a_n\left(\int_0^{\cdot} X_1^n dX_2^n - \int_0^{\cdot} X_1 dX_2\right) \xrightarrow{\mathcal{D}} \int_0^{\cdot} Z_1 dX_2 + \int_0^{\cdot} X_1 dZ_2 \qquad (A.42)$$

on $D[0,\tau]$. If Z_2 is not of bounded variation then the last integral is defined by

$$\int_0^t X_1(u) dZ_2(u) = X_1(t)Z_2(t) - X_1(0)Z_2(0) - \int_0^t Z_2(u) dX_1(u).$$

Theorem A.12. *Suppose that*

1) $x \in D[0,\tau]$ is a nondecreasing function, differentiable at the point

$$x^{-1}(p) = \inf\{t : x(t) \geq p\} \in (0,\tau),$$

where $p \in \mathbf{R}$ is a fixed number.

2) $\{X^{(n)} \in D[0,\tau]\}$ is a sequence of nondecreasing stochastic processes such that

$$a_n(X^n - x) \xrightarrow{\mathcal{D}} Z$$

on $D[0,\tau]$; here $Z \in D[0,\tau]$ is a nondecreasing process, continuous at the point $x^{-1}(p)$.

Then

$$a_n((X^n)^{-1}(p) - x^{-1}(p)) \xrightarrow{\mathcal{D}} -\frac{Z(x^{-1}(p))}{x'(x^{-1}(p))}. \qquad (A.43)$$

Theorem A.13. *Suppose that*

1) x is a continuously differentiable function on $[0,\tau]$;

2) $\varphi = \varphi(t,\theta) : A \times B_\varepsilon(\theta_0) \to \mathbf{R}$, $B_\varepsilon(\theta_0) \subset \mathbf{R}^s$, $A = [0,\tau_0]$ or $(0,\tau_0)$, is a continuous non-decreasing in t function such that $0 < \varphi(t,\theta_0) < \tau$ for $t \in A$;

3) $\{X^{(n)} \in D[0,\tau]\}$ is a sequence of stochastic processes such that

$$a_n(X^n - x) \xrightarrow{\mathcal{D}} Z$$

on $D[0, \tau]$, *where* Z *is a continuous on* $[0, \tau]$ *stochastic process;*

4) $\{\hat{\theta}^{(n)}\}$ *is a sequence of random variables such that*

$$a_n(\hat{\theta}^{(n)} - \theta_0) \xrightarrow{\mathcal{D}} Y.$$

Then

$$a_n(X^n(\varphi(\cdot, \hat{\theta}_0^{(n)})) - x(\varphi(\cdot, \theta_0)) \xrightarrow{\mathcal{D}} Z(\varphi(\cdot, \theta_0)) + x'(\varphi(\cdot, \theta_0))Y \qquad (A.44)$$

on $D(A)$.

Theorem A.14. *Suppose that* $\{X^{(n)} \in D[0, \tau]\}$ *is a sequence of cadlag stochastic processes and*

$$a_n(X^n - X) \xrightarrow{\mathcal{D}} Z$$

on $D[0, \tau]$, *where* Z *and* X *are continuous on* $[0, \tau]$ *stochastic processes.*

Then

$$a_n \left(\prod_{0 \le s \le t}(1 + dX^n(s)) - \prod_{0 \le s \le t}(1 + dX(s)) \right) \xrightarrow{\mathcal{D}} e^X Z. \qquad (A.45)$$

References

Aalen, O. (1978) Nonparametric inference for the family of counting processes. *Ann. Statist.*, **6**, 701-726.

Aalen, O. (1980) A model for nonparametric regression analysis of counting processes. In *Mathematical Statistics and Probability Theory*, Lecture Notes in Statistics, **2**, (Eds. W. Klonecki, A. Kozek and J. Rosinski), New York: Springer-Verlag, 1-25.

Altman, D.G. and Andersen, P.K. (1986) A note on the uncertainty of a survival probability estimated from Cox's regression model. *Biometrika*, **73**, 722-724.

Andersen, P.K. and Gill, R.D. (1982) Cox's regression model for counting processes: A large sample study. *Ann. Statist*, **10**, 1100-1120.

Andersen,P.K., Borgan, O.,Gill, R.D. and Keiding, N. (1993) *Statistical Models Based on Counting Processes*. New York: Springer.

Aranda-Ordaz, F.J. (1983) An Extension of the Proportional-Hazards Model for Grouped Data, *Biometrics*, **39**, 109-117.

Arjas, E. (1988) A graphical method for assessing goodness of fit in Cox's proportional hazards model. *JASA*, **83**, 204-212.

Bagdonavičius, V. (1978) Testing the hyphothesis of the additive accumulation of damages. *Probab. Theory and its Appl.*, **23**, # 2, 403-408.

Bagdonavičius, V. (1990) Accelerated life models when the stress is not constant. *Kybernetika*, **26**, 289-295.

Bagdonavičius, V. (1993) The modified moment method for multiply censored samples. *Lithuanian Mathematical Journal*, **33**, #4, 295-306.

Bagdonavičius, V. and Nikulin, M. (1994). Stochastic models of accelerated life. In: *Advanced Topics in Stochastic Modelling* (Eds. J.Gutierrez, M.Valderrama). Singapore: World Scientific, 73-87.

Bagdonavičius, V. and Nikulin, M. (1995) *Semiparametric models in accelerated life testing*. Queen's Papers in Pure and Applied Mathematics, **98**. Kingston: Queen's University, Ontario,Canada.

Bagdonavičius, V. and Nikulin, M. (1996) Analyses of generalized additive semiparametric models. *Comptes Rendus, Academie des Sciences de Paris*, **323**, 9, Série I, 1079-1084.

Bagdonavičius, V. and Nikulin, M. (1997a) Analysis of general semiparametric models with random covariates, *Revue Roumaine de mathématiques Pures et Appliquées*, **42**, #5-6, 351-369.

Bagdonavičius, V. and Nikulin, M. (1997b) Statistical analysis of the generalized additive semiparametric survival model with random covariates, *Qüestiió*, **21**, #1-2, 273-291.

Bagdonavičius V. and M. Nikulin. (1997c) Sur l'application des stress en escalier dans les expériences accélérées, *Comptes Rendus, Academie des Sciences de Paris*, **325**, Serie I, 523-526.

Bagdonavičius V. and Nikulin, M. (1997d) Transfer functionals and semiparametric regression models, *Biometrika*, **84**, 2, 365-378.

Bagdonavičius, V. and Nikulin, M. (1997e) Asymptotic analysis of semiparametric models in survival analysis and accelerated life testing, *Statistics*, **29**, 261-281.

Bagdonavičius, V. and Nikulin, M. (1997f) Accelerated life testing when a process of production is unstable, *Statistics and Probabilité Letters*, **35**,#3, 269-275.

Bagdonavičius, V., Nikulin, M. (1997g) Some rank tests for multivariate censored data. In: *Advances in the Theory and Practice of Statistics: A Volume in Honor of Samuel Kotz*, (Eds. N.L. Johnson and N. Balakrishnan), New York: John Wiley and Sons, 193-207.

Bagdonavičius, V., Nikoulina, V. and Nikulin, M. (1997) Bolshev's method of confidence interval construction, *Qüestiió*, **21**, #1-2, 37-58.

Bagdonavičius, V. and Nikoulina, V. (1997) A goodness-of-fit test for Sedyakin's model. *Revue Roumaine de Mathématiques Pures et Appliquées*, **42**, # 1, 5-14.

Bagdonavičius, V. and Nikulin, M. (1998). *Additive and multiplicative semiparametric models in accelerated life testing and survival analysis*. Queen's Papers in Pure and Applied Mathematics, **108**. Kingston: Queen's University, Ontario, Canada.

Bagdonavičius, V., Nikulin, M. (1999a) Model Building in Accelerated Experiments. In: *Statistical and Probabilistic Models in Reliability*, (Eds. D. Ionescu and N. Limnios) Boston: Birkhauser, 51-74.

Bagdonavičius, V., Nikulin, M. (1999b) Semiparametric Estimation of Reliability From Accelerated Life Data. In: *Statistical and Probabilistic Models in Reliability*, (Eds. D.Ionescu and N.Limnios), Boston: Birkhauser, 75-89.

Bagdonavičius, V. and Nikulin, M. (1999c) Generalized Proportional Hazards Model Based on Modified Partial Likelihood, *Lifetime Data Analysis*, **5**, 329-350.

Bagdonavičius, V., Bikelis, A. and Kazakevičius, V. (2000) *On nonparametric estimation in the linear path models when traumatic event intensity depends on degradation*. Bordeaux: Preprint 2002, Université Victor Segalen Bordeaux 2.

Bagdonavičius, V., Gerville-Réache, L., Nikoulina, V. and Nikulin, M. (2000) Expériences accélérées: analyse statistique du modèle standard de vie accélérée. *Revue de Statistique Appliquée*, XLVIII, 3, 5-38.

Bagdonavičius, V. and Nikulin, M. (2000a) Modèle statistique de dégradation avec des covariables dépendant du temps. *Comptes Rendus de l'Académie des Sciences de Paris*, **330**, Série I, # 2, 131-134.

Bagdonavičius, V. and Nikulin, M. (2000b) On nonparametric estimation in accelerated experiments with step-stresses, *Statistics*, **33**, #4, 349-365.

Bagdonavičius, V. and Nikulin, M. (2000c) On goodness-of-fit for the linear transformation and fraility models, *Statistics and Probability Letters*, **47**, #2, 177-188.

Bagdonavičius, V. and Nikulin, M. (2000d) Semiparametric Estimation in Accelerated Life Testing. In: *Recent Advances in Reliability Theory. Methodology, Practice and Inference* (Eds. N. Limnios and M. Nikulin). Boston: Birkhauser, 405-418.

Bagdonavičius, V., Gerville-Réache, L. and Nikulin, M. (2001) On parametric inference for step-stresses models. *IEEE Transaction on Reliability* (to appear).

Bagdonavičius, V., Bikelis, A. and Kazakevičius, V. (2001) Estimation of tire wear and traumatic event intensity. *Tire Science and Technology* (to appear).

Bagdonavičius, V. and Nikulin, M. (2001a) On goodness-of-fit for accelerated life models. *Comptes Rendus de l'Académie des Sciences de Paris*, **332**, Série I, 171-176.

Bagdonavičius, V. and Nikulin, M. (2001b) Mathematical models in the theory of accelerated experiments. In: *Mathematics in the 21st Century*, (Eds. A.A. Ashour and A.-S.F. Obada), Singapore: World Scientific, 271-303.

Bagdonavičius, V. and Nikulin, M. (2001c) Goodness-of-fit tests for Accelerated Life Models. In: *Goodness-of-fit Tests and Validity of Models*, (Eds. C. Huber, N. Balakrishnan, M. Nikulin and M. Mesbah), Boston: Birkhauser (to appear).

Bagdonavičius, V. and Nikulin, M. (2001d) Estimation of Cycling Effect on Reliability. In: *Probability and Statistical Models with Applications*, (Eds. Ch.A. Charalambides, M.V. Koutras, N. Balakrishnan), Boca Raton: Chapman and Hall/CRC, 537-545.

Bagdonavičius, V. and Levulienė, R. (2001) On Goodness-of-fit for Absence of Memory Model. *Kybernetyka* (to appear).

Bagdonavičius, V. and Nikulin, M. (2001) Estimation in Degradation Models with Explanatory Variables. *Lifetime Data Analysis*, **7**, 85-103.

Bain, L.J., and Engelhardt, M. (1991) *Statistical Analysis of Reliability and Life Testing Models. Theory and Methods.* New York: Marcel Dekker.

Bar-Lev, S.K. and Enis, P. (1986) Reproducibility and natural exponential family with power variance functions, *Ann. Statist.*, **14**, 1507-1522.

Barlow, R.E. and Proschan, F. (1975) *Statistical Theory of Reliability and Life Testing*. New York: Holt, Rinehart, and Winston.

Basu, A.P. and Ebrahimi, N. (1982) Nonparametric accelerated life testing. *IEEE Trans. on Reliability*, **31**, 432-435.

Bayer, R.G. (1994) *Mechanical Wear Prediction and Prevention*. New York: Marcel Dekker.

Beichelt, F. and Franken P. (1983) *Zuverlassigkeit und Instanghaltung*. Berlin: VEB Verlag Technik.

Bhattacharyya, G.K. and Stoejoeti, Z. (1989) A tampered failure rate model for step-stress accelerated life test, *Comm. in Statist.*, Theory and Methods, **18**, 1627-1643.

Bickel, P.J., Klaassen, C.A.J., Ritov, Y. and Wellner, J.A. (1993) *Efficient and Adaptive Estimation for Semiparametric Models*. Johns Hopkins University Press, Baltimore.

Billingsley, P. (1968) *Convergence of Probability Measures*. New York: John Wiley and Sons.

Birnbaum, Z.W. and Saunders, S.C. (1969) A new family of life distributions.*Journal of Applied Probability*, **6**, 319-327.

Bogdanoff, J.L. and Kozin, F. (1985) *Probabilistic Models of Cumulative Damage*. New York: John Wiley and Sons.

Bonneuil, N. (1997) *Introduction á la modélisation démographique*, Paris: Armand Colin/Masson.

Bordes, L. (1996) *Inférence Statistique pour des Modèles Paramétriques et Semi-paramétriques : Modèle de l'Exponentielle Multiple, Test du Chi-deux, Modèles de Vie Accélérée*. PhD. Thèse, Bordeaux, France.

Borgan, O. (1984) Maximum likelihood estimation in parametric counting process models, with applications to censored failure time data. *Scand. J. Statist.*, **11**, 1-16.

Boulanger, M., Escobar, L.A.(1994) Experimental design for a class of accelerated degradation tests. *Technometrics*, **36**, 260-272.

Brémaud, P. (1981) *Point Processes and Queues: Martingales Dynamics*. New York: Springer.

Breslow, N.E. (1975a) Covariance analysis of censored survival data. *Biometrics*, **30**, 89-99.

Breslow, N.E. (1975b) Analysis of survival data under the proportional hazards model, *Internat. Statist. Rev.*, **43**, 45-58.

Breslow, N.E., Crowley, J. (1982) A large sample study of the life table and product limit estimates under random censorship. *Ann.Statist.*, **2**, 437-453.

Carey, M.B., Koenig, R.H. (1991) Reliability assessment based on accelerated degradation: a case study. *IEEE Transactions on Reliability*, **40**, 499-506.

Ceci, C., Delattre, S., Hoffmann, M. and Mazliak, L. (2000) Optimal design in semiparametric life testing. Preprint de l'Université Paris YI, 14 p.

Chang, D.S. (1992) Analysis of accelerated degradation data in a two-way design. *Reliability Engineering and System Safety*, **39**, 65-69.

Chappell, R. (1992) A note on linear rank tests and Gill and Schumacher's tests of proportionality, *Biometrika*, **79**, 199-201.

Cheng, S.C., Wei, L.J. and Ying, Z. (1995) Analysis of transformation models with censored data. *Biometrika*, **82**, 835-846.

Chiao, C.H., Hamada, M. (1996) Using Degradation Data from an Experimet to Achive Robust Reliability for Light Emmitining Diodes. *Quality and Reliability Engineering International*, **12**, 89-94.

Ciampi, A. and Etezadi-Amoli, J. (1985) A General Model for Testing the Proportional Hazards and the Accelerated Failure Time Hypotheses in the Analysis of Censored Survival Data With Covariates, *Communications in Statistical Methods and Theory*, **14**, 3, 651-667.

Cinlar, E. (1980) On a generalization of gamma processes. *J. Appl. Probab.*, **17**, 467-480.

Courgeau, D., Lelièvre, E. (1989) *Analyse démographique des biographies*. Paris: INED.

Cox, D.R. and Snell, E.L. (1968) A general definition of residuals (with discussion). *J.R.Statist.Soc.*, **B 30**, 248-275.

Cox, D.R. (1972) Regression models and life tables, *J. R. Statist. Soc.*, B, **34**, 187-220.

Cox, D.R. (1975) Partial likelihood. *Biometrika*, **62**, 269-276.

Cox, D.R. and Oakes, D. (1984) *Analysis of Survival Data*. New York: Methuen (Chapman and Hall).

Cox, D.R. (1999) Some Remarks on Failure-times, Surrogate Markers, Degradation, Wear, and the Quality of Life. *Lifetime Data Analysis*, **5**, 307-314.

Crowder, M.J., Kimber, A.C., Smith, R.L. and Sweeting, T.J. (2000) *Statistical analysis of reliability data*. Boca Raton: Chapman and Hall/CRC.

Dabrowska, D.M. (1997) Smoothed Cox regression, *Annals of Statist.*, **25**, 1510-1540.

Dabrowska, D.M. (1987) Non-parametric regression model with censored survival time data. *Scand. J. Statist.*, **14**, 181-198.

Dabrowska, D.M. and Doksum, K.A. (1988) Partial likelihood in Transformations Models with Censored Data, *Scand. J. Statist.*, **15**, 1-23.

Dabrowska, D.M. and Doksum, K.A. (1988) Estimation and Testing in a Two-Sample Generalized Odds-Rate Model, *JASA*, **83**,# 403, 744-749.

Derringer, G.C. (1982) Considerations in single and multiple stress accelerated life testing. *Journal of Quality Technology*, **14**, 130-134.

Devarajan, K. and Ebrahimi, N. (2000) Inference for a non-proportional hazards Regression model and applications. (Submitted for publication)

Doksum, K.A. (1987) An extension of partial likelihood methods for proportional hazards model to general transformation models. *Ann. Statist.*, **15**, 325-345.

Doksum, K.A., Hoyland, A.(1992) Models for variable-stress accelerated life testing experiment based on Wiener processes and the inverse Gaussian distribution. *Technometrics*, **34**, 74-82.

Doksum, K.A., Normand, S.-L.T. (1995) Gaussian Models for Degradation Processes - Part I: Methods for the Analysis of Biomarker Data. *Lifetime Data Analysis*,**1**, 131-144.

Dowling, N.E. (1993) *Mechanical Behavior of Materials*. Englewood Cliffs: Prentice Hall.

Duchesne, T. (2000) Methods Common to Reliability and Survival Analysis. In: *Recent Advances in Reliability Theory.Methodology, Practice and Inference*, (Eds. N. Limnios and M. Nikulin), Boston: Birkhauser, 279-290.

Duchesne, T. and Lawless, J. (2000) Alternative Time Scale and Failure Time Models. *Lifetime Data Analysis*, **6**, 157-179.

Efron, B. (1988) Logistic Regression, Survival Analysis, and the Kaplan-Meier Curve, *JASA*, **83**, 414-425.

Escobar, L.A. and Meeker, W.Q. (1992) Assessing influence in regression analysis with censored data. *Biometrics*, **48**, 507-528.

Finkelstein, M. (1999) *Models of point stochastic processes for analysis of reliability and safety*. St.Petersburg: Elektropribor.

Fleming, T.R. and Harrington D.P. (1991) *Counting processes and survival analysis*. New York: John Wiley and Sons.

Fukuda, M. (1991) Reliability and Degradation of Semiconductor Laser and LED's. Boston: Artech House.

Gasser, T. and Müller, H.G. (1979) Kernel estimation of regression functions. In: *Smoothing Techniques for Curve Estimation*, Lecture Notes in Mathematics. Berlin: Springer, 23-68.

Gaver, D.P.(1963) Random hazard in reliability problems. *Technometrics*, **5**, 211-226

Gertsbakh, L.B. (1989) *Statistical Reliability Theory*. New York: Marcel Dekker.

Gertsbakh, L.B. and Kordonskiy, K.B. (1969) *Models of Failure*. New York: Springer Verlag.

Gerville-Réache, L. and Nikoulina, V. (1999) Analysis of Reliability Characteristics Estimators in Accelerated Life Testing, In: *Statistical and Probabilistic Models in Reliability*, (Eds. D. Ionescu and N. Limnios), Boston: Birkhauser, 91-100.

Gerville-Réache, L. and Nikoulina, V. (2000), In: *Abstract's book of the Second International Conference on Mathematical Methods in Reliability, MMR'2000*, **1**, Bordeaux: Université Victor Segalen Bordeaux 2,447-450.

Gerville-Réache, L. and Nikulin, M. (2000) Analyse statistique du modèle de Makeham, *Rev. Roumaine Math. Pure and Appl.*, **45**, #6, 947-957.

Gill, R.D. (1989) Non- and semi-parametric maximum likelihood estimators and the von-Mises method, Part I, *Scand. J. Statist.*, **16**, 97-128.

Gill, R.D. and Schumacher, M. (1987) A simple test of the proportional hazards assumption. *Biometrika*, **74**, 289-300.

Glaser, R. E. (1984) Estimation for a Weibull Accelerated Life Testing Model, *Naval Research Logistics Quarterly*, **31**, 4, 559-570.

Gnedenko, B., and Ushakov, I. (1995) *Probabilistic Reliability Engineering.* New York: John Wiley and Sons.

Grambsch, P., Therneau, T.M. (1994) Proportional hazards tests and diagnostics based on weighted residials, *Biometrka*, **81**, #3, 515-526.

Greenwood, P.E. and Nikulin, M. (1996) A Guide to Chi-squared Testing, New York: John Wiley and Sons.

Hafdi, M.A. (2000) Modèle de Cox généralisé: Etude par simulation. In: *Abstract's book of the Second International Conference on Mathematical Methods in Reliability, MMR'2000*, **1**, Bordeaux: Université Victor Segalen Bordeaux 2, 494-497.

Hafdi, M.A., El Himdi, K., Bagdonavičius V. and Nikulin, M. (2001) *Modèle Génŕalisé de Cox: Etude par Simulation.* Bordeaux: Preprint 0102, Université Victor Segalen Bordeaux 2.

Hamada, M.(1995) Analysis of Experiments for Reliability Improvement and Robust Reliability. In: *Recent Advances in Life-Testing and Reliability*, (Ed. N. Balakrishnan), Boca Raton: CRC Press.

Hjort, N.L. (1992) On Inference in Parametric Survival Data Models, *International Statistical Review*, **69**, #3, 355-387.

Hirose, H. (1987) A method to estimate the lifetime of solid electrical insulation, *IEEE Transactions on Electrical Insulation*, **22**, 745-753.

Hirose, H. (1993) Estimation of Threshold Stress in Accelerated Life-Testing, *IEEE Transactions on Reliability*, **42**, 650-657.

Hirose, H. (1997a) Mixture Model on the Power Law, *IEEE Transactions on Reliability*, **46**, 146-153.

Hirose, H. (1997b) Lifetime Assessement by Intermittent Inspection under the Mixture Weibull Power Law Model with Application to XLPE Cable, *Lifetime Data Analysis*, **3**, 179-189.

Hougaard, P. (1986) Survival models for heterogeneous populations derived from stable distributions, *Biometrika*, **73**, #3, 387-396.

Hougaard, P. (1995) Frailty Models for Survival Analysis. *Lifetime Data Analysis*, **1**, 255-273.

Hougaard, P. (2000) *Analysis of multivariate survival Data*. New York: Springer.

Hsieh, F. (2000) Accelerated Failure Time Model with Time-Dependent Covariates for Heterogeneous Populations. *Abstract's book of the Second International Conference on Mathematical Methods in Reliability*,**1**, Bordeaux: University Victor Segalen, 526-528.

Huber, C. (2000) Censored and Truncated Lifetime Data. In: *Recent Advances in Reliability Theory*, (Eds. N. Limnios and M. Nikulin), Boston: Birkhauser, 291-306.

Huffer, F.W. and McKeague, I.W. (1991) Weighted Least Squares Estimation for Aalen's Additive Risk Model. *JASA*, **66**, #413, 114-129.

Hyde, J. (1977) Testing survival under right censoring and left truncation, *Biometrika*, **64**, 2, 225-230.

Iuculano, F. and Zanini, A. (1986) Evaluation of Failure Models Through Step-Stress Tests, *IEEE Transaction on Reliability*, **35**, 409-413.

Kahle, W., Lehmann, A. (1998) Parameter Estimation in damage Processes: dependent observation of damage increments and first passage time. In: *Advances in Stochastic Models for Reliability, Quality and Safety*, (Eds. W. Kahle, E. Collanti, J. Franz and U. Jensen), Boston: Birkhauser, 139-152.

Kahle, W., Wendt, H. (2000) Statistical Analysis of Damage Process. In: *Recent Advances in Reliability Theory.Methodology, Practice and Inference*, (Eds. N. Limnios and M. Nikulin), Boston: Birkhauser, 199-212.

Kalbfleisch, J.D. and Prentice, R.L. (1980) *The Statistical Analysis of Failure Time Data*. New York: John Wiley and Sons.

Kaplan, E.L. and Meier, P. (1958) Nonparametric estimation from incomplete observations. JASA, **53**, 457-481.

Kartashov, G.D. (1979) Methods of Forced (Augmented) Experiments (in Russian). Moscow: *Znaniye Press*.

Kartashov, G.D. and Perrote, A.I. (1968) On the principle of "heredity" in reliability theory. *Engrg. Cybernetics*, **9**, 2, 231-245.

Khamis, I.H., Higgins, J.J. (1998) A New Model for Step-Stress Testing, *IEEE Transactions on Reliability*, **47**, #2, 131-134.

Kim, J., Song, M.S. and Lee, S. (1998) Goodness-of-Fit Tests for the Additive Risk Model with $(p > 2)$−Dimensional Time-Invariant Covariates. *Lifetime Data Analysis*, **4**, 405-416.

Klein, J.P. and Basu, A.P. (1981) Weibull Accelerated Life Tests When There are Competing Causes of Failure, *Communications in Statistical. Methods and Theory*, **10**, 2073-2100.

Klein, J.P. and Basu, A.P. (1982) Accelerated Life Testing under Competing Exponential Failure Distributions, *IAPQR Trans.*, **7**, 1-20.

Klinger, D.J.(1992) Failure time and rate constant of degradation: an argument for the inverse relationship. *Microelectronics and Reliability*, **32**, 987-994.

Lawless, J.F. (1982) *Statistical Models and Methods for Lifetime Data*. New York: John Wiley and Sons.

Lawless, J.F. (1986) A Note on Lifetime Regression Models, *Biometrika*, **73**, 509-512.

Lawless, J., Hu, J. and Cao, J.(1995) Methods for the estimation of failure distributions and rates from automobile warranty data. *Lifetime Data Analysis*, **1**, 227-240.

Lehmann, A. (2000) Statistical inference for failure and degradation data based on Wiener models. In: *Abstract's book of the Second International Conference on Mathematical Methods in Reliability, MMR'2000*, **2**, Bordeaux: University Victor Segalen Bordeaux 2, 691-694.

Lenglart, E. (1977) Relation de domination entre deux processus. *Ann. Inst. Henri Poincaré*, **13**, 171-179.

Lin, D.Y. (1991) Goodness of fit analysis for the Cox regression model based on a class of parameter estimators. *JASA*, **86**, 725-728.

Lin, D.Y., Geyer, C.J. (1992) Computational methods for semiparametric linear regression with censored data. *Journal Comput. and Graph. Statist.*, **1**, 77-90.

Lin, D.Y., Wei, L.J. and Ying, Z. (1993) Checking the Cox model with cumulative sums of martingale-based residuals. *Biometrika*, **80**, 557-72.

Lin, D.Y. and Ying, Z. (1994) Semiparametrical analysis of the additive risk model. *Biometrika* **81**, 61-71.

Lin, D.Y. and Ying, Z. (1995). Semiparametric inference for accelerated life model with time dependent covariates. *Journal of Statistical Planning and Inference*, **44**, 47-63.

Lin, D.Y. and Ying, Z. (1996) Semiparametric analysis of the general additive-multiplicative hazard models for counting processes. *The Annals of Statistics*, **23**, 5, 1712-1734.

Lu, C.J. (1995) Degradation processes and related reliability models. Ph.D. thesis, McGill University, Montreal, Canada.

Lu, C.J., Meeker, W.Q. (1993) Using degradation Measures to Estimate a Time-to-Failure Distribution. *Technometrics*, **35**,161-174.

Lu, C.J., Meeker, W.Q. and Escobar, L.A. (1996) A comparison of degradation and failure-time analysis methods of estimating a time-to-failure distribution. *Statistica Sinica*, **6**, 531-546.

LuValle, M. (2000) A Theoretical Framework for Accelerated Testing. In: *Recent Advances in Reliability Theory. Methodology, Practice and Inference*, (Eds. N.Limnios and M. Nikulin,) Boston: Birkhauser, 419-434.

Mann, N.R., Schafer, R.E. and Singpurwalla, N.D. (1974) *Methods for Statistical Analysis of Reliabitity and Life Data*. New York: John Wiley and Sons.

Mazzuchi, T. and Soyer, R. (1992) A Dynamic General Linear Model for Inference from Accelerated Life Tests, *Naval Research Logistics*, **39**, 757-773.

McKeague, I.W., Utikal, K.J. (1991) Goodness-of-fit tests for additive hazards and proportional hazards models. *Scand. J. Statist.*, **18**, 177-195.

McKeague, I.W., Sasieni, P.D.(1994) A partly parametric additive risk model. *Biometrika*, **81**,#3, 501-514.

Meeker, Jr.W.Q. (1984) A comparison of Accelerated Life Test Plans for Weibull and Lognormal Distributions and Type I Censoring, *Technometrics*, **26**, 157-172.

Meeker, Jr.W.Q. and Hahn, G. J. (1978) A comparison of Accelerated Life Test Plans to Estimate the Survival Probability at a Design Stress. *Technometrics*, **20**, 245-247.

Meeker, Jr.W.Q. and Escobar, L.A. (1993) A review of recent research and current issues in accelerating testing. *International Statistical Review*, **61**, 1, 147-168.

Meeker, W.Q., L.A.Escobar. (1998).*Statistical Methods for Reliability Data*. New York: John Wiley and Sons.

Meeker, W.Q., Escobar,L.A., Lu, C.J. (1998) Accelerated Degradation Tests: Modeling and Analysis. *Technometrics*, **40**, 89-99.

Meeker, W.Q. and LuValle, M.J. (1995) An Accelerated Life Test Model Based on Reliability Kinetics. *Technometrics*, **37**, 133-146.

Meeter, C.A. and Meeker, Jr.W.Q. (1994) Optimum Accelerated Life Tests with a Nonconstant Scale Parameter. *Technometrics*, **36**, 1, 71-83.

Miner, M.A. (1945) Cumulative Damage in Fatigue, *J. of Applied Mechanics*, **12**, A159-A164.

F.Mitsuo.(1991). *Reliability and Degradation of Semiconductor Lasers and LED*. Norwood: Artech House.

Moreau, T., O'Quigley, J. & Mesbah, M. (1985) A global goodness-of-fit statistic for the proportional hazards model. *Biometrics*, **34**, 3, 212-218.

Moreau, T., Maccario, J., Lelouch, J. and Huber, C. (1992) Weighted logrank statistics for comparing two distributions. *Biometrika*, **79**, 195-198.

Mudholkar, G., Srivastava, D. (1995) The Exponentiated Weibull Family: A Reanalysis of the Bus-Motor-Failure Data. *Technometrics*, **37**, 14, 436-445.

Murphy, S.A., Rossini, A.J. and Van der Vaart, A.W. (1997) Maximum Likelihood Estimation in the Proportional Odds Model, *JASA*, **92**, # 439, 968-976.

Murphy, S.A. (1995) Asymptotic theory for the frailty model. *Ann. Statist.*, **32**,182-198.

Nagelkerke, N.J.D., Oosting, J. and Hart, A.A.M. (1984) A simple test for goodness of fit of Cox's proportional hazards model. *Biometrika*, **40**, 483-486.

Nelson, W. (1969) Hazard plotting for incomplete failure data. *Journal of Quality Technology*, **1**, 27-52.

Nelson, W. (1980) Accelerated Life Testing - Step-Stress Models and Data Analysis. *IEEE Transactions on Reliability*, **29**, 103-108.

Nelson, W. (1990) *Accelerated Testing: Statistical Models, Test Plans, and Data Analyses*. New York: John Wiley and Sons.

Nelson, W. and Meeker, W. (1991) Accelerated Testing: Statistical models, test plans, and data analysis. *Technometrics*, **33**, 236-238.

Nelson, W. and Macarthur, E. (1992) Accelerated Testing: Statistical models, test plans, and data analysis. *Applied Statistics*, **41**, 224-225.

Nielsen, G.G., Gill, R.D., Andersen, P.K. and Sorensen, T.I.A. (1992) A counting process approach to maximum likelihood estimation in frailty model. *Scand. J. Statist.*, **19**, 25-43.

Nikulin, M.S. (1991) A chi-squared goodness-of-fit tests for the natural exponential families with power variance function, *Mathematical Reports of the Acad. Sci. Canada*, **XIII**, #6, 265-270.

Nikulin, M.S., and Solev, V.N. (1999) Chi-Squared Goodness-of Test for Doubly Censored Data With Applications in Survival Analysis and Reliability. In: *Statistical and Probabilistic Models in Reliability*, (Eds. D.Ionescu and N.Limnios), Boston: Birkhauser, 101-111.

Nikulin, M.S. and Voinov, V.G. (2000) Unbiased Estimation in Reliability and Similar Problems. In: *Recent Advances in Reliability Theory.Methodology, Practice and Inference*, (Eds. N. Limnios and M. Nikulin), Boston: Birkhauser, 435-448.

Nikulin, M.S., and Solev, V.N. (2001) Testing Problem for Increasing Function in a Model with Infinite Dimensional Nuisance Parameter. In: *Goodness-of-fit Tests and Validity of Models*, (Eds. C. Huber, N. Balakrishnan, M. Nikulin and M. Mesbah), Boston: Birkhauser.

Oakes D. (2001) Biometrika Centenary: Survival analysis. *Biometrika*, **88**, #1, 99-142.

Parner, E. (1998) Asymptotic theory for the correlated gamma-frailty model. *Ann. Statist.*, **26**, 183-214.

Pieper, V. and Tiedge, J. (1983). Zuverlässigkeitsmodelle auf der Grundlage stochastischer Modelle von Verschleißprozessen. *Math. Operationsforschung und Statistik, Series Statistics*, **14**, 485-502.

Pieruschka, E. (1961) Relation between lifetime distribution and the stress level causing failures, *LMSD-800440*, Lockhead Missiles and Space Division, Sunnyvale, California.

Pons, O. and Huber, C. (2000) *Survie et Processus Ponctuels*. Paris: l'Université René Decartes.

Prentice, R.L. (1974) A log gamma model and its maximum likelihood estimation, *Biometrika*, **61**, 3, 539-544.

Prentice, R.L. (1975) Discrimination among some parametric models, *Biometrika*, **62**, #3, 607-614.

Quantin, C., Moreau, T., Asselain, B., Maccario, J and Lelouch, J. (1996) A regression model for testing the proportional hazards hypothesis, *Biometrika*, **52**, 874-885.

Rebolledo, R. (1980) Central limit theorem for local martingales. *Z. Wahrsch. Verw. Gebiete*, **51**, 269-286.

Robins, J.M., and Tsiatis, A.A. (1992) Semiparametric estimation of an accelerated failure time model with time dependent covariates. *Biometrika*, **79**, #2, 311-319.

Rodrigues, J., Bolfarine, H. and Lourada-Neto, F. (1993) Comparing Several Accelerated Life Models, *Communications in Statistical Methods and Theory*, **22**, 2297-2308.

Rukhin, A.L., and Hsieh, H.K. (1987) Survey of Soviet work in reliability. *Statistical Science*, **2**, 484-503.

Schaebe, H. (1998) Accelerated Life Models for Nonhomogeneous Poisson Processes, *Statistical Papers*, **39**, 291-312.

Schaebe, W., Viertl, R. (1995) An Axiomatic Approach to Model of Accelerated Life Testing, Engineering Fracture. *Mechanics*, **50**, #2, 203-217.

Schoenfeld, D. (1980) Chi-squared goodness-of-fit tests for the proportional hazards regression model. *Biometrika*, **67**, 145-153.

Schmoyer, R. (1986) An exact distribution-free analysis for accelerated life testing at several levels of a single stress. *Technometrics*, **28**, 2, 165-175.

Schmoyer, R. (1991) Nonparametric analyses for two-level single-stress accelerated life tests. Technometrics, **33**, 175-186.

Sedyakin, N.M. (1966) On one physical principle in reliability theory.(in Russian). *Techn. Cybernetics*, **3**, 80-87.

Sethuraman, J. and Singpurwalla, N.D. (1982) Testing of hypotheses for distributions in accelerated life tests. *JASA*,**77**, 204-208.

Schmoyer, R. (1991) Nonparametric Analyses for Two-Level Single-Stress Accelerated Life Tests, *Technometrics*, **33**, 175-186.

Shaked, M. and Singpurwalla, N.D. (1983) Inference for step-stress accelerated life tests. *J. Statist. Plann. Inference*, **7**, 295-306.

Singpurwalla, N.D. (1971) Inference from Accelerated Life Tests When Observations Are Obtained from Censored Samples, *Technometrics*, **13**, 161-170.

Singpurwalla, N.D.(1987) Comment on Survey of Soviet work in reliability. *Statistical Science*, **2**, 497-499.

Singpurwalla, N.D. (1995) Survival in Dynamic Environments. *Statistical Science*,l,**10**, 86-103.

Singpurwalla, N.D.(1997) Gamma processes and their generalizations: an overview. In *Engineering Probabilistic Design and Maintenance for Flood Protection*, (Eds. R.Cook, M.Mendel and H.Vrijling), Kluwer Acad. Publishers, 67-73.

Singpurwalla, N.D., Castellino, V.C. and Goldschen, D.Y. (1975) Inference from accelerated life tests using Eyring type re-parametrization. *Naval Research Logistics Quarterly*, **22**, 289-296.

Singpurwalla, N.D., Wilson, S.P. (1999) *Statistical Methods in Software Engineering*, New York: Springer.

Singpurwalla, N.D, Youngren, M.A. (1998) Multivariate distributions induced by dynamic environments. *Scand.J. of Statist.*, **20**, 251-261.

Spiekerman, C.F. and Lin, D.Y. (1996) Checking the Cox model for correlated failure time data. *Biometrika*, **83**, 143-56.

Stacy, E.W. (1962) A generalization of the gamma distribution, *Ann. Math. Statist.*, **33**, 3, 1187-1192.

Suzuki, K., Maki, K., Yokogawa, S. (1993) An analysis of degradation data of a carbon film and properties of the estimators. In: *Statistical Sciences and Data Analysis*, (Eds. K. Matusita, M. Puri and T. Hayakawa,) Utrecht: VSP, 501-511.

Therneau, T.M., Grambsch, P.M. and Fleming, T.(1990) Martingale-based residuals for survival models. *Biometrika*, **77**,#1, 147-160.

Therneau, T.M. and Grambsch, P.M. (2000) *Modeling Survival Data. Extending the Cox Model*. New York: Springer.

Tibshirani, R.J. and Ciampi A. (1983) A family of proportional- and additive-hazards models for survival survival data, *Biometrics*, **39**, 141-147.

Tseng, S.T., Hamada, M.S., and Chiao, C.H. (1994) *Using degradation data from a fractional experiment to improve fluorescent lamp reliability*. Research Report RR-94-05. Waterloo: The Institute for Improvement in Quality and Productivity, University of Waterloo, Canada.

Tseng, S.T. and Wen, Z.C. (1977) Step-stress accelerated degradation analysis for highly reliable products. *Institute of Statistics*, National Tsing-Hua University.

Tseng, S.T. and Yu, H.F. (1977) A termination rule for degradation experiment. *IEEE Transaction on Reliability*, **46**, 130-133.

Tsiatis, A.A. (1981) A large sample study of Cox's regression model. *Ann. Statist.*, **9**, 93-108.

Tsiatis, A.A. (1990) Estimating regression parameters using linear rank tests for censored data. *Ann. Statist.*, **18**, 354-72.

Van der Vaart, A.W. and Wellner, J.A. (1996) *Weak Convergence and Empirical Processes*, New York: Springer.

Van der Wiel, S.A. and Meeker, W.Q. (1990) Accuracy of approximate confidence bounds using censored Weibull regression data from accelerated life tests. *IEEE Transactions on Reliability*, **R-39**, 346-351.

Vaupel, J.W., Manton, K.G. and Stallard, E. (1979) The impact of heterogeneity in individual frailty on the dynamic of mortality. *Demography*, **16**, 439-454.

Viertl, R. (1988) *Statistical Methods in Accelerated Life Testing*. Göttingen: Vandenhoeck and Ruprecht.

Viertl, R., Gurker, W. (1988) On filtering methods in accelerated life testing using infinitesimal characteristics. In: *Proceedings of Electronics 88*, **1**, Budapest, 60-68.

Viertl, R., Gurker, W. (1995) Reliability Estimation based on Fuzzy Life Time Data. In: *Reliability and Safety Analysis under Fuzziness*, (T. Onisawa and J. Kacprzyk, eds.) Heidelberg: Physica-Verlag, **50**, #2, 153-168.

Viertl, R. and Spencer, F. (1991) Statistical Methods in Accelerated Life Testing. *Technometrics*, **33**, 360-362.

Voinov, V.G. and Nikulin, M.S. (1993) *Unbiased Estimators and Their Applications*, **Vol.1**: Univariate Case. Dordrecht: Kluwer Academic Publisher.

Voinov, V.G. and Nikulin, M.S. (1996) *Unbiased Estimators and Their Applications*, **Vol.2**: Multivariate Case. Dordrecht: Kluwer Academic Publisher.

Vonta, F. (2000) A Computational Method for Efficient Estimation in Parametric Generalization of the Cox Model. In: *Abstract's book of the Second International Conference on Mathematical Methods in Reliability, MMR'2000*, **2**, Bordeaux: Universityé Victor Segalen Bordeaux 2, 1026-1029.

Wei, L.J. (1984) Testing goodness-of-fit for the proportional hazards model with censored observations. *JASA*, **79**, 649-652.

Wei, L.J., Ying, Z. and Lin, D.Y. (1990) Linear regression analysis of censored survival data based on rank tests. *Biometrika* **77**, 845-51.

Wendt, H. (1999) *Parameterschatzungen fur eine Klasse doppelt-stochastischer Poisson Prozesse bei unterschiedlichen Beobachtungsinfmationen*. PhD. Thesis, University of Magdeburg.

Wilson, S.P. (2000) Failure Models Indexed by Time and Usage. In: *Recent Advances in Reliability Theory.Methodology, Practice and Inference*, (Eds. N. Limnios and M. Nikulin,) Boston: Birkhauser, 213- 228.

Whitmore, G.A.(1995). Estimating Degradation By a Wiener Diffusion Process Subject to Measurement Error. *Lifetime Data Analysis*, **1**, 307-319.

Whitmore, G.A., Schenkelberg, F. (1997) Modelling Accelerated Degradation data Using Wiener Diffusion With a Time Scale Transformation. *Lifetime Data Analysis*, **3**, 27-45.

Whitmore, G.A., Crowder, M.I. and Lawless, J.F.(1998) Failure inference from a marker process based on bivariate model. *Lifetime Data Analysis*, **4**, 229-251.

Xie, M. (2000) Software Reliability Model - Past, Present and Future. In: *Recent Advances in Reliability Theory*, (Eds. N. Limnios and M.Nikulin,) Boston: Birkhauser, 325-340.

Yanagisava, T. (1997) Estimation of the degradation of amorphous silicon cells. *Microelectronics and Reliability*, **37**, 549-554.

Ying, Z. (1993) A large sample study of rank estimation for censored regression data. *Ann. Statist.*, **21**, 76-99.

Yu, H.F. and Tseng, S.T. (1998) On-line procedure for terminating an accelerated degradation test. *Statistica Sinica*, **8**, 207-220.

Zacks, S. (1992) *Introduction to Reliability Analysis*, New York: Springer-Verlag.

Author index

Subject index